Anonymous

Mathews' medical quarterly

Vol. 3

Anonymous

Mathews' medical quarterly
Vol. 3

ISBN/EAN: 9783337713522

Printed in Europe, USA, Canada, Australia, Japan

Cover: Foto ©berggeist007 / pixelio.de

More available books at **www.hansebooks.com**

MATHEWS'
MEDICAL QUARTERLY

A JOURNAL DEVOTED TO

DISEASES OF THE RECTUM AND GASTRO-INTESTINAL DISEASE,

——AND——

RECTAL AND GASTRO-INTESTINAL SURGERY.

VOLUME III, 1896.

JOSEPH M. MATHEWS, M. D.,
Editor and Proprietor.

HENRY E. TULEY, A. B., M. D.,
Associate Editor and Manager.

LOUISVILLE, KY.
PRINTED BY JOHN P. MORTON & COMPANY.
1896

CONTRIBUTORS TO VOLUME III.

ADLER, LEWIS H., JR.
BEACH, WILLIAM M.
BEAVER, DANIEL B. D.
BLECH, GUSTAVUS M.
BUCHANAN, A. D.
COSTON, H. R.
CUMSTON, CHARLES GREENE.
EARLE, SAMUEL T.
GALLANT, A. ERNEST.
GANT, S. G.
GASTON, J. McFADDEN.
GILLETTE, WILLIAM J.
GRANT, H. H.
HOWLETT, K. S.
LAWS, WILLIAM V.
MACLAREN, W. S.
MANLEY, THOMAS H.
MARTIN, THOS. CHAS.
MATHEWS, JOS. M.
ROBERTS, DEERING J.
ROBINSON, BYRON.
TODD, LYMAN BEECHER.
UPSHUR, J. N.
VANCE, AP MORGAN.
VANDER VEER, A.
WALKER, H. O.
WYMAN, HAL C.

GENERAL INDEX.

PAGE

Abdominal Surgery, Seven Cases in which the Murphy Button was Used. 24
Address on Surgery, The 301
ADLER, LEWIS H., JR. 333
American Medical Association 152
An Abdominal Case 149
Anastomosis Buttons, Dr. Chaput's, for Gastro-Intestinal Surgery 129
Ascending Colon, The, in One Hundred and Thirty Autopsies 52
BEACH, WILLIAM M. 1
BEAVER, DAN'L B. D. 340
BLECH, GUSTAVUS M. 358
Bogus Medical Journals (Editorial) 152
Books and Pamphlets Received 95, 195, 295, 405
BUCHANAN, A. P. 36, 121, 264
Button, Murphy 24
Button, Dr. Chaput's 129
BOOK REVIEWS—N. Senn, The Pathology and Surgical Treatment of
   Tumors, 88....Richard C. Norris, An American Text Book of Obstet-
   rics, 89....Dillon Brown, George A. Carpenter, Pediatrics, 91....
   Eugene Fuller, Disorders of the Male Sexual Organs, 91....J. H. Kel-
   logg, The Art of Massage, its Physiological Effects and Therapeutic
   Applications, 92....Theophilus Parvin, The Science and Art of
   Obstetrics, 93.... Rachford, B. K., Some Physiological Factors of the
   Neuroses of Childhood, 93....James Nevins Hyde, Frank H. Mont-
   gomery, A Manual of Syphilis and the Venereal Diseases, 94....
   Charles Denison, Exercise and Food for Pulmonary Invalids, 94....
   Thomas Morgan Rotch, Pediatrics, 184....A. Jacobi, Therapeutics of
   Infancy and Childhood, 184....Charles E. Sajous, Annual of the Uni-
   versal Medical Sciences, 187....Egbert H. Grandin, George W. Jar-
   man, Pregnancy, Labor, and the Puerperal State, 188 ...George M.
   Gould and a corps of collaborators, The American Year-Book of
   Medicine and Surgery, 189....Max Nardau, A Comedy of Sentiment,
   190....T. M. Rotch, Herbert L. Burrell, Medical and Surgical Report
   of the Children's Hospital, Boston, 191....L. M. Phillips, Miskel, 191
   ....George Roe Lockwood, A Manual of the Practice of Medicine,
   192....Transactions of the New York Academy of Medicine for 1893,
   192....Henry C. Chapman, A Manual of Medical Jurisprudence and
   Toxicology, 193....Joseph McFarland, A Text Book upon the Patho-

Books Reviews—Continued— PAGE

genic Bacteria, for Students of Medicine and Physicians, 193....
Gwilym G. Davis, The Principles and Practice of Bandaging, 193....
William Osler, The Principles and Practice of Medicine, 193....
Transactions of the Medical Society of the State of North Carolina,
194 ...Willis P. King, Stories of a Country Doctor, 194....S. G.
Gant, Diagnosis and Treatment of Diseases of the Rectum, Anus, and
Contiguous Parts, 285....The Annual Report of the Supervising Sur-
geon-General of the Marine Hospital Service for the Years 1894 and
1895, 287....Sidney Martin, Functional and Organic Diseases of the
Stomach, 288....Frederick S. Dennis and a corps of collaborators,
System of Surgery, 289....J. H. Kellogg, The Stomach, Its Disorders,
and How to Cure Them, 290....A. Brothers, Infantile Mortality
During Child-birth and Its Prevention, 291....Louis Starr, Diets for
Infants and Children, 292....George M. Gould, Borderland Studies,
292....Irving S. Haynes, A Manual of Anatomy, 293....Edwin J.
Houston, A. E. Kennelly, Electricity in Electro-Therapeutics, 293....
Charles Wilson Ingraham, Dont's for Consumptives, or the Scientific
Management of Pulmonary Tuberculosis, 293....Marcus P. Hatfield,
A Compend of the Diseases of Children, 294....L. Ch. Boisiliniere,
Obstetric Accidents, Emergencies and Operations, 294....William H.
Wells, A Compend of Gynecology, 295....J. C. Wilson, M. D.,
Augustus A. Eshner, M. D., An American Text-Book of Applied
Therapeutics, for the Use of Practitioners and Students, 398....
Emily A. M. Stoney, Practical Points in Nursing for Nurses in Private
Practice, with an Appendix, 399....Frederick S. Dennis, M. D., John
S. Billings, M. D., System of Surgery, 400....Charles Marchand, The
Therapeutical Applications of Peroxide of Hydrogen (medicinal), Gly-
cozone, Hydrozone, and Eye Balsam, 400....Roswell Park, A. M., M.
D., A Treatise on Surgery, by American Authors, for Students and
Practitioners of Surgery and Medicine, 401....William Allingham,
F. R. C. S., Herbert W. Allingham, F. R. C. S., The Diagnosis and
Treatment of Diseases of the Rectum, 402....Louise E. Hogan, How
to Feed Children, 402....John B. Deaver, M. D., A Treatise on
Appendicitis, 403....W. A. Newman Dorland, A. M., M. D., A
Manual of Obstetrics, 404....Miss E. Hibbard, Mrs. Emma Drant,
Diet for the Sick, 404....C. Henri Leonard, M. A., M. D., The
Multum in Parvo Reference and Dose Book, 404.

Cancer of the Rectum, Treatment of............. 383
Change of Name (Editorial)..... 383
Chaput's New Anastomosis Button................................... 129
Clamp Forceps for Suturing Excised Hemorrhoids 21
Colon, The Ascending, in One Hundred and Thirty Autopsies........... 52
Constipation 121

PAGE

Correspondence 73, 154
COSTON, H. R. 225
CUMSTON, CHARLES GREENE 129
Dilatation of the Stomach 264
Disease, The Relation of Intestinal Fermentation to 36
Diseases of the Pylorus 140
Disorders, Surgical Interference in Rectal 10
Dyspepsia, Chronic Gastric 254
EARLE, SAMUEL T. 21
Electro-Cautery in Rectal Ulcers 1
Fistula in Ano, The Relation of to Phthisis 16
Flexure, The Sigmoid and its Meso-Sigmoid 239
GALLANT, A. ERNEST 229
GANT, S. G. 16
GASTON, J. McFADDEN 10
Gastric Dyspepsia, Chronic 254
Gastric Hemorrhage 106
GILLETTE, WILLIAM J. 234
GRANT, H. H. 301
Growths of the Rectum; Presentation of Specimens With Micro-Photo-
    graphs, and Report of Cases 342
Hemorrhoids, A New Clamp Forceps for Suturing Excised 21
Hemorrhoids and Their Treatment 220
Hemorrhoids, Treatment of 225
HOWLETT, K. S. 254
Index, Sixty-nine " Medicus," The Treatment of Post-Operative ... 229
Inflammatory Diseases of the Stomach, Treatment of 358
Intestinal Fermentation, The Relation of, to Disease 36
Intestinal Hemorrhage 106
Intestinal Indigestion 348
Intestinal Obstruction, The Localization of 42
July Issue 273
Kentucky State Medical Society 269
LAWS, WILLIAM V. 220
Localization of Intestinal Obstruction 42
MACLAREN, W. S 42
MANLEY, THOS. H 106
MARTIN, THOS. CHAS 201, 310
MATHEWS, JOS. M 110
Medical and Surgical History of the War of the Rebellion 70
Medical Association, The American (Editorial) 153
Medical Journalism, Bogus 152
Medical Society, Kentucky State (Editorial) 269

PAGE

Memoir of the late Dr. Orrin D. Todd............................... 274
Meso-Sigmoid of the Sigmoid Flexure 239
Method, A New, of Procto-Colonoscopy 201
Murphy Button............................. 24
New Evidence That the Rectal Valve Is an Anatomical Fact........... 810
New York Letter..73, 154
Notes and Queries 98, 199, 278, 407
Obstruction of the Bowels.. 376
Obstruction, The Localization of Intestinal........................ 42
Obstruction, Treatment of Post-Operative Intestinal................ 229
Perforative Ulcer of the Stomach.................................. 234
Phthisis, The Relation of, to Fistula in Ano 16
Procto-Colonoscopy and its Possibilities.......................... 201
Post-Operative Intestinal Obstruction............................. 229
Public Health (Editorial)... 70
Pylorus, Diseases of.. 140
Rectal Disorders, Surgical Interference in........................ 10
Rectal Peri-Phlebitis Treated With Galvanism...................... 340
Rectal Hemorrhage 106
Rectal Surgery, Some Late Suggestions in.......................... 101
Rectal Ulcers and the Electro-Cautery............................. 1
Relation of Intestinal Fermentation to Disease.................... 36
Relation of Phthisis to Fistula in Ano 16
Responsibility of Anesthetists (Editorial)........................ 382
ROBINSON, BYRON 52, 239, 360
ROBERTS, DEERING J.. 140
Sigmoid Flexure and its Meso-Sigmoid 239, 360
Specialties and the General Practitioner (Editorial) 270
Stomach, Dilatation of the 264
Stomach, Perforative Ulcer of..................................... 233
Surgery, Some Late Suggestions in Rectal.......................... 101
Surgical Interference in Rectal Disorders......................... 10
TODD, LYMAN BEECHER... 274
Ulcer of Stomach, Perforative 234
UPSHUR, J. N.. 348
VANCE, AP MORGAN 149
VANDER VEER, A 24
WALKER, H. O.. 342
Word to the Busy Doctor... 79
WYMAN, HAL C. 376

WITH OUR EXCHANGES—*Rectal Disease—*   PAGE

Anal Pruritus, Treatment of 390

ARMSTRONG 83

BEACH, WM. M... 385

BROWN, WARREN 282

Clinical Features and Treatment of External Piles................. 170

DE BUCK 82

Diseases of the Rectum, Relation of Constipation to................ 385

DRENNEN, C. T... 387

Extirpation of the Rectum...................................... 82

Fecal Impaction in the Ano-Rectal Space 168

Forceps 172

FREEMAN, LEONARD 386

GUYON 172

HOUSTON, FRANCIS T... 384

Hydrogen Peroxide 282

Impaction, Fecal, in the Ano-Rectal Space....................... 168

KELSEY, CHAS. B... 84

MANLEY, THOMAS H.. 168

MORAIN.. 390

OTIS, WALTER J.. 170

Peroxide of Hydrogen... 282

Piles, Clinical Features and Treatment of External 170

Pruritus Ani, The Etiology and Treatment of..................... 387

Rectal Irrigator, Improved 389

Rectum, Excision of the ; A Method of Operating................. 384

Rectum, Carcinoma of.. 83

Rectum, Total Extirpation of.................................... 82

Rectum, The Relation between Uterine Diseases and Diseases of the, 84

SMITH, Q. CINCINNATUS... 389

Vulvo-Vaginal Anus.. 386

Gastro-Intestinal Disease—

A Method of Temporarily Closing the Opening after Gastrostomy or
    Enterotomy 395

A Rapid Method of Performing Enterectomy without the Aid of
    Any Special Apparatus....................................... 393

BURNS, T. M... 85

CABOT, A. T... 175

Chlorosalol in Diarrhea... 391

Chronic Gastritis of Long Standing with Periodic Attacks of Migraine, 392

Closure of Fecal Fistula Extra-Peritoneally 176

Colon Irrigation 85

Colotomy.. 181

Gastro-Intestinal Disease—Continued.   PAGE

CRIPPS, HARRISON 395
CURRIDEN, GEO. A.. 392
DEANSLEY, EDWARD 181
Diagnosis and Treatment of Gastric Neuroses.................... 175
Endoxine in Gastric Disorders................................. 87
Femoral Hernia, Operations for the Cure of.................... 178
Fistula, Fecal, due to Patency of the Meckle's Diverticulum 87
Fistula, Fecal, Closure by Extra-Peritoneal Method 176
GIRARD 391
Gastric Neuroses, Their Diagnosis and Treatment 175
Gastro-enteritis and Cholera Infantum, The Anatomical Lesions of.. 391
HEATON, GEORGE 87
HEMETER, JNO. C.. 173
HEUBNER, PROF.. 391
HOUCHON.. 183
Hernia, Inguinal and Femoral................................. 178
Indications for the Tube in Stomach Diseases 283
Inguinal Hernia 178
Intestinal Needle with Spring................................ 183
Irrigation of the Colon 85
KIRBY, N. H.. 284
MARTINS.. 283
Meckle's Diverticulum.. 87
Neuroses, Gastric and their Treatment 175
Needle, Intestinal with Spring 183
PEARSE, HERMAN... 175
Peritoneal, Extra Closure of Fecal Fistula 176
Protonuclein, Ulcer of the Stomach Treated with 284
ROGERS, LEONARD 393
SMITH, GREIG 176
Stomach, An Apparatus for Washing Out 173
Stomach, Indications for the Tube in Diseases of................ 83
Stomach, Ulcer of, Treated with Protonuclein 84
Tube, A Return, for Washing Out the Stomach, Rectum, and Sigmoid 178
Tube, Indications for the Use of the Stomach.................... 283
Umbilical Fecal Fistula....................................... 87

MATHEWS'
MEDICAL QUARTERLY.
"ALIS VOLAT PROPRIIS."

| Vol. III. | JANUARY, 1896. | No. 1. |

Original Contributions.

RECTAL ULCERS AND THE ELECTRO-CAUTERY.

BY WILLIAM M. BEACH, A. M., M. D.,

Surgeon to the Presbyterian Hospital.

PITTSBURGH, PA.

[Written for MATHEWS' MEDICAL QUARTERLY.]

This disease manifests itself in many forms; simple ulceraation, or complicated by fistulæ, stricture, growths, hemorrhoids, singly or combined, may occur in the rectum, and in point of frequency in the order named. No disease is so apt to escape the notice of patient or physician till the ulcer is sufficiently advanced to beget obvious symptoms. It is remarkable how free from pain are any such diseases, especially when located above the internal sphincter, also during the incipient stage. Even in the advanced stages pain is often the least prominent of symptoms.

In the light of modern research and invention of new appliances to aid us in diagnosis, together with a thorough knowledge of pathology, the doctrine of palliative and irrational measures in the treatment of diseases of the rectum is relegated to the past, and there opens to the physician a new and interesting field of study about an organ upon whose function depend much of the ills or comforts in human life. Patients suffering from rectal disease no longer knock in vain at the door of the regular practitioner to obtain relief, as they have done hitherto, thereby aiding the charlatan to fill his coffers, but rectal surgery in recent years is receiving the attention its importance deserves, and places it on a dignified plane with other specialties in America.

Rectal ulcers may be acute or chronic and pathologically benign or malignant. The character of an ulcer may depend upon its location ; if within the grasp of the sphincter the lesion, though' usually benign, is very painful and irritable ; hence the term irritable ulcer. 'Should solution of continuity occur at any point above, it is usually regarded as malignant, and the physician should be guarded in his prognosis.

That the reader may have a clear notion of the pathology of rectal ulcers, it is well to mention a few points in the physiological anatomy of the rectum :

The rectum of the adult lies wholly in the true pelvis, with a lateral curve above and two antero-posterior curves below. It is about twenty centimeters long and cylindrical, beginning at the left sacro-iliac junction, curving to the middle line about nine centimeters, then follows the curve of the sacrum and coccyx for about eight centimeters to the tip of the latter, when it curves backward about three centimeters, terminating at the anus. The upper division is entirely covered with peritoneum, being held in position by the meso-rectum. Anteriorly the peritoneum is about seven centimeters, and posteriorly about nine centimeters from the anus. The tube is narrowest at its upper portion, and gradually increases in size toward the anus, immediately above which it presents a dilatation, the ampulla, capable of being enormously distended.

The muscular coat, consisting of circular and longitudinal fibers, is well developed; the latter fibers terminate in the anal connective tissue, while the former becomes thickened about six millimeters from the anus to form the internal sphincter muscle. The mucous membrane is thick and vascular, and forms Houston's valves, semilunar in shape, one anterior, opposite the prostate gland, another posterior, opposite the sacrum, and a third at the upper end of the bowel on left side ; these folds overlap each other when the bowel is in repose, so that it is sometimes difficult to insinuate the finger or a bougie into the organ for any purpose. When the rectum is empty, the mucous membrane appears folded longitudinally, and at the verge of the anus is gathered into looped folds called the valvulæ morgagni (velvet).

The blood supply is derived from the superior, middle, and inferior rectal arteries, from the inferior mesenteric, anterior root of internal iliac, and internal pudic arteries respectively. In the lower segment of the rectum the arteries pass parallel toward the anus, freely communicating with transverse branches. The veins are similarly arranged, forming hemorrhoidal venous plexuses at lower end of rectum; thus, while any portion of the rectum may ulcerate, the lower segment is more predisposed owing to its peculiar blood supply. The portal and general venous circulation communicate through the superior, middle, and inferior rectal veins, the first being tributary to the inferior mesenteric, the two latter terminating in the internal iliac veins, an important factor predisposing the anal veins to become vertical and to the formation of piles.

The nerves of the rectum are derived from the inferior mesenteric, hypogastric, and sacral plexuses, and follow the course of the blood-vessels. At the anus and within the grasp of the sphincter the anal branch of the pudic nerve terminates, which gives rise to the great pain accompanying anal fissure, which pain is almost pathognomonic of ulcers in that portion of the rectum.

The normal mucous membrane is of a pale pink color and of the consistence of velvet to the sense of touch. To the trained finger existence of ulcers can often readily be detected, thereby saving the patient some pain and annoyance by the use of a speculum.

The depth of the perineum varies considerably in different bodies, depending chiefly upon the amount of fat in the ischio-rectal fossæ, being about three inches in relation to the lateral wall of the rectum. It should be noted in this connection that, owing to the attachment of the levator ani muscle and of the anal fascia to the bowel, the opening into the latter in an ordinary case of fistula is nearly always within half an inch of the anus.

For convenience of discussion I make the following classification of rectal ulcers:

Traumatic. { 1. Fissure. 2. Simple ulcer. 3. Gonorrheal.   Constitutional. { 1. Tubercular. 2. Syphilitic. 3. Cancer.

1. This order is very simple, but will be found to cover all forms of ulcers for the most part. It is based on etiology. Agents producing traumatism of the rectum and ending in ulcer-

ation are external and internal. The former may include the use of surgical instruments, as speculums, dilators, applicators, accidental or criminal introduction of indifferent objects.

2. Application of strong chemicals, as the use of carbolic acid in the cure of internal piles.

3. The gonococcus, which may infect the part directly or indirectly from the genitals.

The internal agents are, (1) constipation, (2) lodgment of fecal matter.

Ulcers produced by constipation are generally in the sigmoid. Disease of this part, once the result of constipation, now induces the latter. In other words, constipation, once the *propter hoc*, becomes the *post hoc* in the productive relations.

Small, hard scybala may become permanently lodged in the rectal pouch between the columns and burrow its way through the mucous membrane, terminating in ulcer and fistula.

The origin of tubercular, syphilitic, and malignant ulcers is a specific cachexia, and so familiar that mere mention of them is sufficient. These conditions will be considered more fully in other directions.

The irritable ulcer, which is the most painful of the series, is almost invariably accompanied with a sort of "hang-nail" or an appended skin growth at the anal margin, and indicates the location of the fissure. A knowledge of this fact will enable the surgeon to save the patient considerable pain during an examination. Of course a skin tag is often without a fissure; pain must be associated with the external growth. The margin of the ulcer is indurated and the lesion is usually pyriform in shape, with the base above extending downward between the columns of Morgagni and terminating at the muco-cutaneous surface. The floor of the ulcer presents an inflamed appearance. The pain accompanying and following the act of defecation is excruciating and further excites the spasm of the sphincter, thus predisposing to constipation, the patient deferring the call of nature to avoid misery.

As noted in the anatomy, the cause of pain is the exposure of the terminal nerves and blood-vessels, especially the twigs having their sources in the internal pudic nerve. The slightest epilation exposing a single nerve terminal may cause an alarming coterie of symptoms.

Simple ulcers may be complicated by a similar lesion higher up the bowel through a submucous sinus; or, what is really the case, the former is the result of the latter. This complex pathological condition was recently illustrated most forcibly by a patient that applied to me for examination of some rectal trouble. Digital search revealed a well-formed fissure only; inserting a small fenestrated speculum, an application by electro-cautery was made; this was repeated a third time, but while the fissure improved the patient still complained of constant aching in the sacral region and thighs, with periodical discharge of matter from the rectum. During the fourth *séance* I noticed an oozing of purulent material at upper end of speculum and posteriorly. Connection of the diseased parts occurred to me at once, which was quickly demonstrated by the use of peroxide of hydrogen and the probe. I subsequently operated and found lodged at the upper end of the sinus or closed ulcer a very hard mass of fecal matter, the size of an almond, which being removed and the ulcer laid open and trimmed, an uninterrupted recovery ensued. The upper end had a well-defined margin with its edges ragged and undermined.

In certain cases of painful reflexes at the sphincteric area no trace of an ulcer can be found, and only a slight depression may appear at a point of denuded epithelium. Secondary ulceration may occur in the " hemorrhoidal inch," complicating the varices, as in varicose veins of the legs. Reference to the anatomical distribution of the veins makes this obvious.

Ulcers of gonorrheal origin may appear at the anus and lower segment of rectum, either by direct inoculation or by extension of the disease from the genital organs. This occurs mostly in females, owing to the proximity of the vagina and its discharge.

Tubercular ulcers may occur at any point in the rectum, but usually reside above the sphincters. They are usually oval, smooth edges, but raggedly undermined, the long axis parallel with the bowel; base of ulcer covered with a glairy coat of mucopurulent matter, the product of inflammation. Molecular death of the mucous membrane may extend to the entire surface of the bowel, general proctitis, a condition expressing itself by chronic dysentery, the diarrheal discharges occurring chiefly in the early morning. It is really not a diarrhea, but the product of the pathogenic membrane which collects while the patient is in repose. The sigmoid flexure is most likely diseased, thus intensifying the already emaciating symptoms. The bacillus tuberculosis is in evidence, and is the *sine qua non* in the diagnosis of tubercular ulcer. The tubercle often invades the tissues below the ischio-rectal fossa, and breaks down the rectum and anus with multiple sinuses, with indurations, tumefactions, and discolorations. These occur in phthisical subjects.

Syphilitic ulcer is commonly associated with strong bands of new formation, causing an obstinate stricture of the bowel, generally located just above the internal sphincter. The ulcerated area is above the stricture and often breaks down in the horseshoe fistula. The stricture may become so marked that the excrement is voided in the shape of a ribbon of tape by the sufferer. The stricture fibers run in all directions. In the lumen of the bowel there may be a linear stricture, or the rectal wall may be thickened for several inches. The discoloration is a dark gray, and can not be mistaken. Syphilitic ulcer without stricture may be difficult to differentiate, but clinical history will clear the case; besides, it is often seen with an eczematous eruption about the anus.

Another cause of stricture due to new formation is cancer. Epithelioma of the rectum differs from the tubercular ulcer chiefly in its tendency to bleed. The malignant neoplasm invades the surrounding tissues, obstructs the lumen, and produces symptoms that are insidious in their inception, but certain in their final issue.

It not infrequently happens that the symptoms of rectal ulcer are vague and perplexing. Dysenteric phenomena, very often a marked symptom, are frequently wanting. Physical examination of the rectum may not reveal any inflammatory loss of tissue; the the diseased condition is then most likely in the colon, the sigmoid flexure being a common point for pathogenesis. Indeed, from recent investigations, more success has been achieved in the relief and cure of diarrhea and constipation by treating the sigmoid by direct measures than by the hitherto palliative and expectant plans. Patients suffering years of agony and discomfiture can now be reasonably assured of material benefit, if not a permanent cure.

The physical signs of rectal ulcers are plain, and of the first importance to bring about a successful issue. Therefore the methods of examination become a matter of interest. While it is possible to rely upon the educated finger for purposes of diagnosis, I think the speculum should be used in every case, for "seeing is believing," and the combined faculties of touch and sight will enable the surgeon to calculate more definitely as to diagnosis, prognosis, and treatment. The armamentarium consists of a plain table, tubular speculum, dressing forceps, and good daylight. Some object to the frequent use of the speculum, in that it causes no little pain, but a cautious use of the instrument will obviate much of the alleged annoyance. For instance, having learned the site of the ulcer, pressure can be made toward the sound wall of the gut during the introduction. For illumination, I prefer an electric lamp, which amply reveals the ulcerated area through the speculum, the patient lying on his left side.

The rational signs of rectal ulcers are pain in loins, hips, thighs, and abdomen, morning diarrhea, especially in tubercular ulcer; frequent micturition is not uncommon, and nervousness. It need hardly be mentioned that general emaciation follows in the train. The diarrheal discharges contain muco-purulent material, occasionally blood, which appears in the first portion of the stool.

Treatment. Patients presenting themselves with the ordinary symptoms of diarrhea, pain, etc., frequently baffle the skill of the practitioner with the exhibition of various drugs internally administered. What has once puzzled the enthusiastic therapeutist now stands forth in clear light, through his knowledge of pathology and approved methods. Knowing the structure, pathology, and relative symptoms of rectal ulcers, the physician is now ready to treat intelligently and, if possible, to cure most of such lesions. To arrest molecular death of the part is the first consideration, and cleanliness assumes the head of the list of agents. Rest is another factor of prime importance. To rest and cleanliness are added such measures as will restore the part *cito, cuto, et jucunde.* This applies to disease of the sigmoid as well.

The special object of this essay is to present the claims of the electro-cautery as a remedial agent in certain cases of rectal ulcer, and I can best serve the reader by citing two selected cases :

CASE 1. C. F., male, age twenty-six, occupation, salesman. Had suffered pain in rectum for three years. Was constipated, which grew worse on account of deferring the stool to avoid the great pain connected with it. General health fair, but had a worried appearance. There was always a slight discharge of muco-pus, with blood, from rectum. On inspection an external skin tag was found on left side of anus, which would swell and subside, giving him great annoyance. Opening the anal folds, a well-marked, irritable ulcer was deeply burrowed just above the tag. His sphincters had been divided twice by reputable surgeons, with only temporary relief. Patient was averse to a third anesthetic, if it were possible to avoid it. The plan of treatment by the electro-cautery was begun. The speculum was inserted, the fissure cauterized by a four-per-cent. solution and mopped dry with cotton. The cautery was applied over the entire surface of the ulcer and a furrow was made through its long axis with the cautery knife. Very little pain attended the application, and only slight soreness was felt the following day. Three such applications were made in this case in as many weeks, with complete recovery.

Lunar caustic is highly commended in such cases, but the great pain thus produced is a serious objection; besides I have never succeeded with that agent in fissures.

In addition to the electro-cautery producing healthy granulations rapidly, its trophic influence reaches beyond the point of impact. This patient was not only cured of his fissure, but his constipation was improved on account of the relief to the spasms of the sphincter ani.

The following case will show the efficacy of this subtle agent in the treatment of chronic ulcer of the bowel:

CASE 2. H. K., female, age twenty-two, anemic, poor appetite, nervous, and irritable; had morning diarrhea, passing copious amounts of slime; emaciated, and appeared careworn and in very poor health; had aching in the back and thighs. Examination per rectum with an incandescent light revealed a large ulcer above the sphincters with scalloped edges and an unhealthy base covered with a whitish substance. There was little or no pain to speak of. The ulcer was cleansed with Thiersch's solution, and the rectum irrigated daily with that remedy. It was apparent that we were dealing with a chronic ulcer, possibly tubercular, and treatment was cautiously begun. Arsenauro was given in ten-drop doses after meals, and a tonic laxative pill before meals, consisting of iron, strychnia, and blue mass.

The electro-cautery was applied to the ulcer, which was done twice a week. The number of stools decreased one half after the first treatment, and the patient felt much better. Cauterization was continued twice a week for a month, and once a week the second month, which, combined with the daily rectal douche and constitutional treatment, caused the surface to granulate and the ulcer with its symptoms to disappear.

The cautery knife used has a platinum point adjusted to an insulated handle; knives of different shapes and sizes are made to suit the emergency. The instrument is attached to the electric light current of the office, as is also the head-light for purposes of illumination.

The Physicians' Medical Supply Company, of Boston, placed these instruments in my office at a very reasonable cost.

It need hardly be mentioned that the cautery can be used in any portion of the rectum, even as high as the sigmoid, with the aid of Kelly's speculums.

It is also understood that its use is limited to benign ulcers and fissures, hemorrhoids and certain forms of chronic ulcers; that its use will prove valueless if not hurtful in malignant ulcers, syphilitic or cancerous, the treatment of which we will not discuss here.

It is argued that operation for fistulæ of tubercular origin is justifiable; then is it not equally proper to attempt to heal the tubercular ulcer? We have in the electro-cautery a remedy that gives much promise in these affections.

HORNE OFFICE BUILDING.

SURGICAL INTERFERENCE IN RECTAL DISORDERS.*

BY J. M'FADDEN GASTON, M. D.,

ATLANTA, GA.

The rectum is an ovoidal canal extending from the sigmoid flexure of the colon to the anus. It is narrowest above and becomes more expanded below, presenting a pouch-like shape, capable of great expansion, and serving as a reservoir for the debris, which results from the completion of the process of alimentation. Its walls are thicker than those of the colon and capable of great contraction in expelling the excrement. The serous coat invests its upper part and forms the meso-rectum and extends on the sides sparingly, but more in front. The rectal fascia of fibro-connective tissue, with its blood-vessels, surrounds the lower portion of the rectum.

"There are two sets of muscular fibers encircling the lower part of the rectum, known as the outer and inner sphincters, by which it is automatically closed except when dilated voluntarily for the release of flatus or the discharge of fecal matter. There is also another distribution of circular contractile muscular fibers at the upper limit of the rectum, which constitutes the division between it and the sigmoid flexure of the colon. This may be appropriately designated as the recto-colic sphincter, and forms an effective barrier ordinarily to the descent of the excrement into the rectum. This annular muscle has not received from anatomists or physiologists the consideration which its rôle in the intestinal functions warrants. It is a veritable constriction from muscular contraction, by which the colon is normally closed against the descent of the fecal mass into the rectum." Before the excrementitious matter reaches this constrictor it is deprived of its nutriment, and after passing this annular division, it is deposited in the reservoir below until a convenient opportunity is afforded for its expulsion through the anal outlet.

The multiform departures of the tissues of the rectum from their normal condition induces various pathological modifications and functional derangements. There are not only changes of

* Read before the Southern Surgical and Gynecological Association, at the meeting in Washington City, November 12, 1895.

structure from ordinary inflammatory processes, but also material alterations in the constituent elements of the organism from a deterioration of vitality induced by syphilitic and tuberculous degeneration, as well as a profound depravity of structure from the development of carcinomatous neoplasms.

Simple fissure of the anus, involving the mucous and sub-mucous, tissues of the rectum, may prove a source of reflex disturbance to the nervous system so as to require operative measures for relief. Hemorrhoidal developments, whether confined within or protruding from the anal opening, may be complicated with sanguineous exudation, demanding a surgical procedure. Strictures of the rectum from fibrinous depositions in its walls call for division or excision of the structures involved.

Fistula in ano, resulting from perirectal abscess, may be complete or incomplete, and admits of no radical cure without operation, even when associated with tubercular disease. Papillomatous nodules, located in the walls of the rectum, whether limited to a small area or extending over a considerable portion of the mucous membrane, should be removed, with a view to obviate their degeneration into a malignant form of disease. When carcinomatous induration of the rectal tissues is detected early there is encouragement to undertake an operation, but after the breaking down of the neoplasm with infiltration of surrounding structures no benefit is derived from excision of the parts involved.

It is a mooted point in regard to the practicability of eradicating rectal troubles of syphilitic origin by medication, and with the present light on the subject it seems justifiable to resort to such a surgical measure as the condition indicates, while constitutional treatment is being carried out in the case.

There are instances of supposed development of specific disease in the form of stricture of the rectum, after the lapse of many years subsequent to any syphilitic contamination, and some authors claim their ability to diagnosticate specific stricture even without a previous history of primary syphilis. But the medical profession should be very guarded in coming to such a conclusion upon a general consideration of the nature of the case, and without sufficient evidence of the initial lesion in a patient of respectable position in society no theoretic bias should lead to a diagnosis of syphilis.

There is a rectal relaxation consisting in prolapse of the intestinal wall, which in young subjects is often relieved without resorting to any operative measure. But again this occurs in adults or even in old subjects, when nothing short of active surgical interference can be relied upon for relief, and even with the most vigorous measures prolapse of the rectum proves in some cases intractable.

There is not infrequently an ulceration extending around and within the lower part of the rectum, involving the glandular structure near the anal outlet, which is only amenable to excision. This has been mistaken for the breaking down of hemorrhoids, but is quite distinct in its pathological condition, and should not be confounded with other ulcerative degeneration of the rectal tissues.

While there are some rectal disorders besides those enumerated which call for the surgeon's attention, it is not requisite to enter into their consideration in this general summary of the conditions demanding operative interference.

It will be perceived that the rectum affords material for surgical work of the most important character, and it should not be relegated to those professing to deal with so-called orificial surgery.

The readers of the Annual of the Universal Medical Sciences, for 1895, will note that Dr. C. B. Kelsey limits his contribution to syphilitic and cancerous disease of the rectum. I have noticed, however, that another specialist in this department, Dr. Joseph M. Mathews, is arousing the dormant energies of the profession to the various phases of rectal troubles through his publications, and especially in his QUARTERLY. While the operation of Kraske and others for the extirpation of the rectum does not receive his sanction, yet there are desperate cases in which some form of removal is warranted, rather than leave a patient to die without any operation.

This matter of rectal surgery is not viewed from the standpoint of the specialist, as my attention has not been directed to it except as a part of my work in the rather extended field of a general surgeon. But I am fully impressed with the conviction that many cases find their way into the hands of quacks which ought to be treated by members of the regular medical profession, and preferably by those who have made a special study of rectal diseases and are prepared to treat properly all the surgical disorders of the rectum. In the paper by Dr. Arpad G. Gerster, before the American Surgical Association, upon the surgery of the rectum, in 1893, there is a full résumé of cases treated during four years in the Mount Sinai Hospital, of New York. There were five hundred and fifty-seven patients suffering from rectal ailments admitted within this period; of these two hundred and eighty were classified as hemorrhoids; one hundred and sixty-seven were cases of fistula, including the more acute forms of ischio-rectal trouble; seventeen cases of carcinoma; eleven cases of prolapse; six cases of cicatricial stricture; six cases of chronic ulcers; seven cases of polypus; one case of multiple adenoma; two cases of congenital atresia of the anus, and one of the rectum, with four cases of anal fissure.

It will be observed that hemorrhoids and fistula make up four-fifths of the entire number of cases involving the rectum, and it seems remarkable that only four cases of fissure should have appeared, making an average of one to each year, when this is generally regarded as of frequent occurrence. Dr. Gerster performed extirpation of the rectum in five cases, four times for carcinoma and once for strictures caused by ulcerative proctitis. Two of these died of acute anemia in consequence of the operation; one suffering from carcinoma, the other from ulcerative proctitis. In both the operation of Kraske was performed. Two other cases in which about six inches of the gut were removed by Kraske's operation were successful. " In the case of a woman fifty years old, whose very wide pelvic aperture permitted easy access without extirpation of the sacrum, the coccyx alone being excised, four and a half inches of the rectum were removed according to the old-fashioned perineal method. She made an easy and rapid recovery." By Gerster a preference is given to the radical operation over colotomy, where the condition of the patient permits it, but the patients should be carefully selected, in view of the general powers of resistance in the patient rather than in the extent of the local disease which is encountered. It is urged that only such cases should be selected as have a good circulation and whose heart and blood supply are fairly preserved. Preliminary colotomy is enjoined where much fecal distress and more or less fever exists; while preparatory feeding and general regimen are held to be essential for success.

In the discussion of this paper Dr. Lewis S. Pilcher, of Brooklyn, presents some interesting data from the service of Dr. Fowler and himself in the Methodist Episcopal Hospital, of Brooklyn, N. Y., relative to carcinoma of the rectum. Of ten cases under observation in the last five years, three presented such an extent of local disease and general cachexia that operative interference by extirpation was inadvisable. In a fourth case, in which the disease extended four inches above the anus, operation was refused. In the other six cases operations were done, with two deaths as the immediate result, and without a radical cure in either of the other four cases, though life was prolonged for some time. One of these patients was twenty-nine years old and another only twenty-three years.

Dr. H. H. Mudd, of St. Louis, states that he has removed the upper portion of the rectum and the lower portion of the colon through the abdominal cavity with success.

Dr. L. McLane Tiffany, of Baltimore, says that by proctotomy he has utterly failed to give relief in rectal troubles, and he believes that colotomy gives the best results.

Dr. T. F. Prewitt, of St. Louis, claims that a great many cases of cancer of the rectum are not suitable for excision, and that inguinal colotomy is the proper course in such cases. He thinks it is better to make a complete section of the colon and bring both ends of the bowel out. In this way you avoid the passage of any matter through the lower part and do not need any spur.

In closing the discussion Dr. Gerster remarked that he had not mentioned closure of the colon after extirpation of the rectum, but he had reclosed the opening into the colon when the cause for which the colotomy was done had been removed. His section being transverse, including almost the entire circumference of the bowel, both apertures protrude through the external wound. He believes that the spur is essential in colotomy.

In an editorial review of operative measures for cancerous affection of the rectum, in the *Annals of Surgery*, by Dr. J. P. Warbasse, a full report of those having large experience in the treatment of this class of cases is presented. I would refer those interested in deciding upon the practicability of affording relief in carcinoma of the rectum to this concise and impartial record without entering into minute details of the observations. It may be stated, however, in general terms, that conflicting views are presented as to the stage at which operative measures of relief are admissible, and no definite conclusion can be arrived at in regard to the precise steps which are indicated or as to the mode of proceeding in different cases. An operation adapted only to women is detailed with some confidence in its advantage, which, so far as my observation goes, has not been practiced in this country. It consists in dividing the whole perineal body into the rectum and thus secure an ample field of operation, which is subsequently to be reunited by sutures.

The burning and urgent appeal to the surgeon to-day is for a definite settlement of the issue as to active interference in cases of pronounced cancer of the rectum. Shall we content ourselves with the mere palliative measure of inguinal colotomy and leave the diseased structures untouched, as urged by Dr. Mathews in his paper before the American Medical Association, or shall we endeavor to remove all the tissues involved by extirpation, as recommended by Dr. Gerster in his paper presented to the American Surgical Association. The full statistics of results in the hands of skilled operators ought to be collected and a fair analysis made before a final adjudication of the question can be reached. The materials for such a comparison should be obtained from cancer hospitals in this country and in other countries, as well as from general hospitals receiving and treating this class of patients, and, being grouped together, a fair inference may be drawn as to the feasibility of active interference in any case of carcinoma of the rectum.

THE RELATION OF PHTHISIS TO FISTULA IN ANO, AND THE RATIONAL TREATMENT OF THE LATTER.*

BY S. G. GANT, M. D.,

Professor of Rectal and Anal Surgery in the University Medical College, of Intestinal and Rectal Surgery in the Woman's Medical College.

KANSAS CITY, MO.

The subject selected by the essayist is one that should be of equal interest to the physician and the surgeon. The former will usually see the case first, to determine if an operation is indicated, and, if such proves to be the case, then the surgeon is called in; hence both should have a practical idea of the subject.

Those of my audience who have operated on many cases of fistula in ano will appreciate the importance of this subject when they stop to think of the frequency of phthisis as a complication of fistula, and, on the other hand, the frequency of fistula as a complication of phthisis. There has been considerable discussion both at home and abroad as to the relative frequency of the one to the other. I have given this subject much thought, and have looked up the statistics, and found that from four to six per cent. of all phthisical patients have fistula; on the other hand, a much larger per cent. of people who suffer from fistula have lung trouble, namely, from twelve to fifteen per cent. I have been in the habit of teaching my class at the University Medical College that we have two varieties of tubercular fistula about the anus, viz:

One is caused by the local deposit of the tubercle bacilli in the rectum, which break down, ulcerate, and finally terminate in abscess and fistula.

The second variety consists of a simple fistula in a person very much debilitated as the result of a pre-existing lung trouble. In this variety there are no bacilli in the rectum.

There are many physical signs that enable us to make a diagnosis between a tubercular fistula and the ordinary fistula which we are called upon to treat so frequently.

The only difference between a patient who is suffering from simple fistula and lung trouble, and the ordinary case of fistula

* Read before the Grand River Medical Society, at Breckinridge, Mo., December 5, 1895.

is their emaciated condition and low vitality, which necessarily delays healing after an incision has been made.

On the other hand, when a patient is suffering from a fistula, the result of the local deposit about the rectum of the bacilli of Koch, there will be a marked difference in both the local appearance of the parts and general condition of the patient. These patients are run down in general health, have a sallow complexion, and may or may not be annoyed by a cough. The anus will be patulous, surrounded and almost hidden by an abundance of long, white, silky hairs. The ischio-rectal fossæ are apparently drawn in, owing to the absorption of fat. The external opening of the fistula is large, triangular in shape, and the skin surrounding it is of a bluish tint, the edges of which droop down into the opening. If the finger or a probe is introduced into the opening, it can be made to sweep around in every direction beneath the undermined skin. At the same time it will be observed that the sinus is not a deep one, but of the subcutaneous variety. On passing the finger into the bowel, it will be found that the internal opening is large, angular, and is situated within an inch-and-a-half of the anus. In the majority of cases it will be found at the junction of the external and internal sphincter muscles. Not infrequently the sinus will be so large that the finger can be passed through it and into the bowel. In such cases patients are constantly annoyed by the discharge of feces through the opening.

What a contrast are the physical appearances in this variety of fistula with the ordinary fistula as found in robust persons, with its small opening, tight sphincter, rounded buttocks, increased pain, etc.

Having mentioned a few of the diagnostic features, I will now proceed to give the treatment, which in a large measure should be operative, though now and then we are forced to content ourselves with palliative measures.

The question now arises, should we operate on patients who have lung trouble and a simple fistula? Again, do we accomplish any thing if we operate on fistulæ due to local tuberculosis of the rectum? These questions have perhaps been discussed more than any other in connection with rectal surgery. All the older and a great many of our recent text-books on surgery tell

2

us that we should not operate on persons suffering from a fistula who have lung trouble, or even a predisposition to the same, for, if we do, the sinus will not heal, and, further, in case it should, the lung trouble will be increased as a result of the arrest of the discharge.

Now I wish to state emphatically that such is not the case, for I have cured many patients of fistula who had lung trouble in a marked degree. I have cured others where the fistula was unquestionably due to the local deposit of the bacilli, and many times have these operations not only resulted in the cure of the fistula but a great improvement in the lung trouble, due, as I believe, to the fact that since we have arrested one destructive process nature is more capable of arresting the other. I operate on all cases of fistula complicated with lung trouble with one exception.

If a patient is brought to me suffering from acute lung trouble, greatly emaciated, has constant cough, night-sweats, hemorrhages, etc., and after my examination I am convinced that the lung trouble *alone* will certainly kill him within the next two or three months, I refuse to operate, and simply dilate the opening to allow a free discharge and relieve the pain.

It would certainly be useless to subject such a patient to the dangers of anesthesia and loss of blood without any prospects of doing him any permanent good. On the other hand, I operate on all cases where the disease is of a chronic nature, and the results which I have obtained have been gratifying to my patients as well as to myself.

When it has been decided that the case under advisement is one suited for an operation, there are many precautions we must take before it is made. In the first place we must improve our patient's general health as much as possible by giving cod-liver oil, creosote, tonics, and a nourishing diet, such as milk in large quantities, eggs, pure beef juice, etc. Not infrequently a trip to the seaside, or to some reputable mineral spring, will prove beneficial from the sea breeze, water, and change of scenery, for it is a well-known fact that the mental anxiety of these patients is pitiable. Just before operating the rectum should be washed out and the parts shaved. As regards the anesthetic, chloroform should *always* be selected when lung trouble is present, unless there is some marked contra-indication, for ether irritates the air-passages and causes an increased secretion of mucus.

I have not the slightest doubt that many of the deaths following operations in this variety of fistula are the result of an inflammation of the lungs induced by the ether, and not as the result of the operation.

The operation for tubercular fistula differs slightly from that of the ordinary fistula, in that we should do it more quickly and use every precaution to lose the least amount of blood, for these patients are anemic and can ill afford the loss of even a slight amount. Again, the sphincter muscle must be incised but once, and then at a right angle, for when it is cut two or three times incontinence usually follows, owing to delayed healing and low vitality.

After the sinus has been laid open in the ordinary way it should be curetted and the overhanging edges of the skin trimmed off. The wound should then be irrigated, packed with iodoform gauze, and the patient put to bed surrounded by hot bottles. Stimulants should be given if there is any tendency to shock.

The after-treatment in these cases is very important. It is essential that they should have the sunlight, hence they should not be allowed to remain in bed more than two or three days, but should get up and lie on a lounge near a window. And just so soon as their condition will permit they must be instructed to spend the major portion of their time out in the fresh air. If these precautions are not taken, the lung trouble when present may be increased. Tonics and tissue builders that were beneficial before the operation should be continued afterward as long as they prove beneficial.

There is one more point in this connection that I wish to mention especially, and that is, that it is not necessary for the bowels to move more than once in every three or four days. Many physicians give a purgative daily, which often starts up a diarrhea that is difficult to arrest, and which exhausts these patients very much. It also delays healing, as the result of the irritating discharges passing over the wound.

Ligature Operation. Some of these patients are unable to take.

an anesthetic, others are too enfeebled to submit to the shock following a surgical operation. In such cases I would commend to you what is known as the ligature operation, which is especially valuable in such cases, from the fact that it can be done, and the patient is not confined to his room, but can roam about in the fresh air. The operation consists in passing into the rectum a medium-sized rubber catheter or solid piece of india-rubber which has been threaded upon a probe. The latter is brought out at the anus, followed by the catheter, which is then tied tightly, after which the ends are fastened by a bullet mashed over them to prevent their slipping. In time it cuts its way out as a result of the uniform pressure exerted.

Dr. Allingham, of London, reports several cases successfully treated in this manner. The following are some of its advantages :

1. It does away with the knife.

2. It can be done without an anesthetic.

3. It is comparatively painless.

4. It permits of the patient's spending all his time in the fresh air.

And lastly there is no bleeding.

Understand me, that I do not recommend the ligature method to be used in cases of ordinary fistula, but those cases only who refuse to be operated upon by the knife, where for some reason an anesthetic is contra-indicated, and where it is absolutely essential that the patient must not be confined to his room for reasons already mentioned.

In conclusion, I wish to reiterate that it is not well to operate on cases suffering from acute phthisis who would only live, at best, a few weeks, but that we are justified in operating on all cases of localized tubercular fistulæ uncomplicated by lung trouble, and also all cases of lung trouble of a chronic nature complicated by fistula. By so doing we will offer relief and happiness to a large number of these sufferers who in former years were told that nothing could be done for them, and were allowed to linger on in a melancholy way until death claimed them, because of the careless advice or ingrained ignorance of some physician or surgeon in whose hands they had placed themselves.

A NEW CLAMP FORCEPS FOR SUTURING EXCISED HEMORRHOIDS.

BY SAMUEL T. EARLE, M. D.,

Professor of Physiology and Diseases of the Rectum in the Baltimore Medical College.

BALTIMORE, MD.

[Written for MATHEWS' MEDICAL QUARTERLY.]

In the first volume of MATHEWS' MEDICAL QUARTERLY, page 326, there appeared a recommendation from Dr. Robert Jones, of Liverpool, for " A simple method of treating the wound after excising hemorrhoids," which consisted " in cutting off the hemorrhoid, after being clamped by an ordinary pile clamp, an eighth of an inch above the clamp, and sewing the cut edges together with a continuous catgut suture, after which the clamp is removed and the operation is complete." This is a step in the right direction, but as will occur to most operators the one-eighth-of-an-inch pedicle is scarcely sufficient to depend upon, when one is relying upon the sutures passed through it to control a possible hemorrhage, nor would they be likely to hold sufficiently long for primary union to be complete. The publication of this method brought out an announcement from Dr. A. Ernest Gallant, of New York, in the second volume of the same QUARTERLY, page 518, " that Dr. Outerbridge had abandoned the use of the ligature, clamp, and cautery since 1888, and treated all hemorrhoids by excision and uniting the cut edges by a continuous catgut suture," using the clamp to hold the edges together while being sewed. He relied upon the continuous suture to control all bleeding.

This is certainly a rational method of dealing with the wound of an excised hemorrhoid in the light of our present knowledge and experience, and one which probably many of us occasionally used before we knew of Dr. Outerbridge's practice, but to him is due the systematic adoption of it in all cases of hemorrhoids, and up to the time this report was made, October, 1894, had treated by this method from one hundred and twenty-five to one hundred and fifty cases of all degrees of hemorrhoids. From the time I first saw this statement I also began the systematic use of this method, and have used it with few exceptions in all the cases of hemorrhoids on which I have operated since that time. I soon found, however, that on account of the constant disposition of the adjoining folds of mucous membrane to turn in over the cut surface, I would frequently sew one of them through mistake to the opposite raw surface; the ease with which such a mistake can be made can scarcely be appreciated until it has been tried. I used different kinds of forceps to hold the cut edges together until the sutures could be taken, but not any of them met the conditions. I finally devised a pair myself, which has more than met my expectations, and indeed seemed to be all that could be desired for the purpose. As will be seen from the accompanying cut, the beak is set on to the blade at an angle of

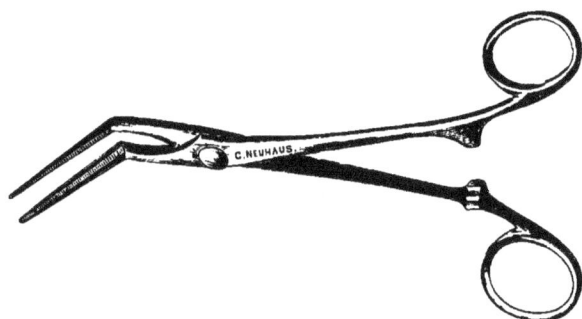

about forty degrees, is one and two thirds inches long, is much smaller at its point than at its junction with the blades; each beak has a serrated edge on the upper part of its inner flat surface to prevent the cut edges of the wound from slipping, at the same time it brings the parts that may be unduly compressed by the serrated edges directly adjoining the cut edges; the blades, six inches in length, are joined together by the most approved lock, so as to be easily taken apart at will, and have three catches near the handle for holding them together when clamped.

Mode of application: The pile is caught with catch-forceps at its most prominent point, pulled out and down, and then the clamp forceps are applied as near the base of the tumor as may be thought proper; after being closed as tightly as possible, the part of the tumor above the beak of the forceps is cut off close, then the suturing is begun at the distal end of the clamp (the end of the suture is caught with a pair of catch-forceps, instead of being tied, in order that the running suture when complete may be drawn from both ends), and is continued over and under the clamp until the whole cut surface is included, but is not drawn tight; the clamp is now loosened, when it can be easily slipped out from between the suture, and the two ends of the suture drawn sufficiently tight to bring the cut edges in nice apposition and control all hemorrhage ; the two ends are now made fast by a knot in each, which should be made close down to the mucous surface.

While this clamp is best adapted to the removal of individual hemorrhoids, either external or internal, yet I have recently used it in a typical case for Whitehead's operation with great satisfaction, where I removed internal hemorrhoids from the entire circumference of the rectum, and external varicosities from two sides, and on the anterior surface. In this case I used the clamp at right angles to the long axis of the rectum, whereas it is generally used parallel to it.

What is claimed for the instrument? It will very much facilitate the time needed for the operation generally ; it adapts the cut surfaces more evenly than can be done without them ; it exposes the least possible amount of raw surfaces to inspection, and it produces very little contusion of the compressed surfaces, which is rendered still less injurious by its occurring just at the edge of the cut surfaces.

GASTRO-INTESTINAL DISEASE.

REPORT OF SEVEN CASES OF ABDOMINAL SURGERY IN WHICH THE MURPHY BUTTON WAS APPLIED. *

BY A. VANDER VEER, M. D.,

Professor of Didactic, Abdominal and Clinical Surgery, Albany Medical College, etc.

ALBANY, NEW YORK.

[Published exclusively in MATHEWS' MEDICAL QUARTERLY.]

Mr. President and gentlemen : I am not unmindful of the fact that the reporting of clinical cases is not always the most interesting material to present at such a meeting as this. On the other hand, the seven cases here presented have a bearing upon the use of the Murphy button that is now receiving much attention not only in this country but abroad, and as a method of intestinal anastomosis is being placed thoroughly on its merits. It is difficult to understand some of the unfavorable reports made by English and German surgeons, when we contrast the very successful results indicated by so many of our American operators in the application in a practical way of this mechanical contrivance. Perhaps there is no part of surgery that, within the past quarter of a century, has presented so much in theory and in which there has been so much disappointment, when practical use has been made of the suggestions, as in the field of abdominal work with all its complications. In other words, how much we have changed from time to time our methods of treatment of many complications, and yet, withal, there have come certain reliable advances that have met all requirements for which they were indicated, leaving permanently in our possession the comforting thought that a grand progress in the sum total has been made ; that we can treat all manner of pathological conditions, traumatisms, malformations, etc., of the intestinal tract and abdominal cavity with less embarrassment than perhaps in any other part of the body, and yet there are very few portions of the human system

---

*Read at the meeting of the Southern Surgical and Gynecological Association, Washington, D. C., November 12, 1895.

upon which we operate where more rapid thought and best judgment is to be employed than in abdominal work. The best
methods for meeting this and that complication must be adopted
at once. There can be no great delay; temporary dressings can
not be applied for the time being; expectant surgery has no field
here. We must meet the emergencies, and I desire to emphasize that much is still wanting in discussions occurring in our
special societies; we need less from a theoretical standpoint and
more of practical experience. Therefore, in presenting the following cases, with such remarks as each one seems to call for, I am
desirous simply of placing on record facts which may assist in
future operations, and aid us in our final determination of certain
procedures when conditions present that require their employment. It is seldom that so small a group of cases cover so wide
a range of pathological conditions.

There can be no doubt but that the consensus of opinion to-
day, among operating surgeons dealing with abdominal cases, is
that when we come to intestinal anastomosis our patient is not
infrequently in a serious condition as regards strength, and all
things being equal, that method which will give the most rapid
and safe manner of procedure is the one that is to claim our
attention. Rapidity of action at such a time is absolutely necessary, and yet with it must be combined thorough safety.

If with carefully reported cases, when some new surgical procedure is on trial, we could have also a report of the same cases
later on as to results, etc., we could then reach a more honest
and clear conclusion than is sometimes accomplished.

Case 1: Gastro-intestinal anastomosis for carcinoma of the
pyloric end of the stomach.

Mrs. J. C., aged sixty-five years; widow; six months' pronounced symptoms of carcinoma of the pyloric end of the stomach. Family history good; pretty continuous vomiting, at times
in large quantities; emaciation very pronounced; much pain and
suffering. Patient very desirous of an operation, although distinctly told that the chances were decidedly against her, as she
had for some time realized. All other organs in a healthy condition.

An anastomosis made by means of the medium-sized Murphy
button, between the upper end of the jejunum and greater curva-

ture of the stomach. Patient was really very comfortable after the operation, but died from exhaustion on the third day. On examination the button was found in excellent position, union quite pronounced, and all surroundings favorable.

These are the desperate cases that present at times to the surgeon for relief, with very little chance for recovery in any operation except they be reached early. It is yet a mooted question whether we ought to operate at this stage or not. I am of the opinion that the weight of argument and results is against doing any operation when the patient is so weak and emaciated.

Case 2 : Carcinoma of the sigmoid flexure, removal and end-to-end anastomosis.

Mr. W. T., aged sixty-two ; native of United States ; machinist by occupation. Family history : mother died of consumption, also two maternal aunts ; father died of some heart trouble, aged seventy-two ; one sister living ; one died in infancy, and one of consumption, aged sixteen.

Personal history : Patient well, save an attack of pleurisy thirty years ago. February, 1893, had attacks of pain and colic after eating, lasting only a few hours. At this time pain occurred at intervals of two and three weeks, but since has increased in frequency and severity until prior to operation when pain was almost constant. Bowels seldom moved voluntarily, injections being necessary, and tube passed to remove gas. Feces passed in small, hardened masses accompanied by pain, occasionally ribbon-shaped and clay-colored. Pain referred to lower portion of bowel in region of sigmoid flexure and up left side of abdomen.

Entered Albany Hospital February 1, 1894, and diagnosis made of stricture of intestine, lower portion descending colon, probably malignant. Operation performed February 3d, consisting of removal of mass in connection with sigmoid flexure, three inches in length, and an anastomosis of the large intestine by means of the Murphy button. Operation revealed cause of constriction to be carcinoma. Mass much adherent to the surrounding parts, which rendered the operation difficult. Wound closed by silk-worm gut sutures, with a tampon of iodoform gauze as drainage in most dependent portions, latter removed on second day and rubber drainage inserted. Patient presented on third day all the marked symptoms of obstruction of the bowels. Up to this time he had been quite comfortable, passing more or less gas per rectum. He now began to vomit, and distension from intestinal gases was so great that I found it necessary to open the abdominal incision, bringing up a portion of the small intestine and attaching it to the edges of the wound. Very free discharge followed, the patient became more comfortable, but finally died from exhaustion on the eleventh day. On examination the ends of the intestine had united and the button was found loose in the lower portion of the rectum, about two inches within the anus. The patient was much exhausted and emaciated previous to the operation.

Cancer of the sigmoid flexure is one of the most difficult conditions in intestinal surgery that we have to deal with. At the time of the operation I was fearful, and said to my assistant that there was great danger of the button impinging too much upon the crest of the sacrum, and I believe that this was really the after-condition that caused the obstruction. I scarcely think I would use the button again when doing the same kind of an operation, and yet I have been equally disappointed in doing an end-to-end anastomosis in similar cases by other methods, and must say that the use of the button saves an immense amount of time and anxiety.

Mathews says that the prognosis in cancer of the sigmoid flexure is very bad, and that there are but two operative procedures : (1) Colotomy, (2) Extirpation. In the former operation he thinks the question of operative procedure should be decided rather by the patient than the physician. In dealing with the subject of excising the sigmoid flexure for cancer he would be inclined to class it as a piece of unjustifiable surgery, except that in several reported cases excision has been practiced successfully, notably the one reported by Dr. W. T. Bull, of New York.

Case 3 : Removal of gall-stones from the gall-bladder, using the long drainage-tube button.

Mrs. H. J. D., aged thirty; excellent history in every respect; had suffered seriously from attacks of gall-stone colic for a period of five years or more. Had had a great variety of treatment from many sources with little benefit. Never jaundiced, and not infrequently received the opinion that her case was probably not one of gall-stones. I had advised an operation in 1892, but patient was reluctant to have it done. Her sufferings became so great, however, that she consented, and operation was done November 13, 1894. Gall-bladder easily reached in the usual manner and several calculi could be felt. The long Murphy button was made use of and the operation was then exceedingly simple. Three gall-stones were removed at once and wound closed without any difficulty whatever. A rise of temperature on the second day to 101°, but on the morning of the third day it became normal and remained so through her entire sickness. The button came away on the eleventh day, wound healed without difficulty, although patient vomited very freely from effects of ether for a period of thirty-six hours. After removal of the button two more gall-stones were removed through the sinus ; wound completely healed at the end of third week, or about twenty-second day.

This patient has been absolutely healthy since, has had no return of her trouble whatever, has gained in flesh, and feels very happy in her recovery.

In all my operations upon the gall-bladder (cholecystotomy and otherwise) in no single instance was the operation done with such absolute ease, and done so quickly as this. When once it is shown that the gall-stones are confined to the gall-bladder, that they are yet outside of the common duct, surely this operation is the easiest of any yet devised. I believe it is much quicker and much easier of performance than the operation of removal of the gall-stones and immediate suture of the gall-bladder, and I have no doubt that the results will prove equally as good.

Case 4 : Removal of eight inches of the small intestine with papillomatous ovarian cyst. End-to-end anastomosis.

Mrs. A. S., aged forty ; widow; history of a rapidly developing abdominal tumor during a period of six months. Patient also gave a history of epileptic seizures extending over a period of ten years, more or less.

Operation January 14, 1895. Multilocular ovarian cyst connected with the left ovary, papillomatous in character, with very many adhesions. Unilocular ovarian cyst on right side removed with very little trouble. On removing the tumor on left side a coil of the small intestine, about eight inches, was so completely imbedded in the growth that it became necessary to do an intestinal resection. More than eight inches of the ileum were removed and the ends brought together by the Murphy button, making the operation exceedingly simple. Patient was very nervous for some time after the operation ; no vomiting; catheter was necessary, and the hypodermic use of morphia, as she had been accustomed to it before. On the third day there was a free movement of gas, and on the eighteenth day a well-formed movement of the bowels. Patient two days after this had two very decided convulsions, and then remained partially delirious for more than a week. Stools were watched carefully for appearance of the button. On the nineteenth day, while in care of her daughter, she had three movements of the bowels, which were thrown away without being examined, and it is very likely that the button passed at that time, as it was never found in stools passed afterward. This patient made a most remarkable recovery and is now in absolute health, having gained more than twenty pounds in flesh. She has had no convulsions since the few immediately after the operation, her abdomen seems soft and in good condition, and she apparently has made a perfect recovery.

When I consider the ease with which this operation was performed compared with some others in which I have done anastomosis by the older methods, I must express myself as feeling exceedingly grateful to Dr. Murphy for his valuable contribution to intestinal anastomosis. There are few more trying positions for the abdominal surgeon than to come in contact with an abdominal tumor that necessitates resection of a portion of the small intestine. The operation is generally long and tedious and patient much exhausted when the point is reached of spending one half or an hour in some other method of bringing together the ends of the intestine. The saving of the strength of the patient is here the great necessity.

Case 5 : Anastomosis of gall-bladder with small intestine.

Mrs. O., aged fifty-four ; good family history, and patient in good health up to the summer of 1894, when, during July and August, she suffered some distress, supposing it to be due to her menopause. Last menstruation August, 1894. In September she had some attacks of gastric disturbance supposed to be simple congestion of the liver. Never had any severe colic ; no attacks like that of passing biliary calculi. The early part of September she developed quite a severe attack of sciatica, left side, from which she gradually recovered, but the second week in October began to develop a condition of jaundice which became very severe. I saw her first April 10, 1895. She had suffered greatly during the winter from itching and nausea, loss of appetite, loss of strength and emaciation. Her pulse was feeble, and she had many ecchymotic spots, the entire body so jaundiced it had almost the appearance of mahogany. Urine was loaded with bile; kidneys otherwise apparently healthy; movements of the bowels were very light in color and had been so for months. She had been recently earnestly advised by her physician not to have any operation, and yet no positive diagnosis given except that of jaundice. I was in doubt as to this being a simple case of catarrhal stenosis of the common duct, or a case of cancer of the liver. Gall-bladder was somewhat distended. I advised an immediate exploration, which was consented to at once by herself and family. Usual incision. Gall-bladder found distended, and twelve ounces of bile carefully withdrawn by means of the aspirator. Thorough search failed to find any nodule, cancer of the liver or stomach, or gall-stones present. An immediate anastomosis was made between the gall-bladder and lower end of the duodenum, or upper portion of jejunum, by small-sized Murphy button. This patient did nicely after the operation. Full movement of the bowels on the third day and afterward by aid of oil enemas. Button passed on twelfth day, after she had had several good movements of the bowels. Soon after this a marked change was apparent in the passages, being yellow and dark in color, her appetite returned, and in every way she continued to improve. At the present time she is in good health, has gained in flesh, is much stronger and able to move about. She is practically well as regards her appetite, digestion, and general condition.

Case 6. Mrs. J. J., aged thirty-four; three children; family history indefinite; patient not strong; general appearance good; menstruated at thirteen; regular but painful. Hernia at umbilicus presented at first confinement; never reduced; pain more or less constant. Two confinements since, natural in every respect. Bowels constipated. October 3, 1895, while straining at stool, suddenly taken ill; tumor increased in size; vomiting and much pain. Was called to see her on Monday, October 7th, at her home some fifty miles distant. Learned that on the previous Saturday she had vomited contents of stomach, and had had a slight movement of the bowels, but none since. Vomited once on Sunday, and once at twelve o'clock on Monday, a yellowish substance, also some portions quite dark in color; had passed some gas from bowels during past forty-eight hours. There was a strong inclination to eructation of gas; able to take very little food; was not thirsty, and drunk but little. Had had three compound cathartic pills without causing a positive movement of bowels; required a moderate amount of morphia to get rest and relief from pain. Her condition of pulse was 88; temperature had been 101°, but now normal; tongue moist and clean; eyes had a sunken appearance, and there was rather an anxious expression of countenance; abdomen soft, except in immediate vicinity of umbilicus. Tumor greatly inflamed and presenting signs not unlike an acute abscess. Amount and condition of urine normal. There was no distension of abdomen nor the appearance of complete obstruction. I ordered the patient removed to Albany Hospital on midnight train on account of her surroundings. She entered the hospital on the morning of the 8th of October, at three o'clock, and operation done at 10 A. M. Tumor, size of cocoanut, in immediate vicinity of umbilicus; portion, size of a silver dollar, implicating umbilicus, in a gangrenous condition. On making an incision there was found a strangulated hernia; many old and firm adhesions. Peritoneum intensely congested, very dark in color. Loop of small intestines included in tumor gangrenous for space of ten inches. Vessels in mesentery secured, and this portion of intestine excised. Murphy button used for end-to-end anastomosis. Two Lembert sutures made use of outside of button. Gangrenous intestine, large portion omentum, some mesentery and peritoneum removed with mass. Wound closed by silk-worm gut sutures; no drainage. After operation patient vomited dark, greenish fluid, and complained of severe pain in the back. Secretion of urine continued normal. Complained of considerable nausea, and continued to vomit at intervals a greenish fluid, from two to six ounces in quantity. At 3:15 P. M. on the 9th she had a large movement from bowels, evidently from that portion below the point of anastomosis. After this vomited a small quantity of clear fluid, still complaining of distress in stomach. At 7:15 P. M. of the 9th had another small movement, and again at 10:20 P. M. No vomiting after this. Was given only hot water and rectal enemata of whisky and hot milk, at intervals of about four hours. At 12:30 A. M. on the 10th had a large fluid movement from bowels, and another at 3 A. M. After this rested very comfortably and in every way feeling nicely. Was allowed some coffee, milk, beef extract, chicken broth, etc., from this time on. At 8 A. M. on the 10th she had another small movement of bowels, and one at 11 A. M., natural in form and color; 2:25 P. M. had a large movement, a small one at 6:15, and then several in succession. Patient now tolerated various kinds of nourishment; her bowels continued to move two, three, and even four times in twenty-four hours, passing gas freely. At 5:30 P. M. the 21st, thirteen days after operation, the button was passed, followed by a large movement of the bowels. On the 24th she had several movements of the bowels, and after this her bowels acted in a normal manner.

This patient made an uninterrupted recovery, and returned home in excellent condition.

This case illustrates the position that the abdominal surgeon is sometimes placed in. One could hardly believe from examination and symptoms that there was present such a gangrenous condition of the small intestines. I really had concluded that the tumor was made up mostly of omentum, and that the gangrenous spot to be observed on the surface would be found to be connected with gangrenous omentum, but, as soon as the sac was opened and the real condition presented, the necessity for prompt, immediate action was upon us.

Most of us will remember that it was but a few years ago when a very able paper was presented at one of our surgical associations in which, regarding the treatment of these cases, it was advocated that the better way was to leave the exposed gangrenous portion of the gut to slough away and a fecal fistula to form, which was opened later if necessary. One can not help feeling grateful for any method that will facilitate and do quickly the work required and save precious time, as was done in this case.

Case 7. Mrs. B., aged sixty-seven; generous, active, cheerful woman, having large interests to look after; much nerve strain at times, and suffering greatly from nervous prostration, being under the care of Dr. Weir Mitchell and other specialists. I was called to see her three years ago by her family physician, Dr. Caverly, when she was suffering from attacks of pain and of colic, in which I made the diagnosis of biliary calculi. A careful, thorough course of treatment was followed out by the use of large doses of olive oil, then smaller doses, also phosphate of soda and other preparations, with some relief, so that for a year she was very much better. However, during the summer and fall of 1894 she suffered several severe attacks of pain, latter increasing in severity during the spring of 1895, and when I saw her again, October 16, 1895, she had suffered two very severe attacks, was slightly jaundiced in one, the only time that she had had this appearance. The pain was so severe that large doses of morphia were required, and at times the administration of chloroform had been necessary to afford relief. Her kidneys were found to be in very good condition, although her general strength was not such as would encourage any one to do a very severe surgical operation; however, the patient was suffering so much that it was decided, if the pain returned, she was to have the benefit of an operation.

The pain continued so severe that on Monday, October 28th, patient having taken ether, I made the usual incision for exploration of the gall-bladder, found it containing about two ounces of bile, and through the walls and down into the cystic duct could be felt a number of small calculi. There were some adhesions, but not serious. Desirous of making the operation as short as possible, the patient's condition being such that we were not warranted in doing a long one if it could be avoided, I made use of the long drainage-tube button to the fundus of the bladder, and closed the wound after a careful examination for any possible cancerous mass, which was not found, then placed patient in bed. She vomited some after the operation, but had no unpleasant complications, aside from the fact that her pulse remained in the neighborhood of 90 to 96, and not very strong. Stomach presented a constant condition of nausea, and was not able to take food readily. Kumyss and other preparations were tried, but she was mostly nourished by rectal enemata. Not very large doses of morphia required to afford relief, in fact it was given up for a couple of nights. Patient's bowels moved thoroughly well on the third day, and each day after that. Tem-

perature, first night morphia was discontinued, reached 101°, after
that it was absolutely normal; wound healed without any disturbance whatever, and button came away on November 6th, eleventh
day, when I visited her for the first time after the operation. I
also removed a small calculus at this time, but refrained from
using the small forceps or scoop until the parts had had a chance
to form more firm adhesions. Secretion of bile had been fairly
free, the dressings at times quite decidedly saturated. Aside from
the nausea and condition of the pulse patient was in every respect
doing well. I believe that this was the best method of treating
this case. The clinical history was in the direction of the gall-
stones being confined entirely to the gall-bladder, and though the
stools had been watched with great care there was no evidence of
her ever having passed any through the common duct. Had I
attempted a more prolonged operation I am sure that the patient's
chances for immediate recovery would have been seriously jeop-
ardized, and that the use of the button in this instance was a
saving of time, leaving the patient in good condition for removal
of the gall-stones later.

On November 7th her physician removed five of the very
irregularly-shaped calculi, which I here present; but at this time
she began to show more marked symptoms of cerebral anemia,
with delirium, which continued, patient finally passing into a
comatose state, and died Friday, November 8th, at 10:30 P. M.,
temperature just before her death going up to 104° and 105.5°.
At no time since the operation had her stomach willingly accepted
nourishment, but during that time she vomited but twice.

Autopsy by Dr. Caverly, twenty-four hours after death.
Stomach in fair condition, possibly somewhat dilated; liver
normal in size but quite fatty. Kidneys also presented a condition of fatty degeneration; but no evidence from gross appearance of nephritis. The doctor states: "The heart was the
most decidedly fatty of any I have ever seen. There was no
color or appearance of muscle. Heart weighed twelve ounces.
Intestines and omentum were loaded with fat; walls of the abdominal cavity were almost entirely fat, the muscular layers very thin.
There did not seem to be any other pathological condition, so far
as the organs of the chest and abdomen were concerned. Brain
not examined. The only remnant of gall-stones found was a

small one about as large as a grape seed.   Gall ducts were patent and unobstructed.   Adhesions left by the button were very nicely observed.   From all appearances we concluded her death was probably due to a fatty heart."

This case is one that puts the surgeon in a most marked position of anxiety: a not very strong patient; a patient evidently suffering from marked cholemia; one who had suffered most severe and agonizing pain, and had not been able to take her usual exercise, carriage drives, etc., in more than six months.   I do not believe that any method of treatment could have been accomplished more rapidly, and yet it was a plain case in which her strength was not equal to the demands made upon it.

At the meeting of the New York State Medical Society, in February, 1894, I gave indorsement to the use of the Murphy button, believing that it was a mechanical contrivance of value, that it would supplant all other devices of its kind for the present, and be a great saving of time to the surgeon in certain operations about the abdomen.   Possibly something still better will be suggested in the near future, yet, for the present, this is certainly worthy our acceptance and use.

Although the cases I have reported are not many, yet they cover the field in which the device may be made use of so readily and easily, and the result so satisfactory, that I have considered them worthy your attention as having a bearing upon statistics. I believe I have given a just criticism of the accumulation of facts so that we can reach and determine definitely as to the value and usefulness of this contrivance.

It will not answer for every lesion about the intestinal tract, but it surely has its sphere of usefulness, being clean, easily handled, and saves the patient from a much longer operation, when time alone is the great desideratum, which can not be secured by some of the other methods.

28 Eagle Street.

THE RELATION OF INTESTINAL FERMENTATION TO DISEASE.

BY A. P. BUCHMAN, A. M., M. D.,

Professor of Diseases of the Digestive System, Fort Wayne College of Medicine,

FORT WAYNE, IND.

In the two preceding papers of this series I have in a limited sense called attention to the processes which are at fault in the digestive apparatus, and in their continued history result in abnormal anatomy and physiology of the organs of digestion. When a pretty thorough pathologic state is once set up, it follows as a logical deduction that the whole organism will come, more or less, under the influence of the products of this faulty mechanism. To more fully illustrate my proposition I will place the following diagram as a schematic view of the normal lines along which food substances are directed. Digestion as a whole is a *unit* of work which really consists of three distinct processes, viz :

1. Digestion proper,
2. Absorption,   } The unit of work = digestion.
3. Assimilation,

The process of digestion proper is not, as is usually taught, a fermentation, but rather a ferment action which consists in the splitting up of food into its various constituents under the action of so-called unorganized ferments, thus preparing them for the further use and nutrition of the body.

The process of absorption is that whereby the food is passed through the second state of its evolutionary journey into the body and passes to other areas. The third number of this trinity need not now be put under discussion, but later on it will be brought to the surface as a concluding part of the argument.

The stomach digestion is comparatively a simple process engaging a simple enzyme and hydrochloric acid, which together change proteid substances. The moment the food reaches the duodenum it is subjected first to the three enzymes from the pancreas, the juices from the duodenal glands, and the liver secretion, thus, at a single plunge the whole process is immensely complicated, and just in proportion to the delicate and complicated mechanism engaged in the operation is there a liability of unbalanced equilibration.

The conditions arising in the stomach and duodenum which result in indigestion, and in the intestines further on, will now be placed in contrast with the normal processes:

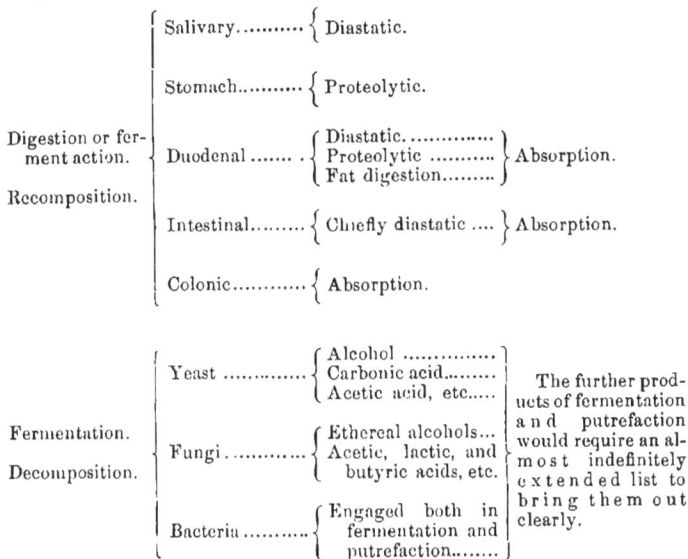

```
                    ┌ Salivary........... { Diastatic.
                    │
                    │ Stomach........... { Proteolytic.
                    │
Digestion or fer-   │                    ┌ Diastatic. ..............  ┐
   ment action.   ⎨  Duodenal ....... . { Proteolytic ...........  ⎬ Absorption.
                    │                    └ Fat digestion.........  ┘
Recomposition.      │
                    │ Intestinal......... { Chiefly diastatic .... } Absorption.
                    │
                    └ Colonic............ { Absorption.
```

```
                    ┌               ┌ Alcohol ...............   ┐
                    │ Yeast ......... { Carbonic acid.........  │ The further prod-
                    │               └ Acetic acid, etc.....   │ ucts of fermentation
Fermentation.       │               ┌ Ethereal alcohols...   │ a n d   putrefaction
                  ⎨  Fungi........... { Acetic, lactic, and   ⎬ would require an al-
Decomposition.      │               └   butyric acids, etc.  │ m o s t   indefinitely
                    │               ┌ Engaged both in        │ e x t e n d e d list to
                    │ Bacteria....... {   fermentation and    │ bring them out
                    └               └   putrefaction........  ┘ clearly.
```

It is intended by the foregoing diagram to present at a glance the essential cause of departure from the health line in the digestive tube. When the digestive enzymes are absolutely mastered and overpowered by the organisms engaged in fermentation, we have to deal with a process which is exactly like that which takes place in a dirty, unclean beer vat. It has been demonstrated that there are yeasts which produce disease in other yeasts; that one kind of fungus will induce degenerative changes in other fungi, but the idea which must be kept fully in mind is, that the process of decomposition is a *going activity* and has for its aim the return of the substances in which it is inducing changes to the inorganic kingdom. The stomach, as well as other areas of digestion, in the struggle against these organisms, soon fails to adjust itself to its surroundings, hence disease. If it were possible now to confine the products of fermentation, the products of the yeasts, the fungi, and the bacteria to the intestinal tube, the quantity of harm to the general organism could be calculated to a mathematical nicety. To run the gamut of all these products would be a work of much time and labor, and perhaps tend to confusion in the minds of many, yet before I enter the domain of absorption I want to accentuate the proposition that the bacteria never initiate a diseased state, that they are simply the handmaids of yeast and fungi, and continue as a gliding continuity the processes inaugurated by yeasts and fungi.

Now, what are the products of fermentation which, when absorbed into the lymph and blood circulation, result in abnormal cell protoplasm, and under what conditions does absorption of such products take place? These two questions are at the foundation of a very wide field of inquiry, and I only regret my inability to go into the subject minutely at this time. However, the object in view will be reached when I call attention to the irritating and paralyzing nature of the carbonic acid gas, acetic, and other acids, and the alcohols which result from the decomposing fermentation which ushers in the first step of the compound unit, which we recognize under the title, indigestion.

Absorption is a process which physicists fail to explain in all its details. It is not enough to say that it falls wholly under the laws of physics, for it does not, and herein lies the missing link which, when once fully recognized, will do much toward clearing up many obscure points in the science of etiology.

The intestinal villi through which *much* of the intestinal contents are taken into the body, do not act simply in a mechanical way, but are under the influence of conscious selective action. This power is paralyzed in part or whole by the products above named, and is therefore now reduced to a mere mechanical action. When this once occurs the imagination does not have to be strained to fully comprehend a source of contamination of the blood and lymph streams.

If I may be permitted to digress for a moment at this point, I will do so only to call attention to the fact that the intestinal canal is the beginning and end of the lymphatic system, just as the heart is the beginning and end of the circulatory system or blood circulation. Each cell being bathed in lymph takes up from the basin its nutriment and discharges its refuse into it. The venous circulation selects out the cell refuse, thus constantly sending on a purified lymph, out of which the nerve structures are resupplied. It is this stream which first becomes contaminated, which when a given point is reached unbalances assimilation. Yeast cells, fungi, and bacteria, together with their products, are now floating freely in the blood and lymph, largely deteriorating both streams. The blood becomes dark and sticky and pasty, with a decided tendency to make capillary circulation slow, thus overfeeding areas that are susceptible on account of previous, or it may be continued, excitation, notably the female reproductive organs. However, the line of least resistance is usually the *mesenteric* system. The lacteals become gorged with yeast cells and fungi, the blood capillaries choked, and very early in intestinal indigestion the mesenteric glands become enlarged, nodulated, and very tender. Consumption of the mesenteric glands is not rare, in fact is a usual sequel of the conditions I have described in the preceding papers of this series. I do not care to leave the inference unguarded, that tuberculosis has its origin in the above named conditions. The guard I wish to throw around the statement is, that the tubercle bacillus finds its soil already prepared for it, finds tissues in a *going* condition, and its work is simply to further the process. This view of the subject applies to the whole "Morgue Brigade;" it is not, nor can it be, limited to any one class or kind of bacteria, and we are therefore put to the necessity of giving some rational sort of analysis of the changed cell protoplasm which fits it for a *food* for so-called pathogenic germs. This I am inclined to call an anhydration process induced by the micro-organisms of fermentation, which now, owing to their environments, no longer act as true ferments. Their action, however, is analogous to that of fermentation, and let me remark, *en passant*, that analogy is the guiding law in nature, the only true Ariadne's thread that can lead through the inextricable paths of her domain. The potential force which changes a peptone to a proteid was supposed to be a characteristic of the blood itself. For a long time I have doubted this theory, and now am convinced that such office is performed wholly by the intestinal epithelium; and this being true, it follows that when the intestinal mucosa is pathologic, the epithelium as a part of its anatomy shall fail in this, its vitally important work, hence, peptones in the blood stream, which I am inclined to look upon as always being pepto-toxines. This view of the office and work of the epithelia of certain areas of the intestinal tube at once negatives the purely mechanical theory of absorption and at the same time places a doubt in front of the leucocyte as a digester. I am convinced that sooner or later the whole mechanical theory of absorption, from the intestinal canal, must give way to the fact that the epithelial cells are conscious entities, and by taking the food products into their interior pass it through a process of final digestion, after which it is cast from their structure either into the blood or lymph streams.

When they are badly fed by material from the intestinal stream it follows as a necessity that what they offer to the blood and lymph streams is imperfect, hence cell digestion, which is now known as assimilation, will be imperfect and result in abnormal cell protoplasm.

As previously remarked, the tubercle bacilli, as well as other so-called pathogenic germs, are utterly incapable of self-support, reproduction, or action, except in the presence of substances having previously been brought under the influence of micro-organisms whose function it is to prepare the way for them.

All the activity of micro-organisms falls under the law of the conservation of energy which, rendered in its simplest definition, means that there is no rest or cessation of motion in nature, and that the whole is a gliding continuity.

This condition now is the same in chemic quality as is produced by a rather prolonged application of high temperature to a like substance.

Now, when we have the blood and lymph streams containing both alcoholic and acid yeasts for a long time, the masses get too large to pass through the pulmonary capillaries, they then are caught up and held until they completely fill the capillary tubes and deprive a given area of blood sufficient to sustain it. Again, they act as nuclei around which tubercles form. When these conditions begin to manifest themselves, we find the whole organism is to a certain extent acetified, so that the breath and perspiration smell distinctly sour. In consumption of both bowels and lungs we find the mucous surface paralyzed and pouring out a tough, ropy mucus, which partially blocks up the follicles, thus preventing a free flow of mucus out of the gland cells, hence, hypernutrition of a low type ensues, which goes on to thickening of the parts. This is characteristic in the early stages of pulmonary phthisis, and as well in the intestinal canal, notably in the colon. A very interesting phase of the colonic condition is the thickened condition of the colon walls, narrowing down the passage often to less than one eighth its normal caliber. This thickening must not be looked upon as the result of inflammation, but is wholly due to a partial paralysis of the parts.

All tissue thickenings brought about in this way have a low state of vitality and are liable to further decomposition changes, wholly under charge of such bacteria as are commissioned with the function of getting such changed tissue into substances which can be discharged from the organism.

Now that the yeast and fungi are floating in the fluids that bathe every tissue in the organism, we may expect fermentation processes to take place as the normal result of their presence, and indeed, disease in general can be reckoned as a fermentation. It seems to me the whole process can be explained fully when we see how rapidly the red blood corpuscles disappear as soon as the blood stream becomes charged with ferments and the products of fermentation. At this point the whole system receives a minimum quantity of oxygen, not enough to sustain life at its normal vigor.

The cells now begin the process of robbing one another of the source of vitality, and the process of ruin is commenced and the whole economy is adapted to the condition in which fermentation becomes the ruling factor. When this condition is established, the micro-organisms, under the title " bacter'a," etc., are a necessary factor to take care and dispose of the products. This is the soil into which the bacteriological tree pushes its roots and from which it derives its substance.

To carry this argument to its ultimate conclusion will necessitate an explanation of the action of typhoid and such other disease-generating entities that produce a specific line of pathologic phenomena. Typhoid fever is legitimately an accident, as is also diphtheria and all other diseases that depend upon a specific germ for their manifestation. The remote factors brought forward furnish the opportunity without which all specific disease-producing germs remain as if they were not. We now have biochemic changes in cell protoplasm rendering it non-resistant and furnishing a medium which intensifies any contagion which may happen to find its way into the organism. Normal living-cell protoplasm attenuates and disinfects contagion.

THE LOCALIZATION OF INTESTINAL OBSTRUC-TION. ILLUSTRATED BY THREE CASES.

BY W. S. MACLAREN, M. D.
LITCHFIELD, CONN.

The three cases which have been selected for this paper have been chosen as illustrating most beautifully what I consider a cardinal point in the localization of intestinal obstruction.

It is, that the *severity* of the symptoms caused by an obstruction will *decrease* as the distance of the lesion from the stomach *increases*.

This is a point which I do not think can be brought out too emphatically.

The symptoms caused directly by an obstruction to the intestine are four in number : pain, generally localized and remittent ; vomiting ; borborygmus ; and constipation or obstipation.

If the pain is very severe or the vomiting is constant, we get the rapid pulse of exhaustion.

If there is strangulation of the intestine the strangulated portion becomes at once a foreign body, and we get a certain amount of peritonitis from its presence. In the early stage of the necrosis, caused by the strangulation, we get some septic absorption ; and later we see the shock and collapse due to perforation. But all of these symptoms are merely concomitant, and are not symptoms of obstruction. I believe this to be a very important distinction. For if we regard these concomitant symptoms as true symptoms of obstruction, we will be misled in our diagnosis in that most important class of cases, the uncomplicated ones. Or we will wait for these symptoms to develop, *i. e.*, we will wait for the development of complications which will largely diminish the chances of cure by operation.

Therefore I say, if we would make a diagnosis in time to have it of any value to our patient, we must not attach any importance to the *severity* of the symptoms.

But having once diagnosed obstruction, then the severity of the symptoms should have the greatest weight in helping us to locate the lesion.

To put it briefly, the diagnosis should hinge altogether on the *character* of the symptoms; the localization largely on their *severity*.

To illustrate what I mean, take one of the most marked and constant symptoms, the vomiting. If the obstruction is high in the small intestine, this will be early, persistent, and severe; if low in the small intestine, it may be early, but will not be persistent, nor, as a rule, severe, for at least twenty-four or forty-eight hours; while, if it is low in the large intestine, there may be no vomiting at all.

My explanation of this fact is very simple. The slight and occasional attacks of vomiting are reflex and due to the irritation from the obstructing cause, and with them nausea is apt to be not very marked. The severe and persistent vomiting does not occur until the intestinal contents have been forced back into the stomach, and then the nausea is great. When the obstruction is high in the small intestine, the proximal portion of the intestine becomes overdistended in a very few minutes, and the first few spasmodic contractions of the intestine are sufficient to force the contents back into the stomach.

If the obstruction is near the ileocecal valve, there are twenty feet of small intestine, which are more or less empty at first and must be filled before the peristaltic action can cause any regurgitation. When the obstruction is in the large intestine, we have a new factor, which exerts a marked influence over our symptoms, the ileocecal valve. If this valve is competent, particularly if the muscular layer of the large intestine is not very active, it will be days instead of hours before the vomiting becomes characteristic. With each contraction of the small intestine enough fluid will be forced through the valve to relieve the tension, and regurgitation does not occur.

If the walls of the large intestine are strong, there will finally come a time when the tension within is so great that the small intestine is powerless to force any thing further through the valve; it distends rapidly and regurgitation into the stomach takes place.

If, on the other hand, the walls of the large intestine are atonic, it distends passively before the powerful contractions of the small intestine, and in such a case I have actually seen rupture of the large intestine occur without any previous vomiting, marked pain, or, in fact, any symptoms other than progressive abdominal distension and complete obstipation of eleven days' standing in a woman whose bowels had previously been perfectly regular.

In regard to pain, this, like the pain of labor, is caused by the effort of a muscular organ to empty itself of its contents against a resistance, and varies with the force of the contractions and amount of resistance to be overcome.

For this reason we do not notice any very great difference in the pain, whether the obstruction is high or low, in the small intestine. The small intestine is strong and active, and hence obstructions here usually cause intense pain. In the large intestine the peristaltic contractions are not so strong as they are in the small intestine, and for this reason pain is not so marked a feature. But if pain and vomiting are not so marked in this region, we have one other symptom which is constant, and that is *obstipation*.

Obstipation, particularly if the bowels have been previously regular, should always make us suspect and look for a possible obstruction.

If, on digital examination, the rectum is found to be *empty*, obstruction should be considered probable. And if high injections at first bring merely small particles, and finally nothing, and if these are followed by high, turpentine or glycerine enemata, and still we get nothing, obstruction is certain. If the obstruction is caused by a fecal impaction, it can generally be exactly located by palpation and percussion.

If by some form of constriction, we can generally map out the colon by these same agencies down to a certain point, at which it becomes lost, and if this point lies near the sigmoid, the portion of the bowel below the obstruction can be distended by high injections, so that going upward we can get a dull percussion note up to the point at which the tympanitic note was lost, and the obstruction will be definitely located.

To illustrate the different pictures we get from obstruction in the three locations, let me briefly give you the histories of three cases :

CASE 1. On February 1, 1895, I was called to see Mrs. H. F., aged twenty-five years. At the time her condition was as follows :

She was suffering intense, constant pain, with frequent marked exacerbations, and referred to a median point midway between the ensiform and the umbilicus. Each exacerbation of the pain was accompanied by loud borborygmus and followed by vomiting. At first the vomited matter was merely the stomach contents ; later it consisted largely of bile. The bowels had been quite regular and moved freely, twice on the day before this attack.

Temperature and pulse normal. Examination of abdomen revealed nothing.

This train of symptoms, together with a history of a probable peritonitis ten years ago, ever since which there have been occasional attacks of pain in this same region, made the diagnosis of intestinal obstruction quite plain.

The fact that the vomiting was so marked from the outset and so soon became intestinal in character, made the high location probable. Large doses of morphine were given hypodermatically to relieve the pain and quiet the exaggerated peristalsis, and glycerine enemata ordered to remove whatever might be in the bowel.

February 2d. There had been no result from the enemata, and the pain, borborygmi, and vomiting still continued.

As the pulse, though still of good quality, was 95 and the patient was beginning to feel the exhaustion due to the prolonged pain and vomiting, an immediate operation was urged as being far less dangerous than any further delay, and with the assistance of Drs. J. L. Buel and J. T. Sedgwick it was performed a few hours later.

The abdomen was rendered aseptic, and as the patient was suffering from a painful ovarian and tubal disease, which it was hoped to relieve at the same operation, a small median incision was made midway between the umbilicus and the pelvis.

The collapsed intestine was soon reached, and upon being

traced up was found adherent high in the abdominal cavity. The incision was prolonged two inches above the umbilicus, and the point of obstruction was reached at once.

The obstructing cause was a dense fibrous band, passing from the mesentery over the intestine and adhering to a neighboring coil of intestine. This band was divided between ligatures and a portion of it removed. The rest of the intestine was searched and six similar bands were found crossing it at various points ; though none of these was tight enough to cause any obstruction, all were removed and their stumps ligated.

In the same region there were a few ordinary peritoneal adhesions. These were broken up with the finger. The right ovary was decidedly enlarged, cystic, and bound down by adhesions. The tube was very much thickened and constricted.

The left appendages were normal. The diseased tube and ovary were removed, after freeing adhesions.

As the stump was dropped back a severe hemorrhage occurred from the slipping of the pedicle ligature. This was secured as speedily as possible.

The abdomen was flushed and left filled with a hot saline solution. The wound was sutured with silk and the usual dressing applied. Time of the operation, two hours. Though the patient lost so much blood that her pulse was 160 at the close of the operation, was between 140 and 150 for two days, and on the third day only got down to 120, she has since made a very good recovery.

CASE 2. On August 18, 1894, I was called to Northfield, Ct., to see Mrs. E. T., in consultation, the diagnosis of acute intestinal obstruction having already been made by the attendant, Dr. Smith, of Thomaston.

I found a patient sixty-six years of age, who gave the following history :

Appendicitis (?) one year ago. Various uterine disorders for several years, otherwise previous history good. Present illness began several days prior to my visit.

At that time she went out in apparently good health to take supper with friends. On returning home was suddenly taken with severe abdominal pain, not definitely located.

She vomited once during the night. Nausea was not marked.

During the next day the bowels moved, the pain continued, and there was occasional vomiting. On the second day the bowels ceased to move, and after that it was not possible to secure a movement by the use of enemata. The physician very properly refrained from the use of any cathartics.

Very little nourishment had been taken during the week. The pain had been constant, and at times quite severe. Vomiting had steadily become more severe, and for the last sixteen hours was occurring every few minutes, was stercoraceous in character, and attended by great nausea. The patient appeared well nourished. The face rather anxious and drawn. Abdomen moderately and evenly distended and tympanitic. Nothing could be determined by palpation.

Temperature normal. Pulse 95 and of good quality.

The diagnosis of intestinal obstruction was concurred in by me, the location was decided to be probably the lower part of the small or upper part of the large intestine.

As Dr. Smith kindly consented to administer the ether, I decided to operate at once without waiting for further assistance.

Operation. After the usual routine of cleansing and disinfecting the abdominal wall, a four-inch incision was made three quarters of an inch to the right of and parallel with the median line. On opening the peritoneal cavity it was found to contain quite a large quantity of bloody serum. The transverse colon was immediately searched for and drawn down into view. It was found to be collapsed. It was traced upward, the caput was passed, and at a point about two feet above the caput the small intestine became distended, and from this point upward it was markedly distended, of a deep purplish color.

At the point of transition from collapse to distension there were no adhesions, nor was there any apparent change in structure. On handling it, however, it was found to contain a hard mass. This mass was so tightly wedged that it could not be moved either up or down.

A one-inch incision was made in the intestine and an enterolith weighing, when fresh, three hundred and four grains, was extruded. The intestinal wound was closed by one row of Lembert sutures. As shock was beginning to be quite manifest, the abdominal cavity was flushed and left filled with hot saline solution. The wound closed in the usual manner.

Owing to the lack of assistance and the great difficulty experienced in returning the distended intestine to the abdominal cavity and keeping it protected, the operation, which should not have taken more than three quarters of an hour, required double that time.

And to this length of time and the exposure of the intestine I think we may attribute the state of shock from which the patient never rallied, and in which she died at the end of twelve hours.

CASE 3. April 13, 1894, I was called to see Mr. C. L., farmer, native born, aged forty-five. Previous history very good, though ever since boyhood he has been liable to attacks of colic following any slight indigestion. They have never been severe, have only lasted a few hours, have always been relieved by pressure, and never have been accompanied by any distention, by vomiting, nor by constipation. Six weeks before my first visit he had an attack with which there was slight vomiting.

A mild cathartic took speedy effect and relieved all the symptoms.

Four weeks later there was a similar attack, which persisted up to the time of my visit.

During this past two weeks pain had been pretty constant, though the patient was able to keep about and do light work. He had taken cathartics quite freely and secured a small movement each day. For the past three days there had been no movement, pain had been more marked, and accompanied by loud borborygmus, and there had been occasional attacks of vomiting. Patient rather poorly nursed. Temperature 98.8°. Pulse 84. Abdomen not distended, percussion tympanitic, except over sigmoid, where there was dullness. Peristalsis not constant, but exaggerated and accompanied by pain and gurgling. Rectum found to be empty on digital examination. As small particles of fecal matter were brought away by high enemata, this line of treatment was persisted in and was soon rewarded by a discharge of flatus.

For ten days this treatment was continued, two enemata being given daily, and mild cathartics were used, as flatus passed freely.

As there was no further improvement, an operation was advised, but owing to the comfortable condition of the patient it was refused.

On May 6th the obstruction became complete. It was not possible to bring away any fecal matter with the enemata, and flatus ceased to be passed. Abdomen slightly distended.

May 7th abdomen much more distended and quite painful.

The colon could be readily palpated to a point just above the left iliac fossa, then it appeared to end by a rounded extremity. On distending the rectum by the injection of water, a fluid tumor was created, which could be readily traced by percussion as high as the rounded end of the tympanitic colon.

The great need of operation and the extreme danger attending delay was again urged, and the patient's consent finally obtained.

With the assistance of Drs. J. L. Buell and J. T. Sedgwick I operated in the usual manner, through a median incision.

As the lesion could be located so exactly, there was no exposure of the intestines. I was able to introduce my fingers and immediately draw the stricture into view. It was found to consist of a dense fibrous construction about an eighth of an inch in width. The bowel above and below the constriction appeared normal externally, but evidently contained a neoplasm.

There were two small hard nodules in the neighboring mesocolon. The lesion was situated in the upper part of the sigmoid, at its junction with the colon. Owing to the short mesocolon, it was difficult to draw the affected portion outside of the abdominal cavity, but by depressing the abdominal wall it was possible to make about two inches of healthy intestine, below the lesion, available for work.

The appearance of the lesion was so suggestive of malignancy that I decided to resect.

This I did, removing about three inches of the colon and sigmoid together, and doing an ordinary end-to-end colon-sigmoidostomy. The nodules in the mesocolon were removed at the same time. The loop of intestine was irrigated and dropped back. The abdomen was not flushed, as it had been fully protected by gauze pads, and there was no shock.

The only point in the after-treatment worthy of note was the introduction of a soft rubber rectal tube, which was inserted its full length. This was kept in constantly for three days, carrying off large quantities of liquid feces and a great deal of gas.

At the end of the third day the rectal tube was removed and there were several large spontaneous discharges of semi-solid fecal matter. The patient made an uninterrupted recovery, and two months later was working in the hay-field.

As the neoplasm proved to be adeno-carcinoma, the probability of ultimate recurrence was of course very great.

After nearly a year of very good health he began to develop symptoms of a new growth in the liver. These symptoms progressed very rapidly, and the patient died just one year after his operation. Through the courtesy of Dr. Buel I was able to be present at the autopsy and secure the specimen, which I show you to-day.

The liver was completely filled with carcinomatous nodules. There was a small recurrent growth in the intestinal scar, not large enough to cause any symptoms. The other organs were perfectly normal.

These three histories are such perfect types of obstruction in the three locations that is is hardly necessary to emphasize the important features.

Obstruction high in the small intestine causes intense pain, each spasm of pain being accompanied by borborygmus, vomiting is severe and persistent from the start, and almost immediately shows intestinal contents. Obstipation may be late in appearing.

When the obstruction is low in the small intestine, the pain and borborygmus are the same. Vomiting at first is apt not to be so constant, and is merely the stomach contents; later it is more constant and is intestinal in character. Obstipation is apt to appear earlier.

Obstruction low in the large intestine gives quite a different picture. Vomiting is apt to be slight, and may be absent altogether. Pain is not quite so severe, and is apt to be decidedly intermittent in character, and may be absent till shortly before the end.

Borborygmus, pronounced or slight, according to the presence or absence of severe pain. Obstipation is apt to be the first symptom, and sometimes the only one.

In obstruction of the small intestine the progress of symptoms is generally so rapid and operation so soon becomes out of the question that there may be some excuse for neglecting it.

But in the case of the large intestine the symptoms progress so slowly and the diagnosis is so readily made that a patient with obstruction in this region never ought to die without being offered relief by operation.

For this reason I believe that too great stress can not be laid upon the urgent necessity for early diagnosis in these cases of low obstruction in the large intestine.

For alarming symptoms are almost certain to be followed very quickly by collapse and death. Hence, if the diagnosis is to lead to operation, it must be made early; if made late, it will merely serve as an interesting preface to the autopsy.

It is a significant fact that as we approach the lower end of the bowel, that is, as the severity of the symptoms decreases, the frequency of operation also decreases.

This is veay clearly shown by such tables as that published by Dr. J. B. Murphy, in the *Medical Record* of May 26, 1894, in which he has collected one hundred and thirty-eight entero-enterostomies.

Of these only thirteen are operations wholly upon the large intestine, and only four involve the sigmoid, a point at which obstructing disease is relatively common.

As an evidence of what becomes of the cases that are not operated upon, and hence not published, let me say, that during my service as a hospital interne three cases of obstruction of the large intestine were admitted, one being of the transverse colon and two of the sigmoid. All of them were in good condition; none of them vomiting and none of them having severe pain.

Alarming symptoms developed in all of them so short a time before the final collapse that all of them died before the condition was relieved.

No other class of obstructions gives us an easier diagnosis, more time to prepare for the operation or greater hope of a successful issue, and yet no class is allowed to die more often unrelieved. And simply, I believe, because we allow ourselves to be misled by the absence of alarming symptoms.

THE ASCENDING COLON IN ONE HUNDRED AND THIRTY AUTOPSIES.

BY BYRON ROBINSON, M. D.,

Professor of Gynecology in Post-Graduate Medical School.

CHICAGO, ILL.

[Written for MATHEWS' MEDICAL QUARTERLY.]

In one hundred and thirty cases there were five cases with an ascending mesocolon. There were two cases of complete mesenterium commune, one of which possessed a mesentery of six inches. In one hundred and thirty subjects there were some ten cases with a slight mesentery at the lower end of the ascending colon. This is, in fact, a partial mesenterium commune existing at the junction of the ascending colon and lower end of the ileum. The average length of the ascending colon in one hundred and thirty subjects is five and three fourths inches. But by taking as a standard of measurement the entering ileum and the ligamentum hepato-colicum, the ascending colon measures slightly over six inches. Heretofore all authors within command have measured the ascending colon from the entering ileum to the so-called flexura coli hepatis, but I have discarded this standard and assumed one which I think not only more anatomical, but more accurate. I have, however, seen no records of any other investigator adopting it. The entering ileum is a common standard assumed by all. But the one I assume, the ligamentum hepato-colicum, I have nowhere seen noted. The ligamentum hepato-colicum is not always present, but, if it is not, the edge of the ligamentum hepato-duodenale is always present and can easily be found. The ligamentum hepato-colicum is a distinct extension of the ligamentum hepato-duodenale at its right border. It extends from the neck or body of the gall-bladder to the flexura coli hepatis. I have quite a number of times observed the ligamentum hepato-colicum extend to the fundus or top of the gall-bladder and inclose it in its blades. By putting the colon on a stretch near the liver it will be easy to observe the ligamentum hepato-colicum. By taking the entering ileum and the ligamentum hepato-colicum as a standard it will make the ascending colon nearly one inch longer, *i. e.*, than by the old standard of measurement. The ligamentum hepato-colicum is the extended right border of the so-called omentum Halleri. Another anatomical peculiarity I mention, which will show that the standard of measurement of the ascending colon is anatomical, *i. e.*, from the entering ileum to the ligamentum coli hepatis. In Figure 2 it will be seen that the

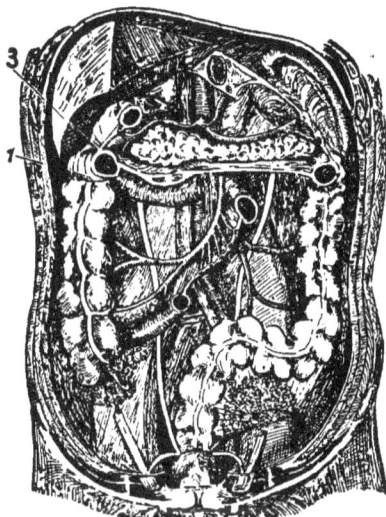

FIG. 1.

Figure 1 (after the author) represents an ascending colon. The bowel is divided exactly at the point where the ligamentum hepato-colicum (3) stretches from it to the neck of the gall-bladder, not here shown. 1 points to the so-called flexura hepato-colicum, which shows that it bends downward at a point one inch from the ligament (3). The ligamentum hepato-colicum (3) is what I have assumed is the upper end of the ascending colon. Figure 1 was sketched, with granulations, over the psoas muscles to show that peritonitic adhesions occurred at these points, produced by the activity of the muscles on the bowel at times when the bowel contained pathogenic microbes.

hepatic flexure of the colon is entirely disturbed, or even reversed. In fact one can not make it out distinctly. In non-descent or partial descent of the cecum the hepatic flexure of the colon can not be taken as a standard of measurement because the normal relations are disturbed. Again, in certain cases the hepatic flexure of the colon is so obtuse or so indefinite that no real point can be fixed as the flexure. But when the ligamentum hepato-colicum or the right edge of the ligamentum hepato-duodenale is taken as a standard, it can always be found with an anatomical certainty.

With the above standard of measurement we may consider that the ascending colon measured on an average of six inches in length in one hundred and thirty adult subjects. Treves gives five and three fourths inches for the length of ascending colon. Otherwise no measurements of this portion of the bowel are within command. So far as my examinations are concerned, which include nearly two hundred and forty subjects (with one hundred and thirty accurate adult records), the ascending colon is the most variable in length and position of any portion of the large bowel. The longest ascending colon was twelve inches and the shortest was three inches. There were quite a number of both males and females which measured only three inches. The twelve-inch ascending colon was from a male. The variation in the length of the descending colon depends on (a) non-descent of the colon, (b) the partial descent, and (c) the excessive descent. In regard to the partial or non-descent of the cecum defining the length of the ascending colon, it may be noted that with the author's standard of measurements the length of the colon will really be about uniform. For it may be frequently observed that when the cecum is fixed under the liver the ascending colon measures just as long as if it descended. But its bend, as in Figure 2, is reversed from the normal. I am quite convinced that non-descent or partial descent of the cecum does not materially lessen the length of the ascending colon. One can frequently observe bends of the colon at the hepatic flexure, as in Figure 2. These same bends can be found in squirrels, gophers, and horses. It is a kind of atavism. Similar bends occasionally occur at the splenic bend of the colon; some may be congenital.

By a mesenterium commune I mean that the ascending colon shares in a mesocolon with the small bowel and the transverse colon. I do not think it is absolutely necessary that a mesenterium commune must include a mesocolon descendens, though it frequently does. The condition which I shall designate as mesenterium commune is where the ascending colon has a complete mesocolon, and that it joins distinctly and is continuous anatomically with both mesocolon transversum and the mesentery. A mesenterium commune in a man is in almost the exact condition of an adult dog, with the exception that man's navel loop is completely rotated in mesenterium commune. In mesenterium commune one can note that the mesentery of the

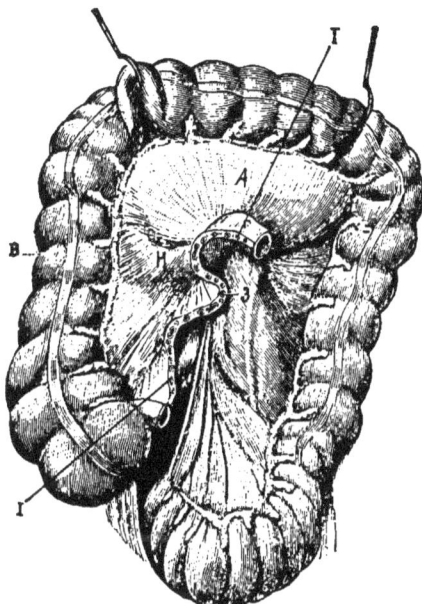

FIG. 2.

Figure 2 is a beautiful cut (I think) from Sappey. B represents the ascending colon. I and 1 point to the mesentery cut-off near to its root. H and 3 indicate the dorsal peritoneum, whence it merges into the right blade of the mesentery and the left blade of the so-called mesocolon ascendens. Note how the left blade of the ascending mesocolon reflects from the bowel nearer to its anterior surface than it does on the descending colon. A is the transverse mesocolon which merges into the mesenterium in the middle and into the mesocolons laterally.

ileum passes without lessening its length into the mesocolon ascendens, and the mesocolon ascendens, which may be four to six inches from root (dorsal wall) to bowel, passes with but little change into the mesocolon transversum. In mesenterium commune the ascending colon is entirely free from the dorsal wall.

The blades of the peritoneum covering the ascending colon come
in contact between the bowel and the dorsal wall. The ascending colon is perfectly free, as is the small bowel and transverse
colon on their mesenterial pedicles. Two cases of mesenterium
commune occurred in one hundred and thirty subjects examined.
The chief feature in regard to a mesenterium commune is that it
affects the mesentery of the ascending colon by elongating it and
endowing it with an anatomical continuation with the mesentery
and mesocolon transversum.

In the case of the mesenterium commune the duodenum had
its full mesentery. The duodenum had no bowel in front of it,
and the mesoduodenum held the head of the pancreas and part
of its body in its blades. The vasa mesenterica superior did not
pass in front of the lower end of the duodenum as in normal
cases. These vessels passed to the left of the mesoduodenum.
The subject had a mesoduodenum faced on both sides by peritoneal endothelium exactly as the mesoduodenum of the lower
quadrupeds, e. g., the dog. It may be remarked that even an
adult human has a mesoduodenum, though it is devoid of the
peritoneal endothelium on both faces. The membrana mesoduodeni propria exist in all humans. It is the real neuro-vascular visceral pedicle which supports the life of the duodenum.
In this subject there was less peritoneal endothelium on the right
face of the mesoduodenum than on the left. The presence of
the liver had displaced the peritoneal endothelium and appropriated it to cover other organs, as the liver. The membrana mesoduodeni propria is covered on the left side to its full extent by the
peritoneal endothelium, which has never been displaced by any
viscus. Hence, in complete mesenterium commune, the mesoduodenum possesses a complete mesoduodenum, i. e., there is a
membrana mesoduodeni propria covered on both sides by a layer
of peritoneal endothelium. Thus a normal mesentery consists
of a membrana mesenterii propria and a peritoneal endothelium
covering for each side. Thus a mesenterium commune alters
very much the relation of the duodenum to the ascending colon.
The cause of a mesenterium commune, which chiefly affects the
ascending colon, is due to defective rotation of the navel loop
The best animal with which I am familiar to learn the process of
axial rotation of the navel loop is the dog. The cecum did not

perform its circle of travels as it usually does in mesenterium commune. The ascending colon hangs on its embryonal mesocolon. About one per cent. of adult bodies show a mesenterium commune. However, in fetuses which I have examined under a month old I found three to four per cent. of mesenterium commune. Now there may be an explanation to this excessive per cent. of mesenterium commune in young fetuses. All of

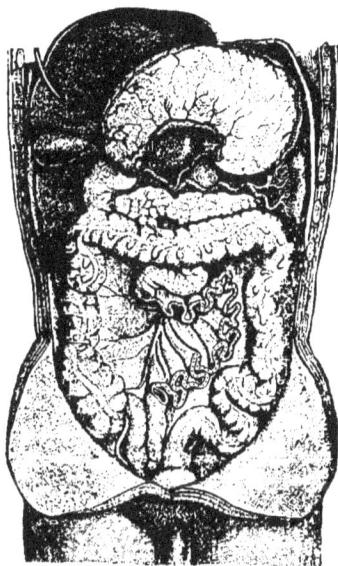

Fig. 3.

Figure 3 (after Sappey) represents an excellent cut of the ascending colon, however without showing the ligamentum hepato-colicum.

this kind of fetuses of course were dead, and it may be that some of the defects, such as mesenterium commune, may have killed the fetuses.

In regard to partial mesenterium commune, we found it in ten cases out of one hundred and thirty subjects. I mean by partial mesenterium commune that the lower end of the ascending colon and the lower end of the ileum are suspended by a free common mesentery. This kind of partial common mesentery does not affect the ascending colon to an extensive degree. The chief effect is on the cecum and appendix, which the common mesentery will allow, from the long pedicle, to lie in almost any part of the abdomen below the navel. The most frequent location of the appendix and cecum when there exists a partial common mesentery is on the lumbar vertebræ among the small intestines, the dangerous ground of peritonitis. This short piece of common mesentery allows the cecum and appendix to occasionally fall into the pelvis and even rest on the pelvic floor. In these cases the fossa ileo-cecalis superior et inferior do not exist as they do when the cecum is in its natural position. The appendix is more apt to lie parallel with the ileum, and also to possess little of the usual spiral condition. In such cases I have not found the bloodless fold or the substituted mesenteriolum of the appendix. It appears that the intra-abdominal pressure, which is muscular tension, forces the cecum and appendix, possessing a partial common mesentery, into the center of the abdomen, which is probably the point of least resistance. I have frequently found the large sigmoid loop with a free, long mesosigmoid also in the middle of the abdomen. Perhaps the vertical colons, had they mesocolons, would be found in the same position, but since the normal vertical colons of man have no mesocolon they are fixed in position.

I would call attention to another point which I have not seen alluded to in writings or heard by lecturers. It is a fact that a part of the ascending colon, situated above the transverse portion of the duodenum and ending at the ligamentum hepatocolicum, occasionally has a short mesentery. It is really a continuation of the mesocolon transversum, which occasionally passes to the transverse portion of the duodenum without being robbed of its original mesentery. I have observed this perhaps six times in one hundred and thirty subjects. The mesentery of the ascending colon, situated between the ligamentum hepatocolicum and the transverse portion of the duodenum, is nearly always short, much shorter than the mesentery found at its lower end.

The middle part of the ascending mesocolon scarcely ever exists normally as such. Excluding the two cases of mesenterium commune in one hundred and thirty subjects, there were four or five cases which possessed a mesocolon ascendens in its middle part.   The reason of the non-existence of a mesocolon ascendens in the middle portion is on account of the kidney appropriating it to cover its anterior and lateral surfaces.   I am now thoroughly convinced that the growing kidney steals away the mesocolon, for I followed the developmental process through quite a number of young embryos, fetuses, and infants.   At the end of the third fetal month the kidney has appropriated nearly all the mesocolon to cover itself.

There is frequently found connected with the ascending colon a fold of peritoneum extending from its right side to the abdominal wall or lower circumference of the diaphragm.   I have

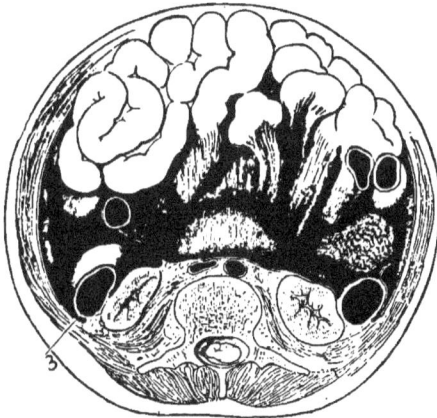

FIG. 4.

Figure 4 (after Henle, 1873,) represents a horizontal section of the abdomen (the upper) between the third and fourth lumbar vertebræ.   3 is the colon dextrum without a mesocolon ascendens.

suggested the name ligamentum phrenico-colicum dextra to this band of peritoneum.   A similar fold exists on the right side, which extends from the upper end of the descending colon, known as the ligamentum phrenico-colicum sinistra.   Treves suggests that we call this fold of peritoneum, sustentaculum heptatis, but I can not see that it had its origin in a design to support the liver, though in adults the right lobe of the liver rests on this fold.   In fact this band, which reaches from the lower end of the upper third of the ascending colon to the level of the

iliac crest, involved my attention for a year before I found what
its origin was. The origin of the ligamentum phrenico-colicum
dextra is the lower pole of the right kidney. I examined a
goodly number of fetuses, and one can note in nearly all of
these that a fold of peritoneum is thrown up by the lower pole
of the right kidney, and that this pole persists in adult life in at
least seventy-five per cent. of subjects. The free border of the
fold is concave from above. Its depth from before backward is
from one to one and one half inches. Vertically it may extend
from one half to three fourths of an inch. This fold serves a
useful purpose in retaining the ascending colon from dropping
into the pelvis. In prolapse of the viscera it clearly demon-
strates that it is useful in holding the ascending colon high up
in the abdominal cavity. In regard to its supporting the liver,
I do not think it plays a very extensive rôle in holding up that
viscus. However, it is easy to note in subjects with enlarged
lobus dextra that the lower border of the liver continually
strikes the band at every downward movement of the diaphragm, and in so doing induces considerable traces of peritonitic scars.

It is stated by one anatomist that the great omentum ceases
where the colon crosses the duodenum. Now the duodenum is
an open spiral ring which fits itself to the uneven right side of
the vertebral column. From its shape and location the colon
ascendens may cross it at varying points. Suffice it to say that
in many fetuses and some infants I found the right border of the
omentum extending frequently to the entering ileum. This
right border of the omentum, omentum Halleri, the part which
descends on the ascending colon, generally ceases at the point
where the ascending colon passes over the transverse portion of
the duodenum, or a little lower. In perhaps forty per cent. of
subjects one can observe Haller's omentum passing below the
transverse portion of the duodenum, and that one can find over
thirty-five per cent. of old peritonitis at this same point. However, the right border of the great omentum, or omentum Halleri, is best demonstrated in the fetus under the fourth month.
I have not infrequently observed the lower end of Haller's
omentum on a level with the cecum or entering ileum. In one
subject, female, twenty-five, the Hallerian omentum descended

below the entering ileum. The lower end of the Hallerian
omentum is very apt to shrink or atrophy on the descending
colon, perhaps on account of the rapid growth of the colon and
cecum dragging on the serous fold, tending to obliterate its
blood-vessels. It may be observed in the fetus that the great
omentum begins to assume relations with the colon at the right
end, and ends by merging into the ligamentum phrenico-colicum

FIG. 5.

Figure 5 (after Henle, 1873,) represents the ascending colon drawn out of its
place a little. 1 points to the right edge of the ligamentum hepato-duodenale,
the standard which I have assumed as the upper end of the ascending colon.
2 points to the peritoneal covering of the right kidney.

sinistra. In the early fetus (ten weeks) it is very plain to occasionally note (1) the lower end of the omentum colicum Halleri
extending down to the cecum. But in adults, from atrophy of
the omental folds, due to dragging of the rapidly expanding
colon and cecum, Haller's omentum is not prominent. I can
scarcely find any mention of it, except in name, in some fifty

anatomies at command. Haller's omentum is simply the right border of the great omentum carried down on the descending colon.

There is also a fold of peritoneum which extends from the kidney to the ascending colon, for which I shall suggest the name ligamentum colico-renale. It passes from the face of the right kidney to the upper end of the ascending colon, and merges into the ligamentum hepato-colicum. The ligamentum colico-renale is a continuation of the ligamentum hepato-renale, which makes the posterior boundary of the inlet of the hiatus Winslowii.

In general the ascending colon extends from the cecum, entering the ileum or right iliac fossa, to the under surface of the liver. The flexura coli hepatis may produce a shallow depression on the under surface of the right lobe of the liver. The flexure is frequently found stained with bile in autopsy. In the vast majority of subjects the ascending colon is covered with peritoneum in its two lateral and anterior surfaces, leaving perhaps an inch of its posterior surface uncovered with serous membrane. The posterior part of the ascending colon not covered with peritoneum lies on the quadratus lumborum muscle, right kidney, and a part of the duodenum, being fixed in position by snowy-white cellular tissue (mesoblastic). The colon on its internal lateral surface rests against the psoas muscle and duodenum.

The ascending colon is concave anteriorly, and its peritoneal attachments are frequently shortened at the point where it crosses the transverse part of the duodenum, due to peritonitis. In a large number of cases it appears to me that the anterior concavity of the bowel is due to peritoneal contractions dorsalward at the point where it crosses the transverse part of the duodenum. It is in contact on the right with the abdominal parietes, and it is separated in front from the abdominal wall by coils of small intestine. The two peritoneal blades which fix the ascending colon to the dorsal wall are short and rigid, and do not come together between the bowel and dorsal wall, even when the bowel is empty and contracted. During bowel distension the two peritoneal blades which fix it to the abdominal wall stand apart, according to the bowel distension.

Hollstein (1872, Berlin, fifth edition,) denies a mesocolon ascendens.

Langer (1885, Vienna, third edition,) says the ascending colon is not covered by peritoneum on its posterior surface.

Simon Rood Pitlord, who wrote the excellent and suggestive article in Told's Encyclopedia on the peritoneum, in 1847, says there is seldom a mesocolon ascendens.

Meyer (1873) says the ascending colon is fixed on the abdominal wall by cellular tissue, hence it could have no mesocolon.

The industrious Cioquet, writing in 1828, says that the ascending colon is connected to the quadratus lumborum muscle and kidney by loose cellular tissue, but sometimes the two blades of peritoneum approach each other sufficiently to form a more or less loose fold, known as the right lumbar mesocolon.

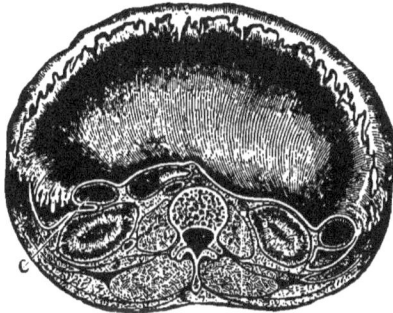

Fig. 6.

Figure 6 (after Abey, 1871,) represents a cross-section of the abdominal cavity at the root of the transverse colon. C points to the cross-section of the ascending colon which shows absence of ascending mesocolon.

Luschka (1861) claims that there is no mesocolon ascendens in the adult. However, he still asserts that in newborn the mesocolon ascendens exists. My own experience is not in accord with Luschka's in regard to infants.

The ever-memorable Huschke (1842) says the ascending colon has "no true mesentery," if we hold the word mesocolon in its proper etymology. He notes that "only occasionally is the posterior surface covered with peritoneum."

I mean by saying that the ascending colon has no mesocolon, because the blades of the peritoneum are so short and stiff and stand so far apart that they do not come in contact between the colon and dorsal wall. The outer mesocolic blade (lamina externa) is a continuation of the peritoneum of the abdominal wall, which, after covering a part of the quadratus lumborum muscle and anterior surface of the kidney, suddenly reflects itself on the ascending colon, covering intimately its two lateral and anterior surfaces. The inner blade (lamina interna) is a continuation of the mesentery of the small bowel, which lines a portion of the right dorsal wall and reflects itself on the colon ascendens to merge into the right blade on the surface of the bowel. It is rare for the two ascending mesocolic blades to approach sufficiently near to each other in the middle of the ascending colon to come in contact and thus form a true mesentery, but it is not rare to find partial mesocolon ascendens at the lower or upper end of the ascending colon.

Cruveilhier (1844) contradicts himself by saying "the peritoneum in the lumbar region does not form a duplicature on the ascending colon." In his anatomy, a little further on, forgetting his previous statement, he says the peritoneum not infrequently "forms a duplicature" behind the colon. No doubt the above statements from this celebrated French anatomist must be considered in the light that as a rule there is no mesocolon ascendens.

Horner (1840, 5th edition,) discredits any mesocolon ascendens.

McClellan (1892) announces there is no right mesocolon.

Allen (1883) says five sixths of the ascending colon is covered by peritoneum, but that "an imperfect mesocolon ascendens" exists.

Hoffman (1863) says there is "an incomplete mesocolon ascendens," hence none exists.

Bischoff, the celebrated embryologist, knowing animal development, said that the fixation of the colon to the dorsal wall removed it from the consideration of having a mesentery.

I would suggest that subjects, so far as a mesocolon ascendens is concerned, should be definitely recorded as to age. So far as my observations are concerned, in two hundred and forty autopsies, visceral prolapse begins at thirty-five years of age, and visceral prolapse is a very variable factor. I know of no rule by which its etiology may be discovered or its degree diagnosed. However, generally when one abdominal organ is prolapsed the remainder are. Lesshaft, the noted Russian anatomist, declares that visceral prolapse rapidly increases with every decade after forty. A natural mesocolon ascendens must not be confounded with a mesocolon ascendens acquired by visceral prolapse, which increases with age after thirty-five. A general statement may be made that females show more visceral prolapse than males, owing doubtless to more extensive and sudden vascular changes during menstrual and reproductive life.

The double-bladed elongated peritoneal supports at the upper and lower end of the ascending colon, or better, colon dextrum, may be termed mesenterium or ligamentum colicum dextrum supremum et inferior.

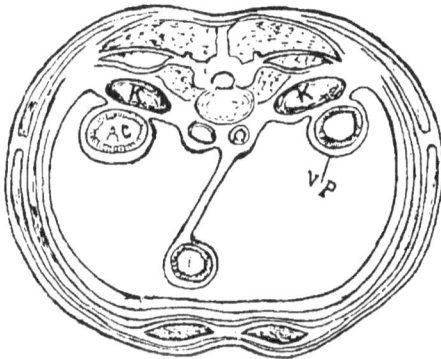

Fig. 7.

Figure 7 (after author) represents a cross-section of the body to illustrate that there is no mesocolon ascendens. V. P., visceral peritoneum; K. K., kidneys; V. C., vena cava; A. C., ascending colon.

The ascending colon crosses the left face of the membrana mesoduodeni propria. It may be remembered that every mesentery consists of a middle portion, called the membrana mesenterii propria, which is the real neuro-vascular visceral pedicle that sustains the life of the bowel. Besides the mesentery is faced on each side by a thin layer of smooth, shining, moist endothelium. In man's mesoduodenum the endothelial layer of each side has been displaced by readjustment, which then only leaves the real mesentery which supports the vessels, nerves, and glands, i. e, the mesoduodeni propria. Now, when the cecum crossed the left side of the mesoduodeni propria it raised the layer of endothe-

lium up to appropriate it for its own covering and that of the
ascending colon. Thus the mesocolon of the upper end of the
ascending colon has at its root (radix mesenterii) origin from the
left face of the membrana mesoduodeni propria. If one exam-
ines the duodenum in a score of bodies it will become apparent
that it is an open, spiral ring, fitting itself closely to the uneven
right dorsal wall. Its two ends, the pylorus and lower end of
the duodenum frequently meet and form a distinct, open spiral
ring, and not a horseshoe, as often compared in resemblance.
The traveling cecum acquires the left face of the mesoduodenum by passing through the open point of the spiral duodenal
ring, whence the cecum travels almost directly to the descending
portion of the duodenum. By this passing of the cecum over

FIG. 8.

Figure 8 (after author) represents a partially descended cecum with the
ileum passing behind the colon and entering its anterior surface. It is a species
of volvulus, a rotation of the bowel on its mesenterial axis. Thirty per cent.
of volvulus occurs at the junction of the ileum and large bowel.

the face of the mesoduodenum from the open point of the spiral
ring to the duodeni descendens the mesoduodenum is divided
into two segments, an upper right and a lower left. The upper
end of the ascending mesocolon is the septum which divides the
mesoduodenum into two segments. The upper right belongs to
the gall-bladder region, and is closely connected with the pylorus and ascending part of the duodenum, the lower left is intimately connected with the transverse portion of the duodenum.
Now this open, spiral, duodenal ring rests on two muscles, viz.,
the lower part rests on the psoas muscle, and the upper right
part rests on the right crest of the diaphragm. Where the
ascending colon crosses the transverse portion of the duodenum
the psoas muscle lies immediately under, and exactly at this

point is found considerable traces of old peritonitis bands, adhesions, and scars, which adhesions, I think, are brought about by the action of the psoas muscle on the duodenum at times when the bowel contains excessive or virulent pathogenic microbes. The most perfect type of animal I know of is the squirrel to show that as the cecum travels over the left face of the duodenum the bowel raises the thin epithelial layer to appropriate the membrane for the covering of the cecum and the ascending colon. The squirrel has two loops of ascending colon, which, with their mesocolons, divide the left face of the duodenum into three distinct segments. (The squirrel possesses a free

Fig. 9.

Figure 9 (after author) represents a non-descended cecum. The cecum and appendix lie immediately under the liver. The ascending colon forms a loop which touches the right brim of the pelvis. Observe the ileum passes behind both branches of the loop to enter the bowel on its anterior surface. It represents a kind of volvulus.

mesoduodenum, i. e., the membrana mesoduodeni propria is faced on both sides by endothelium.) The animals which demonstrate very nicely that as the cecum travels over the left face of the mesoduodenum it appropriates the left endothelial layer to cover itself and the descending colon are the rodents. To demonstrate the segments of the mesoduodenum made by the upper end of the mesocolon ascendens, and also the open, spiral-shaped ring of the duodenum, the small bowel should be tied at the duodenal-jejunal flexure and air or fluid should be forced into the stomach. Now with the hand force the air and fluid from the stomach through the pylorus, whence the shape of the duodenum and the segments of the mesoduodenum with the dividing septum present in view.

Again, I have taken careful pains to examine the posterior surface of the upper end of the colon ascendens. In some cases, young individuals, the posterior surface of the colon does not rest on the kidney at all, but the entire posterior surface rests on the descending portion of the duodenum. In fact, in some subjects I found that the whole posterior surface of the ascending colon not covered by peritoneum, and, lying above the transverse portion of the duodenum, lay entirely on the descending portion of the duodenum, distinctly out of range of the kidney. The ascending colon has running along its wall three fibromuscular bands equally distant from each other, which are about one fourth shorter than the bowel itself. These bands (tenia coli) are the longitudinal muscles of the colon aggregated into distinct bundles, ligamentum intestinale, due, it is supposed, to meconium collections in intra-uterine life. The bands being shorter than the bowel, it throws the bowel wall into sacculations, bulgings, cellulæ, haustra, or diverticula. The three bands, one in front, one internally, and one posteriorly, terminate at a single point at the root of the appendix. Running along the colon, i. e., along the ligamentum intestinale, are two rows of little peritoneal pouches, generally filled with fat. They are known as appendicæ epiploicæ. These peritoneal pouches have been compared to the omentum, but their origin is entirely different. Besides the name appendicæ epiploicæ are the appellations of appendices fallopiæ or appendices omentula. These peritoneal pouches are elongations of the blades of the serous membrane which envelop the colon. The ordinary ones are supplied with one artery and depleted by one vein. They stand in two rows, and the inner is much the larger. They lie on the inner side of the ligamentum intestinale. They are found on the transverse colon only in one series, but on the vertical colons and S-romanum, two series. The rows of appendices epiploicæ of the ascending colon lie quite wide apart There is little doubt that the origin of the appendices epiploicæ is in some way connected with the branches of the colonic arteries.

Mathews' Medical Quarterly

"ALIS VOLAT PROPRIIS."

Vol. III. LOUISVILLE, JANUARY, 1896. No. 1.

JOSEPH M. MATHEWS, M. D., - - - • - - Editor and Proprietor.
HENRY E. TULEY, M. D., - - - - - Associate Editor and Manager.

A Journal devoted to Diseases of the Rectum, Gastro-Intestinal Disease, and Rectal and Gastro-Intestinal Surgery.

Articles and letters for publication, books and articles for review, communications to the editors, and advertisements and subscriptions should be addressed to
Editors Mathews' Medical Quarterly, Box 434, Louisville, Ky.

INDEX MEDICUS.

It is with great satisfaction that we announce the successful campaign in behalf of the Index Medicus.

It would have been a lasting disgrace to have let so excellent a work be discontinued, and the profession as well as the indefatigable editors, Drs. John S. Billings and Robert Fletcher, are to be congratulated at the outcome of the effort to continue its publication.

It will be continued from May 1, 1895, the bibliography from May to December being printed in one number, in two parts, as quickly as the large amount of material will permit.

The first current number will appear in January, 1896, and will continue monthly thereafter. Some two hundred and fifty subscribers are on the list, a number of foreign subscriptions being expected at a later date.

It is rather to be regretted that the announcement has been made that, the requisite number of subscriptions having been received, no more orders will be accepted. The usefulness of so valuable a publication is much curtailed by a limited subscription list.

THE MEDICAL AND SURGICAL HISTORY OF THE WAR OF THE REBELLION.

The *New York Medical Journal* in a recent editorial has started a crusade, prompted by a resolution received from the Vigo County (Ind.) Medical Society, which we hope will enlist the support and co-operation of all medical journals throughout the country. As stated by the *Journal*, the Vigo County Medical Society committee drafted the following resolutions:

" Whereas, We believe a demand exists for a revised edition of The Medical and Surgical History of the War of the Rebellion, and especially so among the younger members of the profession; and

" Whereas, We believe the science and art of medicine and surgery would be materially benefited by preserving in suitable form the knowledge thus acquired from the practical field of experience;

" Therefore, We desire to submit the resolution to this society, with the recommendation that the president appoint a committee, to consist of as many members as he deems advisable, to properly bring the matter before the next Congress."

These resolutions were presented to the Indiana State Medical Society and adopted, and an influential committee appointed whose duty it will be to bring the matter to the attention of medical societies and the profession in general throughout the country; and it is to be hoped that by a concerted action on the part of all who are notified or hear of the movement Congress will be persuaded to make the necessary appropriations for the prosecution of the work.

It is impossible to obtain a complete set of this history, and next to impossible to obtain a single volume. Let the good work go on.

THE PUBLIC HEALTH.

Perhaps at no time in the history of medicine has so much attention been given by the medical profession and medical press to the public health as is being given at the present. By this earnest effort a spirit of interest, at least, has been instilled into the secular press and the people are being educated in the cardinal principles which tend to protect the public health, thereby increasing the longevity of the race. This should be among the first duties of the individual and the State, of corporations as well as the private person, but all are familiar with the gross neglect by all concerned. The officers of the law, such as legislators, mayors, councilmen, etc., cater to the prejudice of an illiterate constituency, and many provisions looking to the maintenance of the public health are held in abeyance or not enforced at all. The citizen is too apt to feel that any improvement of such laws is a menace to his rights and a trespass upon his liberty. Even granting that this is true, it is nevertheless necessary to protect those who can not help themselves, against the invasion of epidemic and contagious diseases; just as well say that a man has the right to allow his rabid dog to roam the streets because he happens to own the animal, as to allow him the privilege of free intercourse with the outer world when his family is afflicted with measles, scarlatina, diphtheria, and such like contagious diseases. There is a great hue and cry raised in some communities, if perchance an order is given to close a surface well which is a breeder of disease and a menace to the health of every person that drinks therefrom. Because the ignorant public is not versed in knowledge to know the truth is no reason why the public should not be protected by law from such dangers. Private interests often interfere with the enactment or enforcement of health laws. A corporation interested in railroad traffic would not hesitate to spend much money to protect them against the trouble and expense of having their railroad coaches fumigated, disinfected, and cleansed in order to prevent the contagion of tuberculosis for instance. It is none the less the duty of legislators to enact, and the civil authorities to enforce, laws which look to the prevention of all contagious diseases, and thus protect the traveling public. Not only should these details be looked after by the minor officers of the law, but it is essential that the nation's health should be protected against the invasion by epidemic diseases. This can never be effectually done until a department of public health is established at Washington, and a health officer placed in the Cabinet. Be it said to the credit of the various health boards of the United States, that their long and tedious fight to educate the masses to a proper view of the subject of the public health is now bearing good fruit, opposition is gradually fading away, and the day of enlightenment is fast drawing near. If the people could be made to understand that the mortality rate has been reduced in the last decade ten to twenty per cent., and that the increase of longevity is due to the earnest effort of the health authorities, much will have been accomplished.

To that able and self-sacrificing body of men composing the American Public Health Association much of this good work is due. Going long distances and expending much time at a personal loss, these eminent physicians and scientists have done much to protect the public health.

Let us all work together, the profession, medical colleges, the medical press, etc., going in hand with the health associations, boards of health, and all those interested in public hygiene.

With a long and determined pull we can put to shame those who stand in the path of advanced thought and action for the sake of personal aggrandizement.

Correspondence.

NEW YORK LETTER.

[FROM OUR SPECIAL CORRESPONDENT.]

WESLEY M. CARPENTER LECTURE; THE HEALTH OF NEW
YORK; AGGREGATION OF POPULATION; LIFE PROBABILITY
LENGTHENED; LONDON LOWEST MORTALITY DURING EARLY,
MIDDLE, AND LATER LIFE; LOOKING FORWARD; INSTRUCT-
ING THE PUBLIC; DIPHTHERIA; BOVINE VS. HUMAN TUBER-
CULOSIS; ERADICATION OF BOVINE TUBERCULOSIS; INFECTED
UDDER; WORK OF FOREIGN VETERINARIANS IN TUBERCU-
LOSIS; THE CONTROL OF TUBERCULOSIS IN MASSACHUSETTS
TUBERCULIN AS A DIAGNOSTIC MEASURE; THE BEST METHOD
FOR THE ERADICATION OF TUBERCULOSIS; THE TRANSMISSION
OF TUBERCLE BACILLI IN MILK; INFECTION THROUGH THE
TONSILS IN MAN; MOUTH-WASHING AS A PREVENTIVE; NON-
ABSORBABLE SUTURES IN OPERATIONS FOR INGUINAL HERNIA.

NEW YORK ACADEMY OF MEDICINE.
(Stated meeting, Nov, 7, 1895.)

The honor of delivering the Wesley M. Carpenter Lecture
was accorded Dr. H. M. Biggs, and for his discourse he made a
careful study of "The Health of New York."

Reference to a table of this city's population, by wards,
revealed the unusual fact that aggregation has little to do with
the death-rate. In the so-called "Eastside," with the highest
population to the square mile of any great city, the mortality
was found to be the lowest in the city. This was accounted for
from the fact that the majority of the people are Russian or
Polish Jews, inured to hardship and a changeable climate, while
among the Italians the death-rate is much higher.

The marked improvement in the last forty years, lengthening
the mean life probability from twenty-eight to thirty-seven years,
shows a steady reduction in the rate per thousand, viz., 1859 to
1863, 33.94; 1864 to 1873, 31.11; 1874 to 1883, 26.87; 1884 to
1893, 25.78; and 1894, 22.16.

In London during the same periods the rates were: 1864 to 1873, 33.89; 1874 to 1883, 22.05; 1884 to 1893, 20.48; 1894, 17.80, the lowest rate of any large city.

Of these deaths there were of those under five years in New York, 1864 to 1873, 123.31; 1874 to 1883, 104.69; 1884 to 1893, 95.09, and 1894, 85.32. Over five years, during the decade ending 1873, 17.32; 1883, 16.43; 1893, 16.92, and the year ending December 31, 1894, 14.73.

From a total population of 1,356,522 in 1884, there were 35,034 deaths, or 25.24 per 1,000. Of this number 6,089, or 17.94 per cent., were from tuberculosis; 7,892, or 5.69 per cent., from zymotic diseases, and 4,732, 3.41 per cent., in children under five years from zymotic and diarrheal diseases. In 1894, population 1,808,294, 41,175 deaths, 21.03 per 1,000; 5,720, 13.89 per cent., from tuberculosis, 7,516, 3.84 per cent., from zymotic diseases, and 4,808, or 2.16 per cent., in children under five years from zymotic and diarrheal diseases.

During early life the greatest mortality is caused by the zymotic diseases; middle life the greatest mortality is due to pulmonary disease, especially tuberculosis, pneumonia, and bronchitis; and past middle age diseases of the renal organs predominate.

Looking forward through the decade ending 1903, with careful inspection of the milk and water supply, rigid enforcement of sanitary regulations, and the erection of new buildings to replace the tumble-down tenements, far better results may be looked for.

The instruction of the public as to the infectious nature of tuberculosis and the grave necessity of isolation and care in handling tuberculous relatives and friends must be pushed with vigor.

Diphtheria, during 1894–5, has shown a notable reduction in death-rate, there having been one third less than at any time during the preceding six years.

Section on Public Healh.

"The Relation of Bovine to Human Tuberculosis." Introducing this subject, Prof. H. D. Gill, of the New York College of Veterinary Surgeons, gave the United States Government the credit for having taken the most active part in the prevention and eradication of bovine tuberculosis.

Physical examination often fails to determine the presence of tuberculosis, the slow, insidious invasion of the disease often giving no sign, as the animal is apparently in perfect health. Tuberculin is a sure test in all cases and ought to be used.

Infection of animal to animal not frequent, but usually occurs through the inhalation of the dried sputum. It may occur during coition or during fetal life, and only healthy animals ought to be brought together.

The chief danger to human beings is through the use of milk from infected animals. The meat if used should be well cooked.

Prof. Leonard Pearson gave a résumé of "The Work of Foreign Veterinarians in Tuberculosis," and called attention to the careful inspection of all meat at the public abattoir, where all slaughtering is done. From fifteen to thirty per cent. have been found infected. The udder is the most dangerous seat. Local foci in glands are not as dangerous, yet it may be carried on the knife of the butcher from carcass to carcass.

One cow is a danger to the whole herd, and infection will be rapidly communicated.

Tuberculin is a valuable test, giving accurate indication of the animal's condition, and five years' use has demonstrated its great value.

Make animal inspection general; destroy all infected animals; restrict the disease; public bear the loss.

"The Control of Tuberculosis in Massachusetts," was presented by Prof. F. H. Osgood, of the Harvard Veterinary School. In 1894 the laws of the commonwealth as regards the inspection of animals was so modified as to compel an inspection of all animals coming into the State, and making inspection of domestic herds only at the request of the owner.

Of six thousand animals inspected up to June 5th, only one tenth of one per cent. of errors occurred.

In 1894 examination of 2,095, 24.58 per cent., were condemned as tuberculous.

Since 1894 all animals coming into the State have been exam ined, 594, or 27 per cent., reacted.

During 1894 one half the market value was paid to the owner by the commonwealth, an average of $21, and a total cost of $47,000. Full value has been paid during 1895, at an average value of $36. Since the latter rule has been in vogue requests for examination of domestic herds have accumulated to such an extent that the inspectors now have over one hundred herds to examine.

Tuberculin has no deleterious effects on healthy animals, and of over 24,000 tested during the past twelve months it has been necessary to destroy 3,450.

" The Best Methods for the Eradication of Tuberculosis," the subject assigned to Prof. James Law, of Cornell University.

The advantage which the veterinarian has over the physician arises from the fact that he can kill the animal at once, and put an end to the liability of infecting others.

In breeding he can select only those classes which are insusceptible to tuberculosis and regulate by police control.

Immunity must be acquired by insisting on separate stalls, with only individual feeding and drinking troughs ; the expectoration must be destroyed by sprinkling in front of the stalls a strong solution of lime, and the construction of stables so as to admit a plentiful supply of fresh air and sunshine, as exposure of a few hours to sunshine will destroy the tubercle bacilli. It is fortunate that in order to procure the largest quantity of milk the dairyman finds it profitable to frequently change cattle and keep up the high standard of nourishment requisite for the well-being of his herd.

When a herd is supposed to be infected tuberculin should be used, and if one or more are found tuberculous they should be destroyed, the healthy animals removed, the premises thoroughly disinfected. In over a dozen herds thus treated no reinfection has occurred, though it is now from six to nine months since the original inspection. All additions to a herd should be subjected to the tuberculin test.

Infection does not occur from the attendants, as the tubercle of bovine destroys that from man, and is much more hardy.

While tuberculin is undoubtedly diagnostic, it has the disadvantage of not discriminating between the advanced and early stages.

The milk from cows with tuberculous udders is the chief source of milk infection, and children are the most susceptible and exposed to the greatest danger.

Still it is a well-known fact that the best care, surroundings, and food are no guarantee against infection. Old cows should all be thrown out, new stock added only after being tested by tuberculin.

Tuberculous attendants are a source of danger. The police or sanitary system must be extended and enforced; all animals examined; herds registered; tuberculin test; isolation and destruction of all infected animals, and abattoir inspection. If all herds were registered infected animals could then be traced back and the original herd examined.

" The Transmission of Tubercle Bacilli and Tubercular Products in Milk," by Prof. Harold C. Ernst, of the Harvard Medical School.

From a report of the Royal Commission examination (forty-one tests) of seventeen cows, eight were found to have tuberculosis affecting the udders. Microscopic examination shows in thirty-three per cent. udder disease. Guinea-pigs fed on milk from cows with diseased udders, thirty-three per cent. were infected, and out of twelve pigs five were inoculated.

While tuberculosis of other organs in cows may be a source of danger, yet the udder is undoubtedly the chief factor in producing the high rate of intestinal tuberculosis in infants and young bottle-fed children.

Transmission of this disease through the milk direct is without doubt one means of communication, yet the diseased udder is the most likely source.

During the general discussion which followed, Prof. W. H. Thompson expressed the belief that infection from the tubercle bacilli occurred less often by inhalation than *through the mouth*. This was demonstrated by the report of the London Hospital for Chest Diseases, where dust, the accumulation of years, was injected into one hundred guinea-pigs, and only two were infected. Also shown by the infrequency of pulmonary tuberculosis in rag-pickers.

Fatty and greasy food on its way to the intestines adheres to the lymphoid tissues, the pharyngeal tonsils, and pharynx, and readily enters at minute injured points on the abraded mucous membrane, is readily carried to the mediastinal glands and forms foci, from which the lungs are later on infected. The absorption through the tonsils is a frequent source of introduction of tubercle bacilli in children.

We must insist on thorough mouth-washing at night, and even in adults subject to tonsillitis and pharyngitis the attacks will be avoided.

SECTION ON SURGERY.

"The Disadvantages of Non-absorbable Sutures in Hernia Operations" was the subject of a paper by Dr. W. B. Coley, and was made up of a report of fourteen cases seen at the Hospital for Ruptured and Crippled. In all these cases the use of silk, silk-worm gut, or silver wire was followed, within a period varying from a few days to two years, by suppuration, with the formation of sinuses, extrusion of the buried sutures, and relapse of the hernia. The reappearance of the hernia occurred as early as two months and as long as three and a half years after the original operation.

One case, first operation in 1893, silk suture; sinus in March, 1894, three or four sutures extruded, sinuses scraped; second operation following June, silk-worm gut, again hernia reappeared; third operation, kangaroo tendon, successful.

Two cases of double inguinal hernia, recurrence on both sides within a year.

In three cases of Dr. Coley's relapse followed the use of silk-worm gut, and in none did sinuses heal until all the sutures came away.

Dr. Coley has had no suppuration, no sinuses or relapse in any case where he has used the kangaroo tendon for buried sutures.

The trouble in these cases he believes to be due to the irritation due to the presence of a foreign body.

Kangaroo tendon is found to be absorbed in from eight to ten weeks.   A. E. GALLANT, M. D.

A Word to the Busy Doctor.

I WISH in this issue to discuss the treatment of fistula in ano from a different standpoint from any that has been mentioned before in these columns. We all are familiar with the time-honored surgical maxim that to heal a fistula it is necessary to "lay open the tract freely and allow the wound to heal by granulation from the bottom." Indeed, very few authors at the present day mention any other plan. A few years ago Langé, of New York, suggested that this disease, fistula in ano, might be dealt with in an entirely different manner, claiming that in many cases at least these wounds could be made to heal by *first intention*, thereby saving a long, tedious convalescence. It is to this plan that I wish to refer.

I WISH to say in the beginning that if the surgeon labors under the impression that fistulæ found around or communicating with the rectum are trivial affairs and easy of cure, he will find himself in the majority of cases greatly deceived. It is no wonder, considering the enormous wounds that are sometimes made in these cases, after the ideas of antiseptic surgery were promulgated, that surgeons began to think of the propriety of trying to get union by *first* intention after operations for fistula in ano. So universal was such practice in other wounds that it seemed plausible in these. But it must be remembered that in days past there were good and great surgeons, and no doubt they thought of the same thing. But, be that as it may, certain it is that their attempts were futile. It may be asserted, however, that surgery has been revolutionized, and that with the antiseptic era we have much better success in the healing of wounds than of yore. The etiology and pathology of the disease was well understood in the past, so we must look for some other reason for supposing that they failed to instruct us in a method to get union by *first* intention, knowing as they did how much more desirable it was.

THE only question then to be asked, has the plan suggested by Langé been successful? If so, to what is this success attributable? or, if answered in the negative, to what is the non-success attributable? We quite understand that wounds of every character are not disposed to heal by first intention if particles of dirt, foreign bodies, or an uncleanly condition exists. But we do know that if a thorough aseptic condition is had, together with a perfect apposition of the parts, union by first intention is the rule. Now I shall contend that the plan suggested by Langé has *not* been successful except in the rarest of cases of fistula in ano. What then is the explanation? I would ask. Is the morbid change in this disease different to-day from what it was fifty years ago? Certainly not. Fistula in ano invariably has its beginning in an abscess; pus burrows, new and extensive channels are formed. The whole anal and perineal region may be invaded. The scrotum, labia, sphincter muscles, etc., may be destroyed by the invasion of pus. If the older surgeons had known that which modern research has made so plain, viz., that pus must be evacuated as soon as detected, much of this disease would have been prevented, and to-day no stronger injunction could be given than this important surgical lesson. I must confess that I believe that much more good could be accomplished to-day by insisting upon *free drainage* in this surgical disease than by an attempt to teach that we might get union by first intention after operating.

IF it could be maintained that union could be had by first intention in any thing like a respectable proportion of these cases, it would be worth the trial. But is this true? I certainly must dissent from any such belief. If in the majority of cases of fistula in ano only one main sinus existed, no one would deny but that an attempt should be made to close it by apposing its surfaces after division. But the surgeon who sees much of this affection is well aware of the fact that such cases are the rarest exception. No surgeon but would say in operating, "Lay open freely every channel." Then, admitting that the majority of cases have many channels and that to "lay open, trim edges," etc., would put the wound beyond any condition to get apposition, and without apposition there can be no union by *first* intention, to even attempt such a thing would result in a pus discharge, breaking of stitches, and an ugly condition of affairs generally.

To CONCLUDE, then, I would submit that in these cases where the least doubt exists as to whether apposition can be had, no attempt should be made to obtain it. Much better to lay open all channels, curette all surfaces that need it, divide the indurated bottom, and trim off overlapping edges, and allow the wounds to heal, as did the old masters, by granulation, trusting to your antiseptic methods to prevent all pus. You then will have a perfect drainage and a perfect cure.

DISEASES OF THE RECTUM.

Vanderlinden and De Buck: Total Extirpation of the Rectum. (*La Flandre Med.; British Medical Journal.*)

The authors claim that partial resection, or even total extirpation of the rectum for cancer, is abundantly justified where at all practicable from the point of view both of its immediate and ultimate results. They record two successful cases of this kind.

Case 1. A multipara, aged thirty-one, in August, 1892, gave a history of a year and a half of pain in the lower belly, constipation, difficult defecation, grooved feces. For a year glairy mucus, blood, and yellowish fetid sanious liquid had been passed with the feces; marked loss of appetite and body weight. *Per anum* a growth was felt, ulcerated in places, extending seven cm. from below, and invading the whole circumference of the rectum with its greatest thickness posteriorly. The summit of the growth was easily reached, and the whole tumor could be moved downward. The operation was performed on October 30, 1892. The dorsal position was used, with the pelvis raised and thighs strongly flexed on the abdomen. The anus was surrounded by two short incisions, which joined in front and behind. A posterior median incision was prolonged from these to the coccyx. The anal canal and rectum were dissected out as far as three cm. above the growth, where section of the bowel was made. Suture of the bowel walls to the skin wound completed the operation, which lasted an hour. The patient returned home at the end of four weeks, and three months later had gained ten kilos. in weight, and could already retain firm stools. There has been no recurrence up to the present time. The growth proved microscopically to be a lobulated epithelioma.

Case 2. C. D., aged fifty-two, married, no children. Three years' history, commencing from the climacteric, and in its details is very similar to Case 1. Two indurated ulcerated masses were found in the anal region. The rectum was invaded in its whole girth by a soft, yielding easily-bleeding tumor, whose summit was reached with difficulty, ten to eleven cm. above the anus. Operation February 24, 1895. Left lateral position, thighs strongly flexed, and pelvis raised. The incision ran from two finger's breadth below the posterior superior iliac spine along the groove between the gluteus maximus and sacrum toward the median line as far as the summit of the coccyx, then surrounding the anus. The musculature of the buttock was detached, the insertions of the great and small sacro-sciatic ligaments cut, and the coccyx extirpated. A part of the left side of the sacrum was removed, the abundant bleeding controlled, and the rectum isolated, commencing with the anal aperture. The peritoneum was opened after isolation of the anterior rectal wall. The bowel was cut transversely two cm. above the growth, and its end, slightly twisted on its axis, was sutured to the borders of the skin wound. The operation lasted an hour and three quarters, much blood being lost. A large quantity of NaCl solution was therefore injected, and, except for two days' fever, the patient did well, and at the date of report was convalescent.

ARMSTRONG : CARCINOMA OF THE RECTUM. (*Montreal Medical Journal.*)

At a meeting of the Montreal Medico-Chirurgical Society Dr. Armstrong related the clinical history as follows : Man, forty-five years of age. Disease first noticed in July, 1892; it was a pretty high cancer of the rectum. I removed it by Kraske's method, made an incision on the left side, separated the attachments of the rectum from the sacrum, turned it down, and got a very free entrance into the region of the rectum ; then introduced ' the sound into the urethra, which assisted me very much in separating the growth from the urethra, prostate, and base of the bladder. I found I could get the rectum down very nicely, so that I was enabled to remove the tissues well above and around the disease. There were a few enlarged glands about the sacrum, which I removed. Then I sutured the ends of the bowel together. In Kraske's operation the sphincters are not disturbed ; we go in from the sacrum and divide the rectum about one and one half inches or two inches above the internal sphincter, then bring down the malignant disease, divide the rectum above that, and suture the two ends together. This case, like most cases, did not unite by first intention. There was leakage. A large mass, which will be shown to you, developed from the fistula where I made the entrance from behind. It was very slow in closing, and the fecal discharge which came down for a long time was the cause of the fungoid growth which developed. The bowel, although three years have elapsed, only shows one or two small nodules in the neighborhood of the growth. One year ago obstruction was so great that I did an inguinal colotomy. Of course this operation has been developed very much since 1892, and many improvements have been made. One of the most important is just now under discussion, and it looks very feasible; that is, to do a colotomy first (Schede does the colotomy afterward). It seems to me it would be a good idea to do a colotomy first, establish an artificial anus, get the patient somewhat accustomed to the use of this artificial anus, get the lower rectum thoroughly cleaned out, and by thus rendering the field of operation aseptic, primary union of the cut ends may be obtained, which it is believed lessens the liability to local recurrence.

KELSEY, CHAS. B., NEW YORK : THE RELATION BETWEEN UTERINE AND RECTAL DISEASE. (*New York Post-Graduate.*)

If you expect to gain any reputation as a specialist in the surgery of the rectum, one of the first things you have to learn is diagnosis in gynecology. Then, unless you are satisfied to confine your attention simply to hemorrhoids and other minor surgical affections of the rectum, you must learn to treat these cases. You can not, in practice, first go through a surgical operation on a lady for hemorrhoids and then send her to your friend for another operation for proctocele or retroversion, with the confession that you have done only part of what you knew was necessary to effect a cure. Nor can you say to your patients, "If your pain were caused by a fissure it would be my place to cure it, but since it is caused by retroversion of the uterus you had better go to somebody else." This, however, is only a business side of the question. What I started to impress upon you is, that, if you would treat diseases of the rectum with any success, you must not limit your studies to the lower four inches of the alimentary canal.

GASTRO-INTESTINAL DISEASE.

BURNS, T. M.: COLON IRRIGATION. (*Denver Medical Times.*)
Indications:

1. Constipation from any cause. An enema two to three times a week at bedtime.

2. Cases where the bowels need to be emptied immediately, without having to wait for a cathartic to act, e. g., colic, intestinal obstruction, sick-headache due to a deranged stomach, septic infection, onset of all acute fevers and inflammations, especially peritonitis. An enema every two to four hours until effectual. In cases of fecal impaction use as hot water as can be borne along with strong soapsuds.

3. Diarrhea and dysentery. Repeat once to three times a day, using cool or cold water, as the patient prefers. Use plain water, containing gr. v to x sulphocarbolate of zinc.

4. Chronic stomach trouble, consumption, anemia, etc. Use as directed for constipation.

5. Kidney disease. Use as for constipation, but follow by one of about a pint, to be retained over night. I have had patients tell me that after taking the regular colon irrigation it seemed as if they passed more water from the bladder than they did from the bowels.

6. Uterine disease, especially that of girls. A hot rectal douche may be used once a day in place of the vaginal douche.

7. Hemorrhoids: Use as for constipation, but use only cool or cold water, containing a dram of fl. ext. hamamelis.

8. In fact all forms of disease are benefited by colon flushing. I would suggest its use in skin diseases, as they are generally due to indigestion or lack of assimilation.

Technique: Patient in the Sims position, hips a little higher than the head, a four-quart fountain syringe filled with lukewarm water containing a teaspoonful of table salt. Allow the patient to pass the lubricated rectal tube herself. Tell her that if she bears down as if she wished her bowels to move she can insert it much easier. Hold the bag about a foot or two above the patient's hips and let the water run. Stop the water or lower the bag a little whenever the flow causes pain. As soon as the pain disappears start the water again, and so on until nearly all the water is out of the bag, then stop, so as not to let air into the rectum. Tell the patient to retain the water about ten minutes, if she can do so conveniently. In some cases more force is required; then the bag has to be raised from two to six feet. Soapsuds should only be used when the salt and water fail to act, and then never in hemorrhoids, or diarrhea, or dysentery. A smooth vaginal tube is often better for the rectum than the rectal tube, because it is large, and therefore prevents a return flow; and again, because it has more openings, it is not near as liable to become occluded.

If the tube becomes "stopped up," denoted by the patient being unable to feel the water or a cool sensation, pass the tube in further or pull it out some, then turn it backward or forward, and raise and shake the bag. If these fail, remove the tube and clean it with a pin.

A bulb syringe is almost useless, as it takes so much time, and on account of the unevenness of the flow causes much more pain, and toward the last almost always air enters the bowels, which causes still more pain.

A rectal tube made to pass the sigmoid flexure is theoretically correct, but practically unnecessary except in cases of spasm of the rectum or sigmoid flexure.

Pain during the injection can be greatly relieved by deep breathing. Pain in the left iliac fossa indicates the passage or attempted passage of water through the sigmoid flexure. Pain in the right iliac fossa indicates the same in regard to the ilio-cecal valve. In animals colored water has been made to pass out of the mouth when injected *per rectum*.

After the bowels move, from a large injection, there is often a strong desire to eat.

Danger:

1. Rupture of the gut. Probably theoretical, although it is possible if the injection was not administered carefully.

2. Injury to the mucous membrane, probably from the use of hot water, which would cause desquamation to occur.

3. Weakening of the patient. This is present more often from mental effort than any thing else, but it may be real with the first three or four injections, or if the injections are repeated

too often. This would be much less if the patient would eat a short time after the evacuation.

4. Constipation is caused theoretically, but practice does not demonstrate it. It may be caused if rectal is substituted for colon irrigation, or if the injections are used too close together. Colon flushing should not be blamed when the patient was previously constipated.

HEATON, GEORGE: UMBILICAL FECAL FISTULA DUE TO PATENCY OF MECKEL'S DIVERTICULUM. (*Birmingham Medical Review.*)

The speaker showed the intestines, umbilicus, and bladder of of an infant with a congenital fecal fistula. Feces were first discharged from the umbilicus two days after birth, and from that time till the patient's death, when three weeks old, all the child's motions were passed through the abnormal aperture.

The specimen showed an opening at the umbilicus, lined by mucous membrane and leading by a short, wide canal into the small intestine some six or eight inches from the ileo-cecal valve. The small intestine above the aperture was much dilated. The large intestine was very much contracted. The cecum and vermiform appendix lay deeply in the true pelvis attached to the sacrum by a long meso-cecum, and the ascending colon passed upward to the left side of the spinal column.

ENDOXINE IN GASTRIC DISORDERS.—The *Belgian Medical Press* announces a new compound of iodine, endoxine, which is the bismuth salt of nosophene. It is a reddish-brown powder with neither taste nor smell, insoluble in water, but differs from the other bismuth salts inasmuch as it is soluble in caustic alkalies, producing a violet-blue tint.

Endoxine possesses to a marked degree the power of healing, has no toxic properties, and may be administered internally, even in gastric and intestinal disturbances, up to 1.25 grams a day with impunity, the daily quantum being .75 to 1.20 grams in three doses.

Book Reviews.

The Pathology and Surgical Treatment of Tumors. By N. Senn, M. D.,
Ph. D., LL.D., Professor of Practice of Surgery and Clinical Surgery, Rush
Medical College, etc., Chicago. Illustrated by five hundred and fifteen
engravings, including full-page colored plates. Price, $6, cloth; $7, half
morocco. For sale by subscription only. Philadelphia: W. B. Saunders,
925 Walnut Street. 1895.

Rarely have there appeared in any one year two such ·books
as the one before us for review, by Dr. Senn, and "Surgical
Pathology and Therapeutics," by J. Collins Warren, M. D., a
review of which appeared in a recent issue of the QUARTERLY.

It has been known for some time that Dr. Senn had in preparation a work on tumors, and its advent was awaited by all
interested in the subject. The volume just issued surpasses even
our fondest hopes and gives evidence of great labor on the part ·
of a tireless worker, the author.

The book is begun with a fitting dedication, in the following
words: "To the memory of Samuel David Gross, a master in
surgery; a pioneer in pathological anatomy; a surgeon honored
and revered wherever Hippocratic medicine is taught or practiced; a man whose eminent professional reputation was crowned
by the purity of his private character."

It can not be gainsaid that there has been a crying need for a
treatise on tumors, and just such a treatise as given by Dr. Senn.
The subject has been handled in a masterly way, and the text not
only proves instructive, but is fascinating reading to surgeon and
general practitioner alike.

The illustrations deserve more than special mention, as does
the admirable typographical execution of the work.

The plan of the treatise is best told perhaps in the author's
language in the preface: " For the purpose of simplifying diagnosis a special effort has been made to trace every tumor to its
proper anatomical starting-point and histogenetic source, and to
make a sharp histological and clinical distinction between true
tumors, inflammatory swellings, and retention cysts. . . . The
microbic origin of tumors is briefly disposed of, as it has not been
established by any convincing experimental investigations or
clinical observations."

The structure and character of a tumor is stated to depend upon the stage of the arrested cell-growth and the embryonic layer from which the matrix is derived, and that the tumor cells always correspond in type to the embryonic cells from which they are derived.

In regard to malignant tumors, it is stated that their density is in proportion to their benignity and their softness to their malignancy.

It is gratifying to read the author's opinion of the value of the microscope as an aid in the diagnosis of tumors. There is no surgeon of any experience whatever who has not been misled by the microscopic report of a tumor, and mistakes in diagnosis made in consequence. As an instance of such a mistake, the author cites the case of the late Emperor Frederick of Germany.

In the chapter on the treatment of tumors proper emphasis is made as to the importance of antiseptic precautions in extirpation and other operative procedures.

Space prevents a complete review of the work, for without doubt it is the most thorough one extant. An attractive feature is the use of italicized paragraphs through the text for the purpose of emphasizing salient points in the various chapters.

An American Text-book of Obstetrics. For Practitioners and Students. Edited by RICHARD C. NORRIS, M. D.; Art Editor, ROBERT L. DICKINSON, M. D. With nearly nine hundred colored and half-tone illustrations. Price, cloth, $7; sheep, $8; half russia, $9. For sale by subscription only. Philadelphia: W. B. Saunders. 1895.

The editors have been assisted in this work by the following collaborators: Drs. James C. Cameron, Edward P. Davis, Charles Warrington Earle, James H. Etheridge, Henry J. Garrigues, Barton Cook Hirst, Charles Jewett, Howard A. Kelly, Richard C. Norris, Chauncey D. Palmer, Theophilus Parvin, George A. Piersol, Edward Reynolds, and Henry Schwartz. Such an array of talent announced as authors in one book is sufficient guarantee for a popular and useful volume.

The profusion and general excellence of the illustrations throughout gives evidence of a wonderful amount of work and attention to details, and it has never been our good fortune to see a work its equal in this regard. The illustrations are, without exception, well executed and well chosen. They are drawn to scale, and a uniform scale has been adopted throughout. In some of the illustrations photo-brown proves itself a very attractive color, and the cuts appear very distinctly. The various cuts representing the succeeding steps in normal and instrumental labors will prove themselves valuable assistants to students of the text.

Many of the chapters are worthy of special mention; the pathology of pregnancy; the conduct of normal labor, especially the prominence given the importance of external examination of the pregnant uterus to make the diagnosis of position, etc.; the mechanism of labor; the newborn infant, which in the main are the posthumous writings of Dr. Charles Warrington Earle, and a most excellent condensed statement it is about the newborn, a subject to which little or no attention is paid as a rule.

It is surprising, however, that in this enlightened day, the opium treatment of peritonitis is recommended so highly as is done by Dr. Garrigues in the chapter on Pathology of the Puerperium, though his statements are very candid, and he is sincere in his belief and in his preference for opium to the saline treatment.

Symphysiosotomy (and not -eotomy as has been suggested) is given eleven pages, and its proper position among the obstetrical operations defended. However, one is left in doubt on a point we all thought settled, that is, in regard to embryotomy upon a living child. We are surprised that such a middle course was pursued in practically dismissing the subject by asking, " Is it permissible to destroy the child in order to save the mother?" However, as a whole, these are minor points when taking into consideration the magnitude and general excellence of the book otherwise.

It is to be regretted that so many *errata* should have crept in, as instanced by the slip accompanying each volume; but they all occur in the illustrations, and not in the text, which shows most careful proofreading.

The book is a fit associate to the other "text-books," in fact is more attractive than the others in the excellence of its typographical execution.

Pediatrics. Volume I, No. 1, January 1, 1896. Published semi-monthly, with illustrations. Van Publishing Co., 1432 Broadway, New York. Owner, DILLON BROWN, M. D., New York; Editor, GEORGE A. CARPENTER, M. D., London.

This journal with the above title has reached our exchange table. Its appearance is not entirely unexpected, as it has been rumored in journalistic circles that owing to a dissatisfaction on the part of the owner of the new journal and the owner of the *Archives of Pediatrics,* for a time under the editorship of Dr. Brown, a new journal devoted to diseases of children would be published with Dr. Brown at its head.

It is a surprise to us, however, that an editor had to be sought abroad. We do not mean this as uncomplimentary to Dr. Carpenter, for he is an able writer and a clinician of wide experience, but rather as a reflection on our own pediatrists, than whom there are none better anywhere.

The editorial staff is well chosen, containing in the list the names of such men as A. Jacobi, Henry R. Wharton, Henry Ling Taylor, Morris Manges, Egbert Grandin, and James Nevins Hyde, of this country, and Frederick S. Eve, Frederick R. Fisher, Dawson Williams, Ward Hovell, of England, and J. Boas, of Berlin.

The first number includes among its contributors, Drs. Jacobi, Fruitnight, Dillon Brown, Howard Lilienthal, A. M. Phelps, and G. A. Sutherland, with an editorial by Dr. M. Manges, besides society reports, book reviews, abstracts, and items.

We trust there is a place for the new journal, and that it will meet with success in a field already so well occupied.

Disorders of the Male Sexual Organs. By EUGENE FULLER, M. D., New York, Member of the New York County Medical Society, Academy of Medicine, American Association of Genito-Urinary Surgeons, etc. Philadelphia: Lea Brothers & Co. 1895.

No one has been identified with the descriptions of affections of the seminal vesicles as has Dr. Fuller, the author of the work under consideration. After a careful perusal of its interesting pages, however, we can not but wonder why the sweeping title, "Disorders of the Male Sexual Organs," was used, as but little reference is made anywhere of any condition except "Affections of the Seminal Vesicles," which we consider would have been a better title.

As said above, the profession is indebted more to Dr. Fuller for careful and painstaking investigations and several timely and interesting articles on this subject, the present volume being the result of much personal work, than to any other worker in the same field.

The volume before us plainly shows that it is written by an enthusiast, but his statements are well guarded by ample illustrations and descriptions of typical cases.

There are a number of half-tone engravings of dissected specimens, which do not show up to the best advantage, one having to draw on the imagination to follow the descriptions, so dark are the shadows. It would have been well to have placed the explanation of reference letters, etc., on the half-tone inserts beneath the cut, rather than part of it on one page and part on another with the insert between. In Figure 5, page 31, the description would have been more explicit had the cut been reversed.

In the chapter on "Differential Diagnosis," it seems that a rather exaggerated statement has been made in regard to the part played by a pathological condition of the rectum in vesicular disease. It is useless to say that the removal of any pathological condition of the rectum, such as mentioned by the author, "fissure, ulcer, spasmodic sphincter," will not relieve the symptoms present, or that the cure of a seminal vesiculitis will relieve the symptoms of the rectal condition. It seems that here is an instance of the "enthusiastic surgeon" erring on the other side.

Many interesting cases are reported in the last pages, showing the clinical history of the acute, chronic, and tubercular vesiculitis, and these are a valuable addition to the work.

The Art of Massage, Its Physiological Effects and Therapeutic Applications. By J. H. KELLOGG, M. D., Battle Creek, Mich. Modern Medicine Publishing Co., Battle Creek, Mich.

Massage is a subject universally, one might say, neglected by physicians and medical colleges throughout the country, and this excellent work is most timely. Unfortunately massage has been relegated in the past to "Swedish movements," "magnetic healers," and the like, or, when an experienced masseur could be had, the treatment was left entirely to his discretion, unadvised by the attending physician. This obtains even to-day, and it behooves the profession to look deeper into the subject and to have a clearer understanding of the possibilities of this therapeutic agent.

Dr. Kellogg has given this subject much study and is thoroughly acquainted with it in its practical aspects. Not only is the text thorough and comprehensive, but it is rendered doubly so by the excellent half-tone cuts illustrative of the various steps in the manual part of the work.

The appendix contains a record of fifty-four clinical cases treated by massage, which adds much to the practical value of the work.

We look forward with interest to the appearance of the companion to this book, now in preparation by Dr. Kellogg, "Medical Gymnastics."

The Science and Art of Obstetrics. By THEOPHILUS PARVIN, A. M., M. D., LL.D. Third edition. Carefully revised. Illustrated with two hundred and twenty-nine wood-cuts, and two colored plates. Philadelphia: Lea Brothers & Co. 1895.

Dr. Parvin's work has always been well received, whether in contribution to current literature or in a volume such as is before us. The fact of its having reached its third edition is proof enough of its popularity. There have been a number of additions to the second edition in rewriting this one, many a decided improvement; the frequent additions to the already copious footnotes on nearly every page is certainly not for the best.

Dr. Parvin follows, in the arrangement of this volume, the course followed in his lectures, which for years have been noted for their clearness.

Several chapters have been added to the last edition, symphysiotomy being discussed at length. We consider the chapter on puerperal fever the weakest one in the book, while diseases of the newborn are discussed in two pages or a little over. We predict for this edition as successful a sale as the previous ones.

Some Physiological Factors of the Neuroses of Childhood. By B. K. RACHFORD, M. D., Professor of Physiology and Clinician to Children's Clinic, Medical College of Ohio, Cincinnati. Price, $1. Cincinnati: The Robert Clarke Company. 1895.

To those who have not had access to the *Archives of Pediatrics*, in which the chapters of this book appeared as a serial, an investment in the book will be well made, especially as the author in preparing the articles for publication has revised many of the chapters and introduced a new one on auto-intoxication. The table of contents includes the following chapter headings: Normal Functions of Nerve Cells; Physiological Peculiarities of the Nervous System of Infants and Childhood; Fever and the Variable Temperatures of Childhood; Heat Dissipating Mechanism; Autogenetic and Bacterial Toxines; Venous Condition of the Blood; An Impoverished Condition of the Blood; Reflex Irritation and Excessive Nerve Activity.

Not only is the book essential to pediatrists, but to the general practitioner; and it will prove a great help to those noble workers, kindergarteners, in providing them with a book which covers so much ground, and so well, in one hundred and twenty pages.

A Manual of Syphilis and the Venereal Diseases. By JAMES NEVINS HYDE, A. M., M. D., Professor of Skin and Venereal Diseases, Rush Medical College, etc., and FRANK H. MONTGOMERY, M. D., Lecturer on Dermatology and Genito-Urinary Diseases, and Chief Assistant to the Clinic for Skin and Venereal Diseases, Rush Medical College, Chicago. With forty-four illustrations in the text, and eight full-page plates in colors and tints. Price, $2.50, net. Philadelphia: W. B. Saunders. 1895.

This is an extremely attractive book of six hundred pages and contains a valuable collection of all that is of importance in a consideration of syphilis and venereal diseases. All controversial points have been avoided, but as a text-book and reference book for the student, and the practitioner as well, it will prove valuable.

The chapters on "Syphilis in Relation with the Family and Society," and "Hypochondriasis," are worthy of special mention, as is the table of differential diagnosis of the venereal diseases and syphilis on pages 316 to 323.

The excellent reproduction of a photomicrograph, by Dr. John A. Fordyce, of gonococci in gonorrheal pus is printed in the text, and adds much to the value of the work.

Exercise and Food for Pulmonary Invalids. By CHARLES DENISON, A. M., M. D., Denver, Colorado, Professor of Diseases of the Chest, and of Climatology, University of Denver. Denver: The Chain & Hardy Co. 1895. Price, 35 cents. To physicians and the trade, $3 per dozen.

This is an attactive little volume of seventy pages, containing much of interest to physicians, but written chiefly for the benefit of that great class of pulmonary invalids, who either through ignorance, or perhaps after being instructed, fail to carry out directions in regard to proper exercise and diet.

It will prove a valuable assistant to the long-suffering physician who finds *spoken* words ineffectual in sufficiently emphasizing the importance of the proper care of one's self.

Books and Pamphlets Received.

The North American Review for November opens with a most unique article on "Quick Transit Between New York and London," by Austin Corbin. An amusing essay on "The Plague of Jocularity," by the late Prof. H. H. Boyesen, illustrates the inability of Americans to consider serious things seriously, and in the "Outlook for Republican Success," the Hon. Charles T. Saxton, Lieutenant-Governor of New York, contributes a thoughtful paper on the possibilities of the next presidential election. "What Becomes of College Women," is ably treated by Charles Thwing, LL.D., President of Western Reserve University and Adelbert College. Mr. Edward Atkinson writes upon " Jingoes and Silverites," while Major-General Nelson A. Miles treats in another chapter from his forthcoming book From New England to the Golden Gate on " Our Acquisition of Territory." A contribution of commanding interest is that on the " Industrial Development of the South," by the Hon. W. C. Oates, Governor of Alabama, who speaks clearly and hopefully of the work of commercial recuperation now going on in that section of the country. A most interesting paper is " The Girlhood of an Actress," by Mary Anderson De Navarro, consisting of three chapters from the advance sheets of her reminiscences, which, under the title of "A Few Memories," will be published by Harper & Bros. early in 1896.

The New Bohemian (Cincinnati, O.) for December is the most attractive number of that sprightly magazine that has yet appeared. The leading feature is a splendid review of the life and work of Eugene Field, by his friend and co-worker, LeRoy Armstrong. The frontispiece is a portrait of the dead poet, the latest and one of the best ever published. "Touch Hands and Part," is an intensely dramatic story by Percival Pollard, the brilliant western journalist. It is a strong conception and skillfully told, being both true to life and true to art. "American Artists Abroad," by Harriet Chedie Connor, is an admirable critique on the work of American painters at the Berlin exposition. It is rather spicy, but sympathetically discriminating and just. James Knapp Reeve continues his interesting and instructive series of "Talks with Young Authors," and this month deals with the more practical phases of the writer's trade.

When you are dieting to reduce flesh you must eat stale bread, and give up potatoes, rice, beets, corn, peas, beans, milk, cream, all sweets, cocoa, indeed, any thing which even suggests sugar or starch.   Dry toast without butter, tea without either milk or sugar, rare meat with no fat, and, as far as possible, no vegetables at all should form your diet.   Take all the exercise you can in the way of walking; go twice a week to a Russian bath (where possible) and invariably go to bed hungry.   Anybody brave enough to live up to these laws will certainly lose flesh.—December *Ladies' Home Journal.*

The Blind of Kentucky, Based on the Study of One Hundred and Seventy-five Pupils of the Kentucky Institute for the Education of the Blind.   By J. Morrison Ray, M. D., Louisville. Reprint from the American Practitioner and News.

The Therapeutic Value of Oxygen and Nitrous Monoxide in the Treatment of Pneumonia, Bronchitis, Anemia, and Chronic Diseases.   By J. N. De Hast, M. D., Brooklyn.   Reprint from the Maryland Medical Journal.

Brief Report of all the Abdominal Operations done in my Private Sanitarium up to July 1, 1895.   By Joseph Taber Johnson, M. D., Washington, D. C.   Reprint from the American Journal of Obstetrics.

Author (to critic): Say, old man, I've just had a new book published.   Critic: Well, Penner, you've always been a good friend of mine, and I don't want to queer you.   I won't review it.   There!—*Truth.*

Clinical Notes on Psoriasis with Especial Reference to its Prognosis and Treatment.   By L. Dunkan Bulkley, A. M., M. D. Reprint from the Transactions of the Medical Society of the State of New York.

Favorable Results of Koch's Tuberculin Treatment in Tubercular Affections that are not Pulmonary.   By Charles Denison, A. M., M. D., Denver, Col.   Reprint from the New York Medical Journal.

Removal of Ingrowing Toe-Nail, a Simplified Operation by Means of a New Instrument.   By A. H. Meisenbach, M. D., St. Louis.   Reprint from the St. Louis Medical Review.

The Necessities of a Modern Medical College.   By E. Fletcher Ingalls, A. M., M. D., Chicago.   Reprint from the Bulletin of the American Academy of Medicine.

The Technique of Tenotomy of the Ocular Muscles.   By Leartus Connor, A. M., M. D., Detroit.   Reprint from the Journal of the American Medical Association.

Herniotomy: Osteotomy.   By Samuel E. Milliken, M. D., New York.   Reprint from the International Journal of Surgery.

Excision of the Coccyx for Constant Pain Resulting from an Ununited Fracture. By Lewis H. Adler, jr., M. D., Philadelphia. Reprint from the Medical News.

Traumatic Separation (compound) of the Lower Epiphysis of the Femur. By A. H. Meisenbach, M. D., St. Louis. Reprint from the Medical Record.

Craniectomy: An Improved Technique. By A. H. Meisenbach, M. D., St. Louis. Reprint from the Journal of the American Medical Association.

Chancre of the Tonsil, Report of a Case. By T. C. Evans, M. D., Louisville. Reprint from the Journal of Cutaneous and Genito-Urinary Diseases.

Ready-made Medicines. By Charles E. Warren, M. D., Roslindale, Mass. Reprint from the Journal of the American Medical Association.

Remarks on Chlorosis and its Treatment. By Frederick P. Henry, M. D., Philadelphia. Reprint from the University Medical Magazine.

The Etiology and Treatment of Endometritis. By Louis Frank, M. D., Louisville. Reprint from the Charlotte Medical Journal.

Tumors of the Mammary Gland. By W. L. Rodman, M. D., Louisville. Reprint from Journal of American Medical Association.

The Treatment of Hernia, including the Operative, Mechanical, and Injective Plans. By Charles C. Allison, M. D., Omaha, Neb.

Uric Acid Formation. By William Henry Porter, M. D., New York. Reprint from the American Medico-Surgical Bulletin.

Surgical Treatment of Laryngeal Tuberculosis. By J. W. Gleitzman, M. D. Reprint from the New York Medical Journal.

The Preston Retreat. A personal letter from Dr. Richard C. Norris to the American Gynecological and Obstetrical Journal.

Primary Idiopathic Perichondritis of the Auricle. By T. C. Evans, M. D., Louisville. Reprint from American Therapist.

Adenoid Growths of the Naso-Pharynx. By T. C. Evans, M. D., Louisville. Reprint from the Medical Record.

Peroxide of Hydrogen. By J. P. Parker, Ph. G., M. D., of St. Louis, Mo. Reprint from the Times and Register.

Pyosalpinx. By Louis Frank, M. D., Louisville. Reprint from the International Clinics. Vol. IV. Series IV.

Malignant Disease. By William L. Rodman, M. D., Louis- ville. Reprint from the International Clinics.

Nucleins. By William H. Porter, M. D., New York. Reprint from American Medico-Surgical Bulletin.

Antiphthisine. By Charles Denison, A. M., M. D., Denver, Col. Reprint from the Medical Record.

Syphilis and Its Treatment. By C. Travis Drennen, M. D., Hot Springs, Ark.

Hydrocele in the Female. By J. T. Dunn, M. D., Louisville.

Hemorrhoids. By Eugene F. Hoyt, M. D., New York.

Notes and Queries.

E. B. TREAT, Publisher, New York, has in press for early publication the 1896 International Medical Annual, being the fourteenth yearly issue of this eminently useful work. Since the first issue of this one volume reference work each year has witnessed marked improvements; and the prospectus of the forthcoming volume gives promise that it will surpass any of its predecessors. It will be the conjoint authorship of forty distinguished specialists, selected from the most eminent physicians and surgeons of America, England, and the Continent. It will contain reports of the progress of medical science at home and abroad, together with a large number of original articles and reviews on subjects with which the several authors are especially associated. In short, the design of the book is, while not neglecting the specialist, to bring the general practitioner into direct communication with those who are advancing the science of medicine, so he may be furnished with all that is worthy of preservation, as reliable aids in his daily work. Illustrations in black and colors will be consistently used wherever helpful in elucidating the text. Altogether it makes a most useful, if not absolutely indispensable, investment for the medical practitioner. The price will remain the same as previous issues, $2.75.

THE ARCHIVES OF PEDIATRICS will commence its thirteenth year with the January number, under the business management of E. B. Treat, publisher, of New York, long identified with medical publishing interests. The *Archives* has been for twelve years the only journal in the English language devoted exclusively to "Diseases of Children," and has always maintained a high standard of excellence.

The new management propose several important changes in its make-up; increasing the text fifteen per cent., and enlarging its scope in every way. This will give room for the fuller contributions and additional collaborators who have been secured for the various departments, all of which give promise of a more successful era than has been known even in the already brilliant career of the journal.

The editorial management will be in the hands of Floyd M. Crandall, M. D., Adjunct Professor of Pediatrics, New York Polyclinic, and Chairman of Section on Pediatrics, New York Academy of Medicine.

A PLEASANT LAXATIVE.—There is probably no laxative or cathartic in the materia medica which is more widely known and more generally used, especially as a home remedy, than castor oil. Its only objection has been its taste. Now, however, even this has been removed, and we have "A Pleasant Castor Oil."

Laxol is pure castor oil sweetened with benzoic sulphinide and flavored with oil of peppermint.

By referring to our advertising pages the readers of this journal will learn how they can procure samples and literature without expense.

Laxol is used throughout many of the best hospitals in the East, where it has been known for some time.

BEGINNING with January, 1896, Mr. W. B. Saunders, publisher, of Philadelphia, announces that he will begin the publication of the *American Year-Book of Medicine and Surgery*, to be edited by Dr. George M. Gould, assisted by eminent American physicians and teachers. It is the special purpose of the editor to review the contributions to American journals, as well as the methods and discoveries reported in leading medical journals of

Europe, thus enlarging the survey and making the work characteristically international. We predict for this enterprise a most successful career.

BAILY & FAIRCHILD COMPANY, of New York, take pleasure in announcing to the medical profession the establishment of the Doctor's Story Series, to be issued quarterly at $2 per year, 50 cents a number. Each number will consist of a complete work of fiction by medical authors. Only such works as are of established value will be reproduced in this popular form. King's " Stories of a Country Doctor " will be issued in January, 1896, to be followed in March by Dr. Phillips' wonderful novel, " Miskel," and later by a novel now in preparation by the same author.

THE VIRGINIA MEDICAL MONTHLY announces that with the new annual volume, beginning April, 1896, its title will be changed to the *Virginia Medical Semi-Monthly*. This change is made after twenty-two years of monthly issue.

THE COLLEGE AND CLINICAL RECORD will be hereafter known under the name of *"Dunglison's College and Clinical Record,"* a monthly journal of practical medicine."

IT is with pleasure we announce that we have secured the co-operation of Dr. Florence Brandeis, of this city, who will in the future furnish our exchange columns with matter culled from the foreign journals.

NOTICE TO CONTRIBUTORS.

Articles and letters for publication, books and articles for review, and communications to the editors, advertisements, or subscriptions, should be addressed to EDITORS OF MATHEWS' MEDICAL QUARTERLY, BOX 434, LOUISVILLE, KY.

All necessary illustrations will be furnished free of expense to authors where they send black and white drawings—or negatives—with their MSS.

All articles for publication in the QUARTERLY will be considered only with the distinct understanding that they are contributed to it exclusively.

Alterations in proof-sheets are charged at the rate of 60 cents per hour, which is the printer's rate to the journal.

Reprints may be had at printer's rate if request is made upon proof when it is returned.

All manuscript must be received by the first of the month preceding its publication.

Remittances may be made by check, money order, draft, or registered letter.

The Editors are not responsible for the views of contributors.

MATHEWS'
MEDICAL QUARTERLY.
"ALIS VOLAT PROPRIIS."

Vol. III.	APRIL, 1896.	No. 2.

Original Contributions.

SOME LATE SUGGESTIONS IN RECTAL SURGERY.*

BY J. M. MATHEWS, M. D.,

LOUISVILLE, KY.

Gentlemen, permit me to thank you for the very kind invitation that I received through your committee to address you to-night. Outside of the professional pleasure that it gives me, there is a social feature that gratifies me very much, in that it gives me the opportunity of renewing old friendships that are very dear to me.

As your committee failed to designate a "theme" for me, I have thought that it would be more to the point for the "shoemaker to stick to his last," so I shall ask your indulgence while I shall speak of *"Some late suggestions in rectal surgery."*

I know of no department of surgery in which there has been a more decided advance in the last decade than surgery of the rectum. The authors of the recent works on general surgery devote much space to this department; all the leading medical journals abound with rectal literature; many medical colleges have annexed a lectureship relating to these diseases, and within the last few years several works have been issued relating to this special work. Numerous new instruments are on the market, to be used by those interested in this work, and a number of operations devised to perfect rectal investigation. In many large cities some one or more surgeons of recognized

* Read by invitation to the Marion County Medical Society, Indianapolis, March 3, 1896.

8

ability have devoted their work specially to diseases of the rectum.   Nor has this impetus to rectal investigation been limited to the United States, but the large European centers have done likewise.   In the great city and medical center of Berlin a surgeon of great reputation is now devoting his life-work to the study of these affections, having been stimulated to do so by the lamented Bilroth.   Investigations in this line were dormant indeed up to within a few years ago, and the field was occupied nearly entirely by the advertising charlatan.   I feel, then, that I will not be trespassing on your valuable time if I devote the hour to the consideration of some recent advances that have been made in this work.   Before mentioning any operative work in this line, you will permit me to say that in this, as in every department of surgery, an operation may be ever so brilliant as far as execution is concerned, yet wholly unadvisable.   I maintain that no surgical operation is admissible, or should be thought of that does not offer to the patient some radical relief, or give promise of a successful termination.   It can be honestly questioned whether the positive advances made in the surgical world in many of its departments, have really resulted in a positive good to the majority of patients operated on.   It is a question worthy of our best thought, whether or not the work of inexperienced and incompetent men in surgery to-day does not bring down the balance when weighed with the successful results of trained and experienced operators.   In a word, is there more good or evil being accomplished by our latter-day surgery?

In the rectal line I might mention for your consideration in this vein of thought three operations, viz:

1. Whitehead's operation for hemorrhoids.
2. Total excision of the rectum (Kraské).
3. The "American" operation.

Relating to the first, Whitehead's operation, it can be seriously questioned whether any improvement has ever been made in the method of operating upon internal hemorrhoids over that practiced by Mr. Salmon and so persisted in by the senior Allingham, viz., by the silk ligature.   It has borne the test of time, and is to-day acknowledged by the most eminent and practical surgeons to be the " easiest of execution, safest in result, and the least dangerous of all other methods."   Mr. Erichsen said, "All external piles should be cut off and all internal piles tied ;" a trite saying that really amounts to an aphorism. But in lieu of this simple and effective operation Mr. Whitehead proposes a plan that has received the sanction of a few, and the condemnation of others.

I therefore deem it proper to consider the operation, but will be forced by my convictions to deal with it as an unjustified piece of surgery. I beg pardon for detailing the plan of procedure of this operation, but it is necessary in order to contrast it with simpler ones. Mr. Whitehead describes his operation as follows :

"The anesthetized patient, having been placed in the lithotomy position, and the sphincters paralyzed by stretching with the fingers, by the use of scissors and dissection-forceps the mucous membrane is divided at its juncture with the skin around the entire circumference of the bowel, every irregularity of the skin being carefully followed. The external sphincters and the commencement of the internal sphincters are then exposed by a rapid dissection, thus separated from the submucous bed in which they rested, are pulled down, any individual points of resistance being snipped across and the hemorrhoids brought below the skin. The mucous membrane above the hemorrhoids is now divided transversely in successive stages, and the free margin of the severed membranes above is attached as soon as divided to the free margin of the skin below by a suitable number of sutures. The complete ring of pile-bearing mucous membrane is thus removed."

For numerous reasons I have opposed this operation, viz:

1. It can not be advised except in selected cases.

2. An anesthetic is necessary in every case.

3. Full and complete paralysis of the sphincter muscles is necessary to do the operation.

4. The operation is difficult, tedious, and bloody.

5. If union does not take place by first intention, pus accumulates, and the result must be an ugly one if not dangerous, and invites sepsis.

6. It is recommended in doing the operation to remove the whole of the hemorrhoidal plexus, which is not necessary to the curing of internal piles.

7. It can be maintained that secondary hemorrhage is likely to occur after the operation.

Prof. Edmund Andrews, M. D., an eminent surgeon of Chicago, in a late article gives the opinions of eminent surgeons, both in this country and Europe, and the consensus is against the Whitehead operation.

In contrast to so formidable an operation as Whitehead's for so simple an affection as internal piles, I beg to call your attention to another operation which has been lately proposed for the same condition. I allude to the excision of piles and suturing the mucous membrane for the purpose of getting union by first intention. The idea was first recommended by Dr. Robert Jones, of Liverpool, in an article entitled, "A Simple Method of Treating the Wound After Excising Hemorrhoids." (MATHEWS' MEDICAL QUARTERLY, page 326, January, 1894.) The method consisted "in cutting off the hemorrhoid, after being clamped by an ordinary pile clamp, an eighth of an inch above the clamp, and sewing the cut edges together with a continuous catgut suture, after which the clamp is removed and the operation is complete."

In the second or succeeding volume of the same journal, Dr. A. Ernest Gallant, of New York, said:

"Dr. Outerbridge had abandoned the use of the ligature, clamp, and cautery since 1888, and treated all hemorrhoids by excision and uniting the cut edges by a continuous catgut suture."

Up to October, 1894, Dr. Outerbridge reported one hundred and twenty-five cases treated by this method. As an improvement upon the plan of suturing in these cases, Dr. Samuel T. Earle, of Baltimore, has invented a special clamp, and alters the method considerably, and certainly advantageously. He describes his mode as follows:

"The pile is caught with catch-forceps at its most prominent point, pulled down and out, and then the clamp forceps are applied as near the base of the tumor as may be thought proper; after being closed as tightly as possible the part of the tumor above the beak of the forceps is cut off close, then the suturing is begun at the distal end of the clamp (the end of the suture is caught with a pair of catch-forceps instead of being tied, in order that the running suture when complete may be drawn from both ends), and is continued over and under the clamp until the whole cut surface is included, but is not drawn tight; the clamp is now loosened, when it can be easily slipped out from between the suture, and the two ends of the suture drawn sufficiently tight to bring the cut edges in nice apposition and control all hemorrhage; the two ends are now made fast by a knot in each, which should be made close down to the mucous surface."

Certainly this operation as perfected by Dr. Earle is much simpler and free of all the dangers of the Whitehead operation, and should receive careful consideration from the profession. My assistant, Dr. William V. Laws, suggests that in doing the operation as described by Dr. Earle, that it may be an improvement to apply small bullets on each end of the ligature in lieu of tying a knot in them.

The American Operation. The operation which has been called the American operation consists simply in removing about two inches of the lower rectum in its entire circumference, pulling it down and stitching it to the true skin. Of course the name is a misnomer. It is simply a modified Whitehead operation, the latter being done in a diseased condition. Its necessity never exists except in the brain of a deluded man who works blindly and never thinks. It is claimed by its originator that the rectum is responsible for " the thousand ills that flesh is heir to," and that as a point of reflex it is " worth going miles to see." From this false premise has emanated the doctrine that heart disease, neuralgias, rheumatism, asthma, etc., can and do result by reflex from the rectum. Hence rectums galore are sacrificed that have never known what disease was. This uncalled for piece of surgery has become so prevalent that no town or hamlet but has suffered in consequence. I have witnessed physical wrecks in people of all ages and stations in consequence of having submitted to it. It has its ludicrous side, but its seriousness far outweighs it, and the profession should at all times give it the stamp of disapproval.

Kraské's Operation or Total Excision of the Rectum. It is not the intention of this paper to criticise the operation known as Kraské's, for as a surgical procedure it is above criticism—there is no method comparable to it, as far as removing the rectum is concerned. If it had no other merit but that the sphincter muscles are left undisturbed, it would still outrank any other plan suggested for the removal of the rectum. Therefore it is not to the operation *per se* that I would object, but rather to the *necessity* of doing it. Good and sufficient reasons can be given against removing the entire rectum for malignant disease, the main object that Mr. Kraské designed his operation for. If it is admitted that malignancy has embraced the whole length of the rectum, I would ask, can it be definitely said that extirpation will avail any good? To have reached such a point a period of three to five years must have elapsed. The infiltration process going on for this length of time must have included adjacent structures. There is no surgeon but will admit that no good whatever can be accomplished by removing portions of cancerous structure. It would be a poor and useless piece of surgery to remove a malignant breast and leave infected glands in the axilla. A hysterectomy, with invaded tissue left, would be an operation devoid of reason, and the uterus is a much more independent organ than the rectum. This particular part of the anatomy receives an enormous blood supply (from these sources), and vascularity means much in malignant disease. Its association with lymph channels is distinctly marked, and hence rapid invasion; its close relation by contiguity with other organs, etc., renders total extirpation out of the question after invasion has taken place. No one would question the advisability of removing a portion of the gut, especially when it is in reach, can be circumscribed, and no cachexia exists, but when such are not manifest it is cruel to subject a patient to so formidable an operation with no prospect of a successful result.

923 Fourth Avenue.

GASTRIC, INTESTINAL, AND RECTAL HEMOR-RHAGE.

BY THOMAS H. MANLEY, M. D.,
NEW YORK.

[Written for MATHEWS' MEDICAL QUARTERLY.]

Hemorrhagic discharges from the terminal portal of the alimentary canal are not uncommon after adult age is attained. The quantity and quality of the discharge varies, depending on the extent and location of the pathological condition giving rise to it, and the circumstances under which it occurs.

Hemorrhage, *per ora*, from the lungs or stomach, is always regarded as a serious symptom of a grave internal disorder, and, with unusual exceptions, subsequent events prove that the reasons for fear and apprehension are well founded.

Bleeding through the anus, in a general way, is not viewed with the same degree of alarm, though, when it occurs in a considerable quantity, or is persistent, to view it with indifference or allow it to continue without treatment, may lead to unfortunate results.

Hemorrhage through the anus is altogether more frequent than through the mouth; it is very seldom a case of immediate death, as is hemoptysis, when a large vessel is opened and the torrent asphyxiates the patient by obstructing respiration. Hemorrhage through the rectum, nevertheless, is often the direct cause of death, especially when its source is located in the higher areas of the alimentary canal.

Hemorrhage into any part of the intestinal canal, either unchanged or digested, is ejected by the rectum at various periods after it leaves the blood-vessels. When the leakage is slow at any point between the pylorus and caput coli, in its descent through, it becomes coagulated, decomposed, disintegrated, and blackened by the action of the acids, salts, and gases in the intestine. It provokes an irritation of the mucous lining of the intestine, and is evacuated in a dark, semi-solid or fluid state, when it is recognized as " coffee grounds."

Not infrequently gastric hemorrhage makes its way out of the body by a downward descent, rather than by the shorter route, in a tardy stream.

Vascularity of the Alimentary Canal. The Stomach. The vascularity of the alimentary canal at the cardiac opening of the stomach varies. This being the center of the greatest activity, in reducing the aliment to a fluid state and pouring acids and ferments in great quantity into it, has a large vascular supply. The gastric vessels, though of large caliber, are not as numerous proportionally as we may find them in other situations. The greater thickness of the gastric wall is made up of secreting structure. The peptic glands are composed of very long tubular follicles. It is only after we have penetrated below the glandular layer that we have come into that network of large vessels which course through the submucosa. Only when an ulcer of considerable area opens into this territory is it that a large hemorrhage is liable to follow. One of slow progress and narrow dimensions, or pursuing certain directions, may advance through all the tissues of the stomach into the peritoneal cavity without any noticeable bleeding at all.

The smaller vessels are closed by the astringent action of the gastric juice and a thickening of the edges of the ulcer by inflammatory changes. After it has advanced into the muscular layer hemorrhage is obviated by the interlacing arrangement of the fasciculi, which tend to contract in different directions, and thus close the orifice. It is probable that in nearly all bleedings from the stomach the hemorrhage is chiefly arterial. In hematemesis we will almost invariably notice the bright crimson color of the blood. When it passes downward the discoloration which the blood undergoes renders it impossible to determine its source. Hemorrhage within any part of the intestinal canal in considerable volume quite invariably provokes an irritation. In the stomach it acts as an emetic; in the bowels, as a purgative.

Causes of Gastric Hemorrhage. It is not my purpose here to enter at length on the etiology of the various phases of hemorrhage under consideration. But there are two prolific causes of degenerative lesion here of great interest, because if either one be definitely recognized diagnosis is established, and prognosis may be made with considerable certainty.

Every one knows how comparatively common very large gastric hemorrhage is among females at about puberty, and the menopause particularly, and here it rarely is fatal. And, *per contra*, how unusual, though fatal, gastric hemorrhage is among males. Indeed, except in that type of hematemesis so general among hard drinkers suffering from cirrhotic livers, barring traumatisms, it is very rarely seen in males, though when it is, it is generally of very grave import.

The pathological lesions which open the way for gastric hemorrhage are simple ulceration, tuberculosis, or cancer.

This is true without doubt when we are reasonably assured that there is a local lesion, or breaking down of tissue. If it be tuberculous, then we may assume that large local bleedings are quite a specific remedy for the malady in females, as we rarely see any succumb from them, and, as a matter of fact, few or none of them later sink from a dissemination of the disease.

For some time before those large hemorrhages set in the patient has indigestion, is extremely anemic and reduced in strength. But after recovery, though the quantity of blood lost may be enormous, all those symptoms pass away and better health may be enjoyed than before.

Sarcoma or Cancer of the Stomach. Malignant disease of the stomach is not often attended with large hemorrhage, except in the male. In young men of good habits sudden and repeated gastric hemorrhage is a sign pointing strongly to sarcoma. A young man begins to lose strength, his appetite fails, and anemia sets in. Along with this he has vague pains over the epigastrium. He is now suddenly seized with hemorrhage from the stomach, when the question arises : Does this succeed from a so-called simple ulcer or malignant disease? Many eminent clinicians are strongly inclined to view a large hematemesis in the male as a very grave symptom, inasmuch as they regard it as very suggestive of malignant disease.

In the walls of the cardiac end of the stomach we will find nothing but tubular glands, but as we approach the pyloric orifice we come on the acinous or mucous, which appear here in great numbers. It is in these epithelial structures that we will find schirrus primarily set in. In the female, proliferation and infiltration of the connective tissue elements, or sarcoma, simultaneously induce stenosis and so much tumefaction that inanition is manifest ; and often through the emaciated abdominal walls the neoplastic formation may frequently be recognized on palpation. In this lesion, in the female, we seldom witness large hemorrhages.

Broadbent and other English writers have called attention to the significance of large bleedings from the stomach in the young male, as often pointing to the existence of malignant disease.

The absence of pain, and other symptoms which are often forerunners of simple or tuberculous ulcers of the stomach, is quite conspicuous in the incipient stages of sarcoma, which invade the gastric walls. In this class the most common symptom is a pronounced anemia with rapid wasting of flesh.

If we will eliminate heavy drinkers, we will find that copious gastric hemorrhage in the male adult is generally associated with malignant disease. Two of such cases have come under my observation within the last two years. One in a young man of eighteen years, and another in a patient of thirty years. In both cases, although there were dyspeptic symptoms for a time in advance, nothing alarming was suspected until an explosive and exhaustive hemorrhage set in. In fact, both followed their usual occupations until this event occurred; but profound exsanguination succeeded, and repeated hemorrhages so speedily followed that neither survived six months. Neither had marked gastric tenderness as might be expected, when it is borne in mind that when cancer is limited to the parenchymatous elements of any of the abdominal viscera, no pain is elicited. It is only when the disease advances outward and reaches the peritoneum that sufferings begin. This is why in cancer of the viscera the patient, as a rule, only becomes cognizant of its presence when its invasion is so far advanced as to render operative relief quite out of the question. In sarcoma of the stomach its site is widespread, stretching far in different directions through the lymphoid layer of the mucosa into neighboring parts. It is unlike malignant epithelial infiltration, which primarily almost invariably seizes on the pylorus.

Intestinal Hemorrhage. Hemorrhage from the small intestine, like that from the stomach, concerns the physician rather than the surgeon; nevertheless the presence of a great hemorrhage is always appalling; and to those not accustomed to free bleeding from opened vessels, more positive resources than those afforded by the tedious and uncertain action of internal medicines seem necessary.

It is only recently that surgery essayed to deal with gastric and intestinal ulcers. May we look for safer and more prompt relief in the future, in cases of large hemorrhages in the alimentary canal, through the aid of surgery? It seems scarcely possible. Eroding ulceration of Peyer's patches or Lieberkuhn's crypts is responsible for those large hemorrhages, so serious in enteric fever.

If we could only determine where the leakage commenced, and whether it was from a large or small vessel, we might promptly institute rational treatment. In a considerable hemorrhage it will be a simple matter to open the abdomen and ligate

the principal mesenteric feeders, but the presence of a large
bleeding in the typhoid is only manifest when evidence of mor-
tal exsanguination, or a state close to it, is apparent. Deep
shock, collapse, and syncope all appear in rapid succession.
From such a hemorrhage a woman may survive; but with a
male who stands a large loss of blood badly, his chances
of completely rallying are slender. Tubercular ulceration,
though common enough in the ileum, for some unexplained
reason, seldom gives rise to hemorrhage. Some have attributed
this to the slow progress of the lesion and the styptic action of
the bile. Malignant disease of the small intestine is very unusual
as a primary affection.

*Hemorrhage of the Large Intestine: Cecal, Colic, Rectal, and
Anal.* Hemorrhage, except at the rectal terminus of the large
intestine is uncommon, if we exclude dysenteric ulceration. As
a symptom of a surgical lesion, we find it beginning at the ileo-
cecal junction in infants and young children, when the smaller
bowel in intussusception is telescoped into the larger.

In various types of ulceration, either simple, tubercular, or
neoplastic, bleeding may follow from any segment of the colon,
from the cecum to the rectum. This part of the alimentary canal
being devoid of lacteals or peptic glands, is the least vascular.
The colon is thin-walled, and we find in its mucosa less lymphoid
tissue than abounds in the small intestine, the adenoid layer of
which occupying more than three fifths of the whole thickness
of its walls.

Traumatic Hemorrhage of the colon is exceedingly rare. When
it does occur in considerable volume the blood ejected is undigested and imperfectly coagulated. Its anatomical site, its functions, and its protected position in a marvelous degree preserve
the colic arch against serious traumatism.

Except its transverse segment, it is so protected that in the
event of blows, kicks, falls, or crushes of the abdomen, it usually
escapes, the other neighboring viscera sustaining the most serious
harm.

Thus we encounter, after varying abdominal traumatisms,
hemorrhage in the renal tissue or other organs and structures,
but fresh leakage into the colon in consequence of direct force,
almost never.

Rectal and Anal Hemorrhage may be considered together with advantage, inasmuch as both are from muscular membranous passages; their source of blood supply is chiefly from the same arterial branches. Those lesions which give rise to excessive bleeding are vastly more common in these areas in close vicinity with the verge, than in the mucous membrane of the ampulla or below O'Bierne's sphincter.

General Consideration of Recto-Anal Hemorrhages; Pathological, Operative, and Post-operative. This phase of intestinal hemorrhage is of the highest importance to the surgeon, for it is only in the ano-rectal, a fixed segment of the alimentary canal, that mechanical expedients or hemostatic agents can be utilized with any degree of safety or certainty.

On the cadaver I have discovered that the adult finger of average length can, by pressing up the anus and inverting it somewhat into the ischio-rectal fossa, penetrate as far up the rectum as the insertion of the peritoneum or the beginning of the sigmoid flexure. This marks the utmost limit of direct surgical manipulation.

The vascular supply of the rectum and anus is more abundant and complex than the continuous intestine above.

The arterial sources of the rectum are extensive and varied, as we may suppose, with an organ so lodged in the pelvis, outside the peritoneum in close contact with the organs of generation, the bladder, the vagina, the muscles, and perineal structures.

Above, the arteries to the rectum are mainly directly or indirectly provided from branches derived from the anterior trunk of the internal iliac; but below, the ampulla and anus are largely supplied by arteries from the internal pudic, which pass horizontally across, beneath Colles' fascia and transversus-perinei muscle.

It may be said that the anus secures its principal arterial supply from the pudic, the same vessel which courses through and conveys blood to the external generative organ, the perineum, the scrotum, and penis. The distribution of the pudic artery will partly account for the simultaneous degenerative or atrophic changes in the penis and prostate, and the vagina in the female, in anal disturbances, with a dry mucous membrane and resulting constipation, fibrous induration of the sphincter, and an occasional tendency to prolapse of the rectum, through the weakened outlet sometimes, seen in old people.

The venous circulation in the ano-rectal end of the intestine is in many important particulars quite unique. The smaller veins in the submucosa here are numerous, with thick walls; the muscularis is denser than in the colon; they are devoid of valves, and drain their contents into the hemorrhoidal vessels. In the annular zone of vessels which encircle the verge the capillaries are short, but large, and in various areas the arterioles and venulæ directly anastomose with each other.

The mixed quality of the hemorrhage which arises through various diseased conditions, the aborescent, turgid condition of the varices, suggest an angeiomatous state of the vessels in the submucosa.

The Causes which Lead to Ano-Rectal Hemorrhage. These are predisposing and immediate. Among the first, may be enumerated the mechanical impediments to the circulation peculiar to this situation; the almost vertical direction of the efferent vessels in the standing position, the absence of valves, the irregular habits of life among human beings; for it does not appear that rectal disease is any thing other than very rare, in the lower animals.

The Pathological Conditions leading to any rectal hemorrhage are numerous. The most common is a hemorrhoidal or varicose state of the vessels about the anal verge. This is so generally the case, that to the laity under nearly all circumstances, an anal bleeding is regarded as a self-evident symptom of piles. Many cases of tuberculous or cancerous ulceration through a misinterpretation of this symptom advance to an incurable degree, before their exact nature is recognized.

My own experience has been that tubercular, next to simple ulceration of hemorrhoidal walls, is the most prolific cause of exhaustive hemorrhage from the anus. Cancer ranks third in frequency. Malignant epithelial proliferation and degeneration on any area of the periphery or surface of the body leads to death, generally through loss of blood, by an interrupted exsanguination, infection, and through pain.

Cancer of the rectum, like the visceral type elsewhere, is not very painful in the beginning, and with unusual exceptions large or frequent hemorrhage is not present even when the disease is making most rapid headway and is spreading into contiguous parts.

The immediate causes of hemorrhage from the anus is through straining at stool, when a thin-walled, widely distended tumor ruptures. The blood which issues from such a vent is of a mixed color and drains off in large quantities. Sometimes the hemorrhoidal mass will burst, on any sudden violent strain of the body when the leak is into the lumen of the bowel. When the colon and rectum are widely distended by such an escape, the patient is seized with an imperative desire to evacuate the imprisoned blood ; so that at such times enormous masses of clots and blood are discharged.

Arterial papillomata of the rectum are not an uncommon cause of most exhaustive depletions. Many cases of the most pronounced anemia, with all the symptoms that attend that state, have come under my observation succeeding this pathological condition. In these cases the mucous membrane of the rectum investing the external sphincter is studded with minute raspberry papillæ. They are apparently devoid of an epithelial investment, and bleed on the least irritation. In aggravated cases they often give rise to copious hemorrhage on evacuation of the rectum.

Hepatic congestion, or a disordered state of the chylopoëtic organs and circulation, have been regarded by the older clinicians as predisposing causes of rectal hemorrhage. In the female, as a vicarious discharge, sanguinous emission from this source is not often encountered ; no such case has ever come to my personal knowledge.

Ano-Rectal Hemorrhage in Immediate Connection with Operations on this region, under ordinary circumstances for lesions occupying but limited areas, is, if present at all, only moderate. But in Whitehead's operation for hemorrhoids, or any other for the same purpose, which includes a deep division of structure, the loss of blood may be considerable.

Stahl divided hemorrhoids into idiopathic and symptomatic. (Mollien, Mal. du Rectum et L'Anus, p. 577.) Ball divides them into three classes : the venous, columnar, and circoid. The venous are the same as the external, except that they are covered by mucous membrane. The columnar, according to Hamilton, consists essentially of folds of mucous membrane surrounding the anal opening or the pillars of Glisson. Within each of these, a descending parallel branch of the superior hemorrhoidal artery penetrates.

The above author regarded this the most common variety as well as the most important, because of its great vascularity. The nevoid has been described under the name of " vascular tumor" of the rectum, which Houston, from its surface characters, compared to a strawberry. This bleeds less often than the second variety, but more abundantly.

Mr. Charles R. Ball, of Dublin, Ireland, a distinguished surgeon and author, claims that hemorrhage is the most constant symptom of internal piles. (Diseases of Rectum and Anus, by Charles R. Ball, F. R. C. S.) "The caking and hardening of fecal concretions lead," he adds, "to the laceration of their surface, and consequent hemorrhage."

Under the designation "hemorrhoid" the ancients described most diseases of the rectum. With one it signified hemorrhage from the rectum, a tumor at the extremity of the digestive tube; finally another invented hemorrhoidal diathesis. We observe also the description of what was called nasal and buccal hemorrhoids. Now we restrict the term to anal varices.

In Kraské's operation for rectal cancer, a large mortality follows, with few exceptions, from the extensive mutilation of structure and laceration of highly vascular parts.

The inevitable loss of blood attendant on various operations for hemorrhoids or other lesions in the verge or ampulla, is one of the most serious objections against it in those, whose general health is enfeebled, have become anemic, or have been exsanguinated by previous vascular drains. When large operative hemorrhage arises on division of large, thick-walled veins, and the momentary gush for an instant floods every thing, moderate compression will promptly subdue it. In operating here, as elsewhere, the divided arteries give issue to the greatest loss of blood. If operating within the lumen of the bowel nothing less than a thorough and complete dilatation of the external sphincter will enable one to expose those arteries which ramify through an atmosphere of loose connective tissue and quickly retract far up out of sight. The best way to provide security against dangerous hemorrhage in operative manipulation, is to be well prepared for it, and close every bleeding point as we proceed with each stage of the operation. This will leave a dry, bloodless field for leisurely and thoroughly dealing with the lesion, spare the patient shock and all those dangers so fraught with serious consequences.

Hemorrhage in all operations on the rectum for malignant disease is often quite unmanageable. If the disease is far advanced its destructive action has extended into the blood itself. Our patient is anemic, and the microscope will show that the red corpuscles are deformed, are in diminished numbers, and destitute of that finely reticulated structure readily seen under high power in the healthy blood. In cancerous blood we can best observe the " transitory " group of Norris. The blood corpuscles lose their vitality early, and under the microscope we can see their color rapidly fade in large groups, and, like phantoms, or as " ghosts," rapidly pass into the invisible.

Marchiafava and Celli, in their extended researches on the pathology of the blood, have fully described this phenomenon. Besides, although in early cancer the blood is hyperplastic, now, when the disease is advanced this living principle of coagulation is enfeebled, the fibrogenous ferment is wanting. In opening up through an osseo-ligamentous structure like the sacrum, in posterior rectal resection, we will note that the vessels are thin-walled, and many of them ramify through tortuous canals or paths in cancellous bone tissue, or through the inter-ligamentous spaces, in places very difficult, if not impossible to secure the mouths of spouting vessels. The entire osseo-ligamentous, vertebral chain, from the occiput down, is enormously vascular, as many times I have had an opportunity to demonstrate experimentally as well as clinically, in laminectomies, and no segment of it is more vascular than its sacral terminus.

Post-Operative or Secondary Hemorrhage is not of common occurrence in rectal surgery, though it is well to be always on our guard for it. It most commonly follows annular resection of the hypertrophied adenoid coat of the intestine, and in many operations for extirpation of piles. The scalpel or scissors may have penetrated too deeply and cut through several of the larger arterial trunks which lie deeply imbedded in the adenoid layer, over the ring muscle, before they split up to supply the smooth and striped muscles which invest the external sphincter. By the adoption of such measures as will insure prompt and secure hemostasis, there will be but little danger of a large secondary oozing, though unless all arterial leakage is arrested by ligation, torsion, or the thermo-cautery at the time of operating, dangerous secondary hemorrhage may follow. After the sphincter has contracted and the dressings are applied, the blood, instead of making its way outward, may drain into the empty intestine. The evidence of its presence there is only made manifest by a death-like pallor of the patient, with a thready pulse and impending syncope.

Several cases have come to my knowledge of very grave post-operative recto-anal hemorrhage. In one case, after operation, the patient had been given a grain of opium by suppository. During the night an attentive nurse on duty was startled by the death-like pallor of the patient, who was now soundly sleeping. Surgical aid was promptly summoned, when it was discovered that blood had saturated the bed-clothes, had soaked through the mattress and was steadily dripping on the floor. The patient was now aroused, and with great difficulty the bleeding artery was discovered and closed. Another case came under my care, a gentleman who a year previously had been operated on for inflammatory piles. It was evident that there had been an extensive mutilation of tissue, for he now had a rectal stenosis, for which he came to be treated. What struck me most at the time was his profound enemia, with an enormous hypertrophy of the spleen and anasarca of the lower extremities. He related that twelve hours after an operation for internal hemorrhoids, one year previously, he nearly bled to death, and now though only forty years old, his health was broken with no prospect of regaining it. Dieting, change of climate, and medication had failed to restore his health, which he said was always robust before he was operated on. He finally sank, and died about six months after I saw him.

It should be always borne in mind, that while under the full influence of a narcotic, especially in all protracted operations, vascular tension is much diminished and the heart's force is enfeebled; hence, when reaction sets in a fresh hemorrhage begins. In order to avert this the greatest caution should be observed in closing all the bleeding points before returning the prolapsed bowel within the sphincter.

Treatment. In many of those small angeiomatous papillæ with sessile bases and deeply imbedded vascular rootlets, nothing has served me to such advantage as the thermo-cautery. Its action is prompt and permanent, but, with vessels of large bore, secure ligation constitutes our main reliance. After any extensive mu-

9

tilative operation above the sphincter ani, for several hours, vigilance should be exercised to detect the first onset of secondary hemorrhage. If it should set in, then we may resort to several palliative expedients, which, failing to stop the drain, must be supplemented by again placing the patient on the operating table, reopening the sphincter, and seeking out the bleeding point.

Symptomatic Hemorrhage from the rectum, as from plethora or hepatic engorgement, must not be confounded with that dependent on local lesion alone. With those of full habit, large eaters, without ample bodily exercise, to maintain a healthy equilibrium there is no doubt but Nature may seek an outlet for the overdistended vessels through the rectum. Such moderate exsanguination is salutary and beneficial when it occurs only at considerable intervals, and is not followed by marked disturbance of the system. Proper dieting, with purgative medicine, will tend to arrest the discharge if it has become habitual, by removing the causes which lead to it.

Hemorrhage from Ulceration; Simple, Hemorrhoidal, Tubercular, or Cancerous. The first has been briefly considered, as no anal varix is liable to bleed without surface erosion. In my own experience profuse hemorrhage from an insidious tubercular ulceration within the ampulla has not been very uncommon. The patient subject to this type of hemorrhage is anemic, in reduced health. If he have no fistulous opening to drain off the ichorous secretions of the ulcer it may provoke an irritation and pruritus.

Several cases have come to my notice of supposed "bleeding piles," which on examination proved to be hemorrhage from a deep-seated ulceration. Those ulcers usually lie in the posterior wall of the bowel, though I have met with them in the form of a deep annular furrow extending around the entire circumference of the rectum.

Cancer of the rectum like the esophageal type is seldom, at least in its early stages, attended with bleeding; the proliferation of epithelial elements is chiefly into the mucosa, which acquires a marked density that only breaks down and sloughs in the advanced stages. In all those cases pain is a prominent symptom during the act of defecation, but hemorrhage we seldom see.

During the past autumn a gentleman came under my observation, in consultation, who was suffering from rectal trouble, in whom on account of the absence of bleeding in defecation, two eminent clinicians had excluded cancer. After a careful examination into his general and local condition, it was my conviction that the lesion was malignant, and subsequent events have verified the accuracy of these conclusions.

It is true in this case that a section of the neoplasm would have definitely decided its pathological character, but it was located high up, and when this was suggested the patient refused permission. Hemorrhage from the anus, therefore, is not a uniform symptom of rectal cancer, and hence its absence is a feature of practically no importance in reaching diagnosis, in certain varieties.

Condylomatous vegetations of the verge of the anus or gummatous infiltrations of syphilis are not very vascular. The former are warty outgrowths of the epithelial investment of the papillary layer of the anus, and are devoid of large vessels, except at their roots. Gummatous masses by selection imbed themselves in the non-vascular stratum of lymphoid tissue, which near the outlet of the rectum is of unusual thickness. The hyperplasia provoked by these deposits, though ample to infringe on and produce a stenosis of the intestine, seldom progresses to ulceration or hemorrhage.

Hemorrhage, succeeding tubercular ulceration of the rectum, should be treated by local applications and attention to the general health. The blood itself in many of these cases is also deteriorated, the venous is wanting in plasticity, and is slow to coagulate. To effectually suppress a leakage, dependent on a depraved state of the system, is to intelligently employ hygienic measures simultaneously with internal medication and topical applications.

Summary and Conclusions. First. Hemorrhage from the rectum may be symptomatic, of constitutional or organic disease, as plethora or hepatic congestion.

Second. In consequence of a lesion of some part of the digestive tube, anywhere from the flexure to the cardiac end of the stomach, bleeding may escape, changed or unchanged, through the rectum.

Third. The local lesions, in their order of frequency, as a source of hemorrhage, in the ano-rectal outlet of the intestine, are: (*a*) hemorrhoids, (*b*) simple or tubercular *ulceration*, (*c*) malignant disease.

Treatment Includes Constitutional and Local Measures. Hemorrhage from simple, tubercular, cancerous, dysenteric, or typhoidal ulceration in any part of the digestive tube above the rectum is quite beyond relief from direct surgical methods, and hence its treatment must, for the present, at least, remain within the domain of medicine.

Surgical treatment of hemorrhage of the rectal pouch and anus, when non-malignant, is generally *practicable*, safe, and permanent in results. In order, however, to be rendered effectual and definite, thorough dilatation of the anus and eversion of the rectum are imperative in order that the bleeding points or source of hemorrhage may be brought under the immediate eye for direct and effective treatment.

When bleeding succeeds from hemorrhoids for the first time, or when its quantity is small, moderate catharsis with simple astringents in the form of suppositories, will favor its arrest without recourse to radical or severe methods.

———————————

COOKE, A. B., NASHVILLE: ANAL FISSURE. (*Charlotte Medical Journal.*)

The author objects to the term "irritable ulcer," as irritable expresses merely a more or less adventitious characteristic, and is in no way distinctive either as to origin or location.

The sensory nerve supply of the anus being more extensive than any other superficial structure of the body, an injury or exposure of a nerve filament is not difficult to account for. Constipation is an important element in the causation, overdistension of the anus and rupture more or less extensive of its circumference results. Forcible extrusion of large polypi and hemorrhoids may produce the initial lesion. Setting the parts at rest by thorough divulsion of the sphincters under anesthesia, avoiding rupture of the sphincter muscle, taking plenty of time for the procedure, will effect a cure in all instances.

GASTRO-INTESTINAL DISEASE.

CONSTIPATION.

BY A. P. BUCHMAN, A. M., M. D.,
Professor of Diseases of the Digestive System, Fort Wayne College of Medicine.
FORT WAYNE, IND.

[Written for MATHEWS' MEDICAL QUARTERLY.]

Constipation is a phenomenon which presents itself as an epi-
sode growing out of the history of a prolonged series of pathological states of the digestive tube. The precedent anatomical
lesions, which as an ordinary evolution lead up to constipation,
include every structure which enters into the make-up of the
intestinal canal from the stomach to the rectum. To fully appreciate the etiology of constipation the condition must be studied
in an orderly way from the initial point of departure up to the
state where it becomes a menace to the health and life of the
patient. That the condition is usually a paralysis I think will
require no argument as a demonstration. At first the peristalsis
is only enfeebled, the normal cycle of intestinal discharge is but
slightly interrupted, not as a rule sufficient to attract the attention of the patient; from this to occasional fits of constipation,
lasting only a few days, to the condition which imperatively
demands either therapeutic or mechanical means of relief. The
general history may extend over many years, and it is not infrequently the case during this time that the system becomes
charged with toxic elements from this source, rendering cell
reaction abnormal, and some organ, usually an emunctory, suc-
cumbs to the lethal influence.

A number of years of close observation and inquiry into the
history of a series of cases of chronic interstitial nephritis has
disclosed the fact that chronic constipation was a precedent condition and had been neglected because of its supposed negative
character. While it is true that the kidney is not the only source
of protection against toxines formed in the intestinal canal, yet, if
they are constantly engaged in eliminating abnormal quantities

of such substances, I think it may be fairly assumed that in time the reaction will be all-sufficient to induce the pathological conditions I have named. I do not want to convey the idea that chronic interstitial nephritis is alone the form of nephritic waste that arises from the named source, but, on the other hand, I am fully convinced that any or all of the destructive kidney lesions may be induced in this way. I have thus dwelt upon the chief emunctory of the body because of its work being in such close relation to the toxic substances arising in the intestine.

A complex of other symptoms are referable to this source, and taken together they exhibit a most interesting clinical picture. Constipated people are never healthy. It is a notable fact that the insane are always constipated. The subjective sufferings of hypochondriacs and neurasthenics are due very largely to the sluggish action or failure upon the part of the intestinal tube to evacuate its contents. Vertigo, headache, migraine, are the daily and almost constant companions of constipated subjects.

The symptoms I have here enumerated do not by any means exhaust the list, but are so frequently referred to some cause other than the one under consideration, that I placed them in this relation so as to give them particular emphasis. The general catarrhal condition of the intestinal canal, which always for a longer or shorter time precedes constipation, is essentially a paralysis, due to a certain extent to the formation of carbonic acid gas resulting from the presence of alcoholic and acid yeast in the tube, also lactic and butyric acid ferments are frequently present. With these conditions present the environments are ripe for the production and maintenance of the bacteria which induce putrefactive decomposition in animal matter. The hydrogen and ammonium sulphides, together with many other poisonous elements which there induce a hyperemic and then a hypertrophic condition of the mucous membrane, from which, instead of having the normal digestive and lubricating fluids poured out, we find a heavy, viscid, ropy, sticky fluid, which is tenacious and clings to the membrane, filling the mouths of the mucous follicles, and so completely covering the membrane as to almost entirely preclude the possibility of any thing like normal action. The hypertrophic condition of the enervated mucous tissue,

being accompanied with capillary stasis, soon results in the interstitial deposit of fibrin, which serves to carry this same general
hypernutrition into the submucosa and from thence to the general muscularis. The intestinal nerve supply which is almost
wholly derived from the sympathetic system, is crowded, and soon
its filaments and plexuses are so tightly held in the grasp of these
abnormal deposits as to wholly paralyze them. During the progress and evolution of these different yet perfectly coherent pathological stages the intestinal peristalsis is gradually but surely
lessening in its power to force onward its contents, until finally
it wholly and entirely fails.

Throughout all of these progressive stages which merge into
each other easily and almost imperceptibly, the patient will often
have a discharge from the bowels once every twenty-four or
forty-eight hours, and when told that he is suffering with constipation will certainly deny the statement. Such motions, however, are purely mechanical, due to general fermentation going
on and to the constant further crowding of material into the
tube. The proof of this proposition is easily ascertained by
simply withholding such foods as are subject to alcoholic and
acid fermentation, when at once the true state of affairs presents
itself. During the earlier and formative stages of constipation,
when there are more or less regular daily bowel evacuations, the
fecal mould as it escapes from the rectum is greatly reduced in
diameter, and is often not larger than an ordinary lead pencil.
Its color and consistence varies, and it is frequently accompanied
by yellowish-brown chunks of mucous or colloid substances. At
other times the discharges will consist of a sticky, tarry-like substance that can scarcely be forced past the sphincter. This stage
of constipation, which is so often overlooked, is one of vital
importance, as all throughout the whole time, which is not infrequently a period of several years, the system is being gradually
but surely subjected to an auto-infection which, while it belongs
to another field of investigation, stands out prominently as a
remote factor in the production of specific organic changes which
are classed, wrongly I believe, as incurable diseases.

While this and following stages of constipation are in progress, the epithelial surfaces of the intestinal mucous membrane are
gradually becoming enervated and their physiological functions

paralyzed so that they soon lose all selective action, which alone
and in itself stands at the outer gate as a sentinel protecting the
general organism from harmful invasions. Now, they indiscriminately accept and transmit to the circulation every thing offered
them, thus vitiating every organ and tissue of the body.

After this stage we come to a condition at which it is usual
for authors generally to begin the description of constipation.
The stage at which the fecal matters begin to and do undergo
different degrees of desiccation. The desiccated, or dry and
hardened condition of the bowel contents, is not in any sense the
cause of constipation, but is simply one of its stages, and is due to
and referable to the pathologic condition of the intestinal mucous
membrane and its contained structure, which is the direct antith-
esis of the condition found in the stage immediately preceding
this one, the stage of hypertrophy. In this stage, once well
under headway, we have an atrophic condition and consequently
gradual lessening of the normal secretions until finally the whole
secretory apparatus fails to functionate, except as it is influenced
by stimulating and drastic cathartics, all of which induce only
the momentary and vicarious discharge of fluids into the canal,
to be followed without fail by just the same or a worse condition
than before the drugs were administered.

Mechanical obstructions, such as intussusception, narrowing
and final closure of the lumen of the bowel, when not the result
of violence or accidental injury, are simply episodes in the history of general intestinal disease, and should not be classified as
causes of constipation. Neither is it the result of atonic dyspepsia, nor its cause, but is one of the phenomena which does happen synchronously with it, and therefore can not sustain the
relation of cause and effect. The neurotic is generally constipated, and here, also, we are taught that because the patient is a
neurotic he is constipated. The neurotic family, together with
those suffering with chlorosis, are produced by the same gener-
ating causes, viz., unhealthy and abnormal feeding. The foundation for all their future trouble is often laid in infancy. The
infant is carelessly fed upon foods containing large amounts of
amylaceous substances, and these are not infrequently rendered
more palatable by the addition of sugars. Thus, long before
nature has had time to evolve the structures which shall furnish

the starch and sugar digesting fluids, we find the infant being fed sometimes exclusively upon them.

The long and painful list of suffering resulting from this source can not be discussed here in relation to infancy ; but when, by reason of the inheritance of a rather robust physical constitution and a high degree of resisting power, the periods of infancy and childhood are merged into adult life with little if any attempt at a correction of the alimentation, we have as a result our large and ever-increasing family of neurotics, made so by inheritance, it is said. However true it may be that the parentage was neurotic, I still maintain that if the infant has been correctly and scientifically fed, the inherited tendencies would have been absolutely and entirely negatived. It is true also that the most robust and healthy child can be rendered neurotic or chlorotic by bad and irregular feeding.

As the stages of constipation succeed each other and the intestinal tube becomes more and more paralyzed, we come to a peculiar phase or type which presents itself in alternating fits of constipation and diarrhea. Almost without exception the diarrheal discharges are accompanied by scybala, which are indicative of the fact that the discharges in this form are provoked by masses of hardened fecal matter lodged, usually, in the transverse colon, or at the bend of the colon where it merges into the descending colon. These fits of diarrhea are simply an effort on the part of nature to dislodge and convey out of the colon these desiccated and hardened masses.

It is not unusual in old, long-standing cases to dislodge, by proper and well directed means, from the left bend of the colon comparatively large, old, dry masses of fecal matter, one side of which bears the mark of having been, by fungous growths, attached to the colonic wall. The side so attached will invariably be covered with a streaked, bloody mucus, and is nearly always accompanied with an odor denoting tissue decomposition. A diagnostic sign of this particular condition is a peculiar throbbing which is almost sure to be looked upon as an aortic impulse. These patients are in constant danger of infection from this particular denuded area.

With this stage of constipation, which varies, as a matter of course, from its early incipient stages to a more remote condition which renders any degree of health and comfort utterly impossible, there is always a greater or less degree of complication resulting from hemorrhoids. On this account defecation is always somewhat difficult and painful, and evacuation is often postponed and sometimes avoided for many days.

Finally, these patients never have a natural passage, they resort to purgative medicines and injections in order to obtain temporary relief. They suffer most of the time from headache and *malaisé*, loss of appetite, and occasionally have fever, which may pass into typhoid; the tongue is always coated, they are rarely free from gastric distress, and suffer much from pains in the lumbar region and down the thighs.

The entire length of the small intestine becomes smeared over with a thin coating of fecal matter which is constantly undergoing putrefactive decomposition. This is finally transformed into a fungous growth which gradually encroaches upon the lumen of the bowel, and from its presence utterly precludes digestion or absorption of foods. This fungous growth at intervals ripens and is cast off in the form of shreds and patches. When this occurs it is always accompanied with large watery diarrheal discharges. The abdomen becomes very tender, and usually, just preceding and during the discharges there is sharp, cutting pain. During the attacks of diarrhea these patients become excessively weak and prostrated. A peculiar sense of weakness and distress in the abdomen is always spoken of, and which in female patients is very apt to be referred to as something wrong with the reproductive organs. It has been my fortune to treat many of these patients and entirely cure them by referring the whole course of treatment to the abnormal condition of the digestive tube. One patient discharged at one sitting a muco-fungoid cast of the small bowel that was a little more than six feet in length.

When these patients are under proper dietary restrictions the reaccumulation of this material on the mucous wall of the bowel is prevented, and very soon much of the destroyed epithelial lining is replaced, when digestion and absorption of food again assume a healthy and normal action and the patient recuperates very rapidly. To secure this end all vegetable foods must positively be withdrawn, since without doubt such foods are mainly instrumental in bringing about the condition. So long as the digestive tube is in this pathological state it is a highly nutritious fountain for micro-organisms which at the first opportunity invade the general circulation and eventually produce disease in other organs.

Then too, the chemical disintegration of the accumulation originates toxic elements which serve to paralyze capillary circulation, especially in the cord and brain, inducing perivascular and pericellular exudates in these organs. A further and more remote consequence may be detected in brain and cord lesions which manifest themselves in a series of well defined symptoms, such as numbness of the extremities, uncertainty of gait, ringing in the ears, loss of memory, at times inco-ordination of thought, sleeplessness, irritability of temper, and sometimes a host of morbid fears haunts the patient night and day.

The colon plays a very conspicuous rôle in the history and evolution of constipation. It is not an organ of digestion in any sense, other than that, normally all material passes into it in a liquid state, where at once rapid absorption begins and continues until the residue is reduced to a semi-solid consistency. Being thus constantly engaged in, as it were, selecting material which shall supply the different organs with products to repair waste with, it is essential to the health of the general body that its condition be maintained in a clean and healthful state. Its anatomical structure is exactly adapted to the function it has to perform. The cecal end being provided with the appendix receives the contents of the small intestine and at once begins a series of contractions which partake of the same nature as those observed in the stomach when engaged in active digestion. The peristaltic movement twists the cecum upon itself, first in one direction and then in another, in this way thoroughly churning its contents. At regular intervals it contracts upon its contents, thus lifting them upward and beyond the first series of colonic valves. While it is so engaged the appendix is rhythmically contracting and relaxing both in its long and transverse diameters, forcing through the cecal sphincter a lubricating substance formed in its interior.

The colon is never the primary seat of disease, except as by accident or injury. As I have demonstrated elsewhere, the digestive tube always begins to be abnormal and pathologic from above, and the different stages merge into one another by regular and well defined gradations from the stomach to the rectum.

The mode of treatment to which these patients must be subjected to cure them is simple enough in itself, yet requires to be done with accuracy in every detail. The plan is decidedly more exacting than the time-honored expedient of administering some drug to relieve or palliate the difficulty; but relieving and curing the disease are two totally different processes, and it is only with the latter that I have any concern.

I have already specifically stated the fact that when constipation is once established the whole digestive tube will be found diseased, and if the patient is not handled with all these conditions kept well in mind the process of cure must inevitably result in failure.

The first requisite is to secure the very best hygienic environments. That clothing should be ordered which affords the most comfort and at the same time best protects the body, and the clothing that is worn during the day must not be worn at night. A daily bath is essential; this should be an ordinary sponge or towel bath, the bath is more stimulating if a little ammonia is added to the water. Twice a week, at least, the patient must have a thorough soap bath.

The whole dietary course must be managed and carried out on lines that will essentially stop fermentative processes in the alimentary canal. The albumins are best adapted to this end. The quantity and quality of food must be under the daily supervision of the attending physician. Frequent microscopic examinations of the fecal matter should be made in order to ascertain how well the foods allowed are being digested. This is most essential. No article of food should be continued for a single meal if its digestion is not complete.

The colon should be thoroughly washed out every second or third day. Massage and electricity are often of great service. The therapeutics, so far as drugs are concerned, is of vital importance. It is as impossible as it is illogical to suppose that any one formula will suit more than any one case. Here the physician must adapt his work to the particular case in hand, and also to the various changes the case passes through on the way to health.

The great majority of these cases have been much over-drugged, and it is therefore incumbent upon the physician who undertakes to make a cure to avoid as much as possible the use of drastic and irritative cathartics. A gentle tonic laxative given at regular intervals during the day, and persistently continued, sometimes for several months together, is vastly more beneficial than the mere effort to secure a motion of the bowels by some rather quick and effective method. When the entire causative history is constantly kept in mind and the general course of treatment directed into this channel the results will follow with a degree of certainty that will be gratifying to both physician and patient.

ON DR. CHAPUT'S NEW ANASTOMOSIS BUTTONS FOR GASTRO-INTESTINAL SURGERY.

BY CHARLES GREENE CUMSTON, B. M. S., M. D.,

Instructor in Clinical Gynecology, Tuft's College, Boston ; Member of the Société Francaise d' Elec-trothérapie ; Corresponding Fellow of the Maine Academy of Medicine and Science ; Director of the Gynecological Clinic, Tremont Dispensary, etc.

BOSTON, MASS.

[Written for MATHEWS' MEDICAL QUARTERLY.]

My distinguished *confrère*, Dr. Chaput, Surgeon to the Paris Hospitals, has had the kindness to send me a set of his new anastomosis buttons for use in gastro-intestinal surgery, and as they appear to me of the greatest practical value, not only on account of their perfect simplicity, but the simple technique employed, I feel that a description of the instrument and a concise report of the cases in which they have been used may be of considerable interest to those occupied in abdominal surgery.

Before entering into my subject I desire to thank Dr. Chaput, not only for the buttons but for his kindness in sending the elec-trotypes which illustrate this article. There are in all five buttons, and I will take for demonstration the button shown in Fig. 1, which is the largest size, and is intended for gastro-enterostomy and entero-anastomosis.

Full view shows that it has the form of an elliptical ring. The central orifice measures five millimeters in breadth and thirty in length. This orifice is bordered by a rim, separated into six sections by V-shaped openings.   When seen in profile a central groove will be remarked, which measures seven millimeters in depth and ten in breadth.   The entire length of this button is about forty-five millimeters, in breadth from twenty to twenty-one.   The circumference is sixty-two millimeters when the borders of the groove are not brought together, but when this has been done by pressing them with the fingers, it is only from fifty to fifty-two millimeters.

The buttons are made of pure tin and are extremely malleable.

Button No. 1 has an orifice of 10 x 5 millimeters, and a groove six millimeters deep by ten in breadth.   It is used for cholecyst-enterostomy or for circular anastomosis of the small intestine. (Fig. 2.)

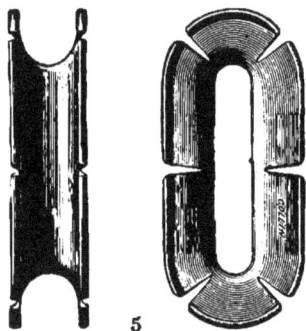

Fig. 1.

Button No. 2 has an orifice of 10 x 5 millimeters, and a groove seven millimeters deep by ten millimeters in breadth, and is intended for circular section of the small intestine.   (Fig. 3.)

Button No. 3 has an orifice of 12 x 5 millimeters, the groove being of the same dimensions as button No. 3.   It is also used in circular section of the small intestine.   (Fig. 4.)

Button No. 4, orifice 15 x 10 millimeters, groove 7 x 10 millimeters.   Used for circular section of the large intestine. (Fig. 5.)

Button No. 5 (Fig. 1) has already been demonstrated, as well as its uses.

To give a general idea of the use of the buttons we will suppose that an entero-anastomosis is to be performed with button No. 5.   A longitudinal incision sufficiently long to insert the button is made on one of the loops of gut. A purse-string suture is then placed in the lips of the orifice of the incision, the button is inserted and the suture drawn tight and tied at the bottom of the groove, so that when this is accomplished one half of the instrument is in the loop of intestine, the other remaining outside.

FIG. 2.   FIG. 3.

The same incision and suture on the other loop of intestine, and the half of the button emerging from the other loop of gut is inserted; the suture of the second incision is tied at the bottom of the groove. *Figure* 6 shows an entero-anastomosis with the button in place, the borders of the gutter still open.

FIG. 4.   FIG. 5.

The borders of the groove are then pressed together from the outside of the intestine, thus saving a plan of sutures. *Figure* 7 shows the borders of the groove brought together by pressure with the fingers. A few sero-serous sutures placed some distance apart may complete operation if necessary.

As is readily remarked, the principle of these buttons is that *the borders of the groove are malleable and may be easily pressed together with the fingers.*

The V-shaped openings in the borders increases the flexibility of the metal and divides the work of pressing them together into several segments which are successively compressed. A fact to be noted is that *it is easier to close the groove than to open it after the borders are pressed together;* and from this it may be seen that as a means of reunion these buttons are firm but in no way dangerous.

The technique that has just been described is good for a demonstration, but as a procedure it has many bad points.

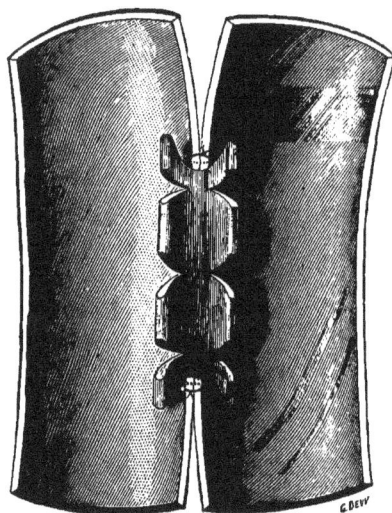

FIG. 6.

The *modus operandi* will now be described:

Entero-anastomosis. First Step. Both loops of the gut are incised; on the posterior lips of the incisions (those the farthest from the operator) a running suture is placed bringing them together. (Fig. 8.)

Second Stage. The posterior groove of the button is placed on the suture, while the upper thread of the suture is brought over into the anterior groove and both ends of the suture are tied, not in the middle of the groove, but at the end. (Fig. 9.)

Third Step. The button is pushed either to the right or left in order to expose the posterior groove, into which a silk suture is

passed and held in place. (Fig. 9. C D, the dotted line rep-
sents the course of the thread.) The button is then put back in its
former position.

Fourth Step. A running suture is then made in the anterior
lips of the incisions with the upper ends (C) of the thread (C D).
In Figure 10 the running suture is not finished, while in Figure
11 it is completed, and it only remains to tie the ends C and D.

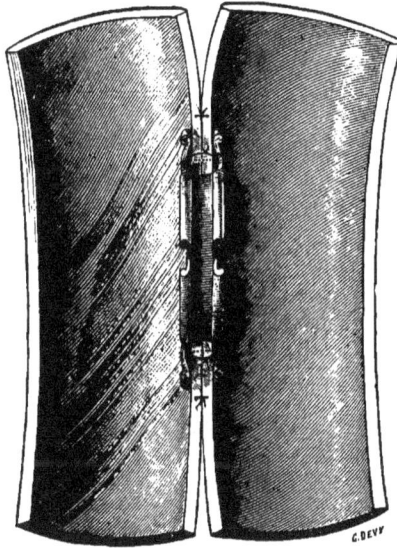

Fig. 7.

Fifth Step. The upper or anterior suture is depressed with a
grooved director, and the borders of the grooves of the button
are pressed together with the fingers through the walls of the gut.
If apposition is perfect, complementary sutures are unnecessary,
but if this is not the case a few sero-serous sutures placed about
one centimeter apart will be sufficient; eight will be found
enough for button No. 4.

The technique of gastro-enterostomy and cholecystenterostomy
is exactly the same as for entero-anastomosis.

Circular Suture. Both ends of the gut are treated in the same
manner as in entero-anastomosis. A purse-string suture is placed

10

in the posterior semi-circumference of both ends of the intestine and then tied on the button. A second thread is stitched into the posterior groove of the button; with one of its ends a running suture will be made in the anterior lips of the intestine, then tied, and the borders of the groove pressed together. (Fig. 12.)

It is absolutely necessary before commencing the operation to make certain that the size of the button selected will be of proper

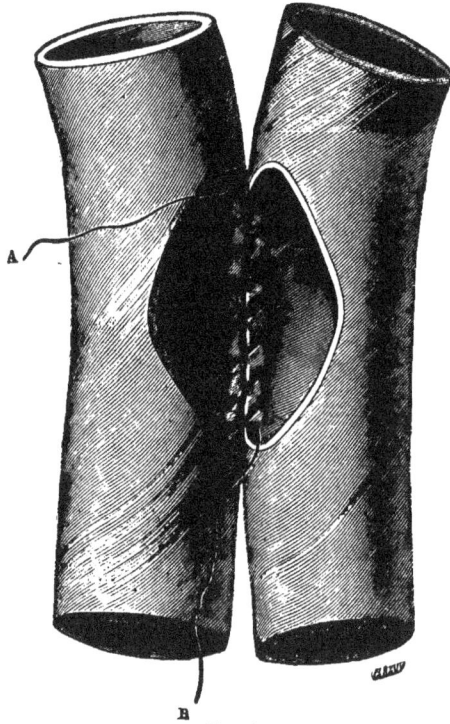

A

B
Fig. 8.

dimensions to keep the instrument *perpendicular to the gut* during its sojourn therein. If one of the orifices of the gut is smaller than the other, it should be increased in size by a longitudinal slit, and the loose triangular flaps thus formed are excised.

Dr. Chaput advises that the sutures employed should be of strong medium-sized silk, as fine silk would break. A straight needle, that of Reverdin in particular is very good.

The operation with Dr. Chaput's button is very quickly done, more so than with Murphy's, because only one plan of sutures is used, while with the latter's instrument two plans are necessary.

Dr. Chaput's largest button is smaller in circumference than the smallest Murphy button; the latter measures 66 millimeters in circumference, while the former, after the groove has been pressed together, hardly exceeds 55 to 60 millimeters.

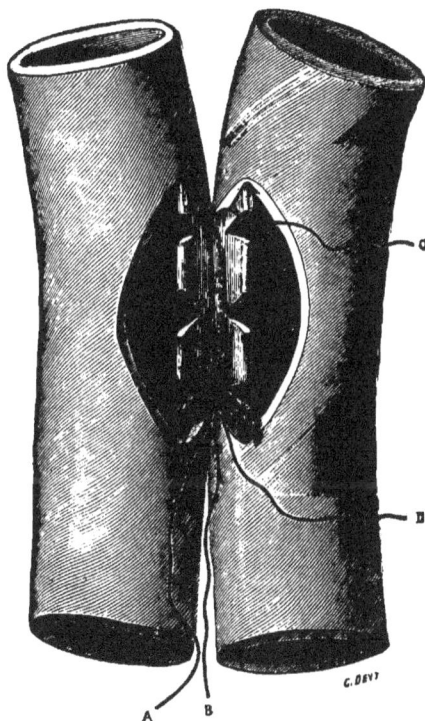

Fɪɢ. 9.

The orifice of each of Chaput's buttons is very much broader than the corresponding Murphy button. With the former an exact idea of the amount of constriction can be obtained, and, what is more, *it is impossible to press too hard*, because the flexibility of the instrument does not allow of a too severe compression.

Now, on the contrary, with the Murphy button there is a tendency to press too tightly, and frequently necrosis or a perforation is the result.

All surgeons have found that when once the Murphy button has been articulated it can not be disarticulated. Now, with Chaput's button, the borders of the groove can easily be separated with a grooved director.

In several cases the orifice of Murphy's button has become blocked up with the feces or alimentary substance; this danger is to be less feared with Chaput's button, whose orifice is very short, only measuring a few millimeters.

FIG. 10.

Murphy's button is a delicate instrument and easily gets out of order; Chaput's button is so simple that there is nothing to get out of order. Murphy's button has too small an orifice for the operation of gastro-enterostomy; Chaput's instrument gives a definitive opening of from six to eight centimeters in length, thus rendering consecutive stenosis far less probable.

We now come to the most important point. Murphy's button *is only liberated by a necrosis of the intestine; Chaput's button causes no necrosis.* And he proves this by his first case, which died forty-eight hours after operation from pulmonary congestion due to the ether, and the second case in whom death occurred on the tenth day from persistent vomiting, the tissues presented no trace of necrosis. In his third case the button came into the rectum on the tenth day, and at the bottom of the groove both purse-string sutures were found, but there was no necrosed tissue between the borders of the instrument.

Mechanism of Elimination. The pressing together of the groove holds the intestine but does not compress it too strongly.

Fig. 11.

After 'the button is applied and pressed together it is easy to extract it from the intestine by slight traction, after having cut the purse-string sutures.

The purse-string sutures which hold, as they do, the borders of a large orifice around the much smaller one of the button, pull strongly on the tissues. After a few days they cut through these structures and the pressure of the groove is not sufficient to prevent the tissues from escaping from their grasp, solicited as they are by their natural elasticity.

The intestinal orifice, strictured as it is by the sutures, tends consequently to return to its former dimensions, which it does by escaping from the groove of the button as soon as the sutures no longer constrain it. When the borders of the anastomotic orifice have become completely free the button falls into the lower end of the intestine because it is pushed by the contents of the alimentary canal.

Clinical Results. The first case was that of a man with cancer of the pylorus, in extreme cachexia, and with tuberculosis as well. Chaput performed a gastro-enterostomy with his button. The upper end of the intestine was surrounded with iodoform gauze and the anastomosis was established between the loop which went

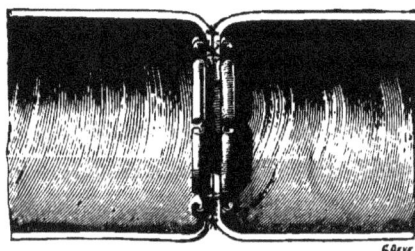

Fig. 12.

toward the stomach and that which went away, in order to re-establish the flow of bile through the intestinal canal. Patient died from pulmonary congestion, due to the ether, forty-eight hours later. The autopsy confirmed the pulmonary lesions and the integrity of the peritoneum and the sutures. There was no necrosis of the tissues comprised in the grasp of the button.

The second patient presented symptoms of serious pyloric stenosis, and an exploratory laparotomy only showed a bend in the pylorus. Gastro-enterostomy was performed. After the operation the patient vomited so repeatedly and with such intensity that the sutures of the abdominal walls gave way, and the intestine had to be replaced within the abdomen and the wall sutured again. Death took place from exhaustion on the tenth day. By going over the history of the patient Chaput arrived at the conclusion that the vomiting was due to the commencement of tabes dorsalis, because he found certain details which allowed of this opinion. However, whatever it might have been, the autopsy showed the integrity of the peritoneum; the gastro-enterostomy was perfect, and it would seem that the method must have been good to have resisted such violent vomiting for ten days. The button was in its place and the tissues in the grasp of the groove were adherent and not necrosed.

The third case reported by Chaput was a man, aged seventy-one, with a carcinoma of the transverse colon. An anastomosis between the small intestine and the sigmoid flexure was performed with the button. The patient supported the operation well, but was subsequently overtaken by intestinal paralysis which had no relation to the operation, but was due to the distension of the large intestine by gas, the feces collecting there without being expelled on account of both the neoplasm and the ilio-cecal valve. A second laparotomy was performed, and showed perfect integrity of the peritoneum and of the sutures, and also that the button had gone. Complete section and obliteration of the end of the small intestine between the anastomosis and the cecum was executed, and the operation terminated by an artificial anus in the transverse colon.

When the operation was completed rectal examination with the fingers revealed the button, which was withdrawn. In its groove were found the purse-string sutures and no necrosed tissue.

The fourth case was that of a woman, aged fifty-eight, with cancer of the pylorus. Gastro-enterostomy, ligature of the upper end of the intestine, and entero-anastomosis, to establish the course of the bile which was cut off by the ligature of the intestine, were executed. The patient was operated upon on September 27, 1895, and was still alive December 11, 1895. Her digestion was quite good, vomiting had ceased, and the gastric dilatation had disappeared. The buttons were not found in the feces.

These four cases show that (1) the employment of these buttons is rational and benign, and that up to the present no objection can be raised against this method; (2 that the buttons are easily expelled; (3) there is no necrosis of the tissues comprised within the grasp of these buttons.

DISEASES OF THE PYLORUS.*

BY DEERING J. ROBERTS, M. D.,

NASHVILLE, TENN.

Gentlemen, this subject having been assigned me by the Committee on Essays, it will be perhaps well enough that a few preparatory words of definition be submitted to you. Taking the stomach as a whole, it is usually divided for the purposes of description into three parts—possibly from its near relationship to the gall, as most of us have been taught in our juvenile days that *omnia Gallia in tres partes divisa est*, etc., or probably from the great prominence given to matters both celestial and terrestrial to the Trinity.

That third lying to the left and into which the esophagus enters being known as the cardiac or greater extremity; that lying to the right and opening into the duodenum is termed the pyloric or upper extremity, the middle division lying between the two.

The stomach is provided with three layers of muscular fibers. (1) External longitudinal, a continuation of those of the esophagus. (2) Middle circular. (3) Internal oblique. The longitudinal fibers are most marked over the lesser curvature and at the pyloric extremity; they are not continued very distinctly over the rest of the stomach. The circular layer is not very distinct to the left of the cardiac opening over the great pouch, being more prominent over the middle portion, and largely developed in the pyloric, and at the junction of the stomach with the duodenum forms a powerful muscular ring. At this point they project considerably into the interior of the organ, almost abruptly ceasing at the opening into the duodenum, so as to form a sort of valve, presenting when contracted a flat surface looking toward the intestine.

The oblique fibers take the place, to a great extent, of the circular fibers over the great pouch, extending obliquely over the fundus from left to right, and moving at a distinct line about the junction of the middle with the pyloric portion of the stomach. At this point the stomach becomes constricted during the

* Read before the Nashville Academy of Medicine.

movements which are incident to digestion, dividing the organ into two tolerably distinct compartments—the cardiac extremity and middle portion forming one, and the pyloric extremity the other.

As a consideration of the diseases of the pyloric portion of the stomach would confine me to that of diseases of the stomach, and would occupy entirely too much time and space for a paper of twenty minutes' reading, I shall confine myself to a brief consideration of the morbid conditions pertaining to the pylorus itself.

In the first place, while the stomach is occasionally absent in a cephalic monster, yet more rarely it is to be found abnormally small in fetuses which are otherwise well developed; and while complete atresia, congenitally, of the pylorus is very rare, yet stenosis or abnormal contraction is more frequent.

While possibly we may have occasionally a deficient contractility of the pylorus, and a too great potency of its lumen, producing a temporary diarrhea by permitting the alimentary masses to pass into the small intestine before gastric digestion is completed, it is to conditions of *stenosis* and obstruction of this orifice that pathological phenomena are almost exclusively attributed.

Obstruction to the passage of the contents of the stomach into the duodenum is not infrequent, and it may arise from very different pathological conditions.

1. Cancer of the pyloric end of the stomach surrounding the opening and gradually spreading to the intestines. The muscular fibers become hypertrophied, the contractile fibers enlarged and increased in number, in some cases becoming condensed into dense fibrous tissue; not only obstructing the passage of the gastric contents by closure of the lumen of the orifice, but by preventing those contractile movements by which they are forced onward.

2. Fibroid thickening of the submucous tissue as an accompaniment of cirrhosis of the stomach occurring independently. This may be confined to the pylorus itself, or may extend some distance back, producing a hard, leathery condition of the gut. The same effect, although to less degree, being produced in cancer. The muscular bundles become hypertrophied, their contrac-

tion being embarrassed by the tough, fibrous tension separating them.

3. Gastric ulcer, located at the pylorus, may produce spasmodic contraction during its existence, or permanent stenosis by reason of cicatricial contraction after healing; or destruction of muscular fibers may prevent the forcing onward of the alimentary mass.

4. Tumors in the vicinity, lipoma, carcinoma, fibroma, an occluded and distended gall-bladder, a floating, dislocated kidney, a fecal impaction of the hepatic flexure, or transverse colon, etc., may so press upon the pylorus as to occlude it.

5. Adhesions may form between the duodenum or pylorus and adjacent parts, and in this way produce a difficulty in the passage of food from the stomach.

The principal effect of any considerable obstruction of the pylorus is a greater or less degree of dilatation of the stomach. This invariably occurs. In fact but few cases indeed of dilatation of the stomach occur independent of pyloric obstruction. The most prominent clinical phenomena are vomiting, occurring at irregular intervals usually, and several hours after taking food, heartburn, pyrosis, other symptoms of indigestion, constipation, and gradual and progressive emaciation.

While gastrectasia, or dilatation of the stomach, may be due to weakness of the muscular tone of the stomach from imperfect nutrition in cases of anemia, chlorosis, exhausting acute or chronic disease, amyloid degeneration of the gastric arteries, chronic gastritis, dyspepsia accompanied by stagnation and fermentation of the gastric contents, polydipsia, polyphagia independent of or attendant upon diabetes, when it results from stenosis of the pylorus it always reaches a more extreme degree than from any other cause, and the clinical features due to other causes are absent.

In regard to vomiting, Flint very correctly says "that it occurs almost without exception when dilatation is due to pyloric obstruction, and in other cases it is a frequent but not a constant symptom."

The characteristic features as regards this symptom are the occurrence of the vomiting after periods varying from a day to several days, the large quantity of matter vomited at a time, absence of nausea and retching, and freedom from bile. The matters vomited are acid and frequently in a state of fermentation, as indicated by froth containing gas. Sarcinæ, yeast fungus, bacteria, and crystals of fatty acid are found. Acetic, lactic, butyric, and other organic acids are usually present. According to Van der Welder hydrochloric acid is persistently absent when the obstruction is due to cancer. Alimentary matters vomited may have been retained in the stomach for several days—this being regarded as somewhat diagnostic. In the unobstructed stomach food partaken of one day is usually digested by the next, or rejected. Repeated attacks of vomiting without a great degree of effort, retching or nausea, and masses of food that have passed into the stomach several days preceding are strong indications that the power of the stomach to propel the food into the duodenum is wanting.

The persistent constipation, unattended by fecal accumulations in the small or large intestine, also adds to the characteristic features of occlusion of the pylorus.

Although the diagnosis of pyloric obstruction is somewhat difficult in its incipiency, yet as the disease progresses it gradually becomes more and more clear. A greater difficulty, that even sometimes projects itself until the end, is a discrimination as to the particular pathological cause producing it. A few brief points will be briefly cited. While all conditions, unless relieved by nature, aided or unaided by art, are as a rule fatal, and while the obstruction from any cause may be followed by fatal results in from a few weeks to a few months, cancerous disease as a rule will rarely, if ever, be protracted beyond one and a half years. Either from cancerous infiltration of adjoining and adjacent areas and viscera, or the gradual but rapid development of a cachectic condition due to maglignant disease, but few cases indeed are protracted over a longer period of time.

In the matters vomited we rarely have indications of hemorrhage other than in cancer or the eroding stage of gastric ulcer.

The absence of the cancerous cachexia, the absence of pain of an acute character, the ability to hold the patient up beyond the time usually observed in malignant disease, together with the development of but a limited degree of tumor beneath the cartilage of the eighth rib on the right, point more toward fibroid thickening. Malignant disease may be more confidently expected after forty-five or fifty years, yet the other conditions are not limited to prior ages.

It is to be regretted that we have no special diagnostic factors; one of some moment has been cited by Van der Welder, viz., the persistent absence of free hydrochloric acid in the matters vomited or removed from the stomach by artificial means. Several practical tests in this are given in Pepper's System of Medicine, Vol. II, page 544.

The measures of treatment to be resorted to are to be found both in the domain of general medicine and surgery. In the early stages, and in fact until the diagnosis of complete or partial occlusion has been thoroughly established, our remedial measures should be limited to the former, leaving the latter as a *dernier ressort*, or held in reserve until we are fully satisfied that other measures are futile, or that the patient is too rapidly succumbing by reason of progressive asthenia.

It is often, indeed, a most difficult matter to decide when to abandon the expectant measures of general medicine and resort to explorative, if not curative operative measures—measures that sometimes show us conditions that we would rather not see. It is a question that can only be decided by a conservative yet thorough judicial investigation of every fact and factor of the case; and it is only he who can weigh well each particular factor individually, and the whole mass of clinical testimony in its entirety, that will be satisfied with the results in his opinion in such cases.

Among the expectant remedies of the therapeutic field first to be considered of paramount importance are dietary measures. By a carefully regulated regimen much suffering may be avoided. And even in those sad cases of malignant disease the bright star of hope may be enabled to shine a little longer. The crippled stomach must be aided by rectal alimentation, and the demands of nutrition placed at low-water mark by a rigid abstinence from all unnecessary mental or physical exertion. As regards alimentation by the stomach it should be limited in quantity, and should consist only of milk, eggs, animal food, and fish. Starchy and saccharine foods should be rigidly excluded. If bread is persistently demanded, let it be limited to gluten bread, or that made from diabetic flour, almonds, etc. The lean of fresh meat, the juice expressed from freshly broiled tenderloin steak, concentrated meat extracts will render up a considerable amount of nutritive matter to the absorbents of the stomach, and aided by the judicious introduction of proper nutritious material into that second stomach, the rectum, life may be maintained for many days with complete occlusion of the pylorus. The character of food introduced into the stomach should be that which would leave but little residue to pass the barred gateway of the pylorus, or be rejected by the cardiac orifice.

In connection with rectal alimentations I will mention that, more than a year ago, having a good deal of trouble to get this portion of the alimentary canal to tolerate the ordinary fluid or semi-fluid material used per enema, I found a most valuable substitute and a far more convenient means in the large size gelatin capsules made by our capsule manufacturers. I found that the Parke, Davis & Co. capsule No. 12 would hold at least half an ounce of any of the handy beef extracts, and was more easy of introduction, and more readily retained than an enema. Since then I have made frequent use of them for introducing both medicine and aliment into the system.

If fermentation and the development of gas persists in spite of the exclusion of starchy and saccharine foods, the anti-fermentative remedies may be used, such as creosote, creosotal, carbolic acid, the salicylates, or that excellent proprietary preparation, listerine.

Lavage of the stomach, introduced by Kussmaul in 1867, marked a new era in the treatment of gastric disorders, and more than one case of apparent permanent occlusion has been saved from the surgeon's knife of later days in the inevitable false result of former days, by the washing out of the stomach with the stomach pump or the more simple siphon. As a rule, washing out the stomach once a day is often enough in any case; in many, every two or three days will suffice. While Kussmaul recommends before breakfast as the most suitable time, I rather agree with Prof. Welch in occupying the half hour previous to the midday meal. It should not be resorted to any oftener than absolutely necessary, as determined by the degree of discomfort and uneasiness produced by gastric distension from decomposition of its contents or other causes.

The contents of the stomach being invariably acid in pyloric obstruction, occasional drinks of alkaline waters may be used, or the administration of carbonate of soda or other simple alkali. If there is much pyrosis and heartburn that is not relieved by anti-fermentation and alkaline measures, an occasional dose of bismuth s. nit., may be given.

Strychnia, from its property of stimulating muscular power, has been recommended and is worthy of consideration, in the hope that at some time the obstruction may be overcome by making an occasional if not a continual pressure onward of the gastric contents; furthermore, it is a most excellent tonic. For the constipation, although it is due partly to the diminution if not total cessation of material coming from beyond the pylorus, and partly to an absence of those reflex peristaltic movements incited by gastric digestion, it will be necessary to resort from time to time to a laxative pill of rhubarb, Carlsbad water, Blue Lick water, the natural or artificial Sprudel salt, or waters of this class. Tonics, stimulants, anodynes, hypnotics, and other drugs have their appropriate places to be used when occasion requires, and only then.

Friction with oil, preferably cod-liver oil, to be followed in three or four hours by a good warm bath, is not only agreeable, but possibly some of the oil may be absorbed by the skin.

In regard to operative measures, as the discussion is to be opened by a gentleman who is more familiar with the glittering sheen of the scalpel of late years than myself, and as more than one member of this academy is under the prevailing influence of the *" cacoethes insecti,"* I shall be brief, and limit myself to a few historical data.

Gastrotomy and forcible divulsion of the contracted pylorus had yielded nine successful cases to Loreta from 1882 to 1884. It is limited to stenosis due to fibriod hypertrophy and cicatricial contraction. It was first attempted by Schede in 1877, who was unsuccessful. Ashhurst, in his fifth edition (1889) gives a list of thirty-four cases, by various operators, nineteen successful, fourteen fatal, and one uncertain. Pyloric gastrectomy, or resection of the pyloric extremity of the stomach, first suggested by the German Merrem, was first put into practice by Pean in 1879. Ashhurst (*op. cit.*) gives a list of one hundred and nineteen cases by different operators. Bilroth leads the lot with twenty-two cases and only eleven deaths. This measure may be resorted to in malignant disease, as a means of prolonging life, and in non-malignant cases where divulsion is impracticable. The younger and stronger patients, and those who were on the operating table the shortest time, bore the operation best. It is necessarily a slow and tedious operation, and according to Druitt (twelfth edition) "no patient recovered when the operation lasted three hours." Bilroth, one of the most expeditious operators, has the best record.

According to Erichsen (eighth edition), V. Winiwarter originated the operation of gastro-enterostomy, as he very properly termed it. He attempted an excision of the pylorus, but he found it impracticable, and very ingeniously drew up the lower part of the duodenum and established a fistulous opening between it and the great curvature of the stomach. Ashhurst gives a tabulated list of sixty-two cases and twenty-four deaths. If resorted to, Senn's decalcified bone plates, the catgut mats of Davis, or similar means will be of no use. In cases where it was found applicable, attachment of some portion of the duodenum to the anterior surface of the pyloric extremity would seem preferable to a more dependent portion of the stomach, or a more direct part of the small intestines. The operation is applicable to all cases of permanent occlusion. Senn regards it questionable in stricture from cancer, the few weeks or months of life not compensating for the immediate risk of life.

In conclusion, I beg leave to state that the history of operative measures in connection with pyloric disease has originated and been developed during the time in which many of the members of this Academy have been connected with medicine, and it is one of the important features of recent—yes, quite recent—developments in the special field of abdominal surgery. And while its results have been brilliant indeed, in some instances absolutely snatching the patient from the very brink of the grave and restoring him to health, in others, relieving suffering and preserving life, yet it is an open and an inviting field that must not be invaded indiscriminately, but only after carefully and well weighing every phenomena connected therewith.

The very interesting case, some weeks ago so graphically

delineated by Drs. Buist, Douglas, and yourself, Mr. President,. while not purely a case of pyloric disease, is quite germane in its relation to pyloric obstruction, and the results of the autopsy in an adjoining State redound as much to the credit of the gentlemen connected with the case as the most brilliant achievement of gastrotomy, pyloric gastrotomy, or gastro-enterostomy.—*The Southern Practitioner*.

CROWLEY, D. D., OAKLAND: PREPARATIONS AND INSTRUMENTS REQUIRED IN EXPERIMENTAL OPERATIONS ON THE DIGESTIVE TRACT OF THE ANIMAL. (*Pacific Medical Journal.*)

The following classification of the details of preparations and instruments is given:

1. Antisepticizing the skin and hair of the animal.

2. Bandages to tie legs and muzzle animal.

3. Chloroform, paper cone for anesthesia, hypodermic syringe, whisky, aromatic spirits ammonia, and digitalis.

4. Razor, soap, carbolized water, and soft cloths.

5. Clean cloths to surround line of incision and a three-percent. solution of carbolized hot water.

6. Plenty of moist carbolized cotton pledgets, and two basins of a half-per-cent. solution of carbolic acid, and a third basin of sterilized warm water.

7. Instruments: Two knives, grooved director, four artery forceps, scissors, four small cambric needles, armed with fine sterilized silk, three large and two small curved needles, armed with corresponding silk.

8. Drainage tubes.

9. Dressings: Iodoform, balsam peru, jute, surgeon's cotton, common cotton, flannel bandage, plaster-of-paris bandage.

CLINICAL MEMORANDA.

AN ABDOMINAL CASE.

BY AP MORGAN VANCE, M. D.,
LOUISVILLE, KY.

On February 8, 1896, Mr. M., aged forty-three, was referred
to me by Dr. Ringo, of South Louisville, Ky., with a note asking me for my opinion in regard to a swelling in the right side
of his abdomen. Mr. M. gave me this history: for three months
he had been sick and had lost forty pounds in weight. The physician who attended him in the beginning called it a case of walking typhoid fever, and gave him, every two hours, a mixture
containing " coperas, strychnine, and murcury." Shortly after
beginning this medicine he " swelled up and vomited a great deal."
During the six weeks prior to his being seen by me he had been
able to be up. In fact he remained in bed very little of the time.
That morning he had ridden two miles in a buggy and about
three miles on the electric cars. For several weeks he had been
passing a great deal of pus by the bowel, as much as " two or three
bucketsful " in all. He had noticed the swelling in right side of
abdomen very early. Sometimes it seemed smaller and at others
larger; was always sore, particularly if he made a misstep or was
jolted in any way. On examination of the man I found he was
very pale with an anxious countenance; pulse 130, temperature
103.5°, still he had walked from the cars two squares to my
office. The abdomen was only slightly distended, his bowels
having acted naturally that morning. In the right side of abdomen was a well-defined, doughy tumor, fully as large as a man's
two fists, moderately tender on handling. The middle of this
swelling corresponded with McBurney's point. I thought the
man was suffering with an old appendicitic abscess, which had a
connection with the bowel, the sac refilling at intervals and keeping up a condition of low septic fever, and in a letter to the
doctor advised an exploratory incision. The next day the man
made the same trip to my office from his home in the country,

11

and I found his condition about the same.   He said when he got home the day before he had eaten a very hearty dinner, as he was very hungry.   I sent him to the Norton Memorial Infirmary with a note to the superintendent, but before the condition of the patient was understood he walked up to the third story, seemingly none the worse for the active exercise.   At three o'clock on Sunday, February 9th, I opened the abdomen as for an appendicitis, and on incising the abdominal wall I came to the tumor so plainly to be mapped out before, firm adhesion being found between the wall and tumor, with evidence of inflammation.   Carefully open-ing the mass I was surprised to find a large cavity comparatively empty.   Sweeping my finger around in search of the cause of the abscess I could discover very little, caseous pus in moderate quantities coming away on my finger.   When just about to close the wound over a packing of gauze I felt what I considered might be the remains of a necrotic appendix deep down in the bottom of this cavity.   Carefully separating adhesions and draw-ing this up to the surface, I found it to be a ragged opening in what turned out to be the small intestine near the cecum.   I was greatly puzzled at first, as the bowel had no peritoneal cov-ering or mesentery and was completely bloodless.   The part between the large opening and the cecum seemed in better condi-tion, but there was no mesentery or peritoneal covering to this. I completed the section at the large perforation and gently drew out this much-narrowed and lifeless tube, coming to new perfora-tions every two or three inches.   After removing about eighteen inches, and no sign of good bowel appearing, the same blanched and necrotic tube was to be traced deep down into the pelvis. All this investigation was done without entering the general cavity.

The part removed seemed to consist of the mucous and sub-mucous coats of the bowel only.   Finding it impossible to bring any good intestine into anastomotic apposition with the cecum, which seemed in fairly good condition, I determined to break up the firm adhesion separating my field of work from the general cavity, and try if possible to untangle the diseased from the healthy bowel, and thus accomplish an anastomosis.   I found a general plastic peritonitis present over the whole area exposed, and the intestines completely matted together.   In attempting to separate some of the coils in an effort to trace down the necrotic tube into the pelvis perforation after perforation was produced. After repairing the tears, the patient was beginning to show marked effect from shock, and I finished the operation by stitching the free ends of the small bowel, or rather the mucous and submucous coats of the small bowel, in each angle of wound, and got the man to bed. I expected him to die within forty-eight hours, of general peritonitis, but, astonishing to state, he lived twenty-five days, finally dying of exhaustion. I failed to say that there was no evidence of any fecal matter in the intestine removed, nor in the cecum, and absolutely no odor connected with the condition.

Dr. J. B. Bullitt made an autopsy, but found very little more to base an opinion upon than we already knew from the discoveries made at the operation. There was no evacuation by the natural channel after the operation, but all came by the artificial opening. At the *post-mortem* the cecum was found full of soft feces and the appendix in a healthy condition. The intestine traced from the fecal fistula at the *post-mortem* went deep down into the pelvis where there was firm attachment to the upper part of rectum and great evidences of inflammation.

This is a curious case. What was the cause of all this poor fellow's trouble? Was it a case of typhoid fever with perforations and abscess, or did the typhoid fever medicine have any thing to do with the remarkable condition found in his abdomen? I rather lean to the opinion that it was primarily a case of typhoid fever. Why the man did not die of peritonitis before I ever saw him I am at a loss to say, and why he lived so long after I did see him I am at a greater loss to explain. The microscopic examination of the specimen removed shows it to be composed of the mucous and submucous coats of the ileum.

218 West Chestnut Street.

MATHEWS' MEDICAL QUARTERLY

"ALIS VOLAT PROPRIIS."

Vol. III.   LOUISVILLE, APRIL, 1896.   No. 2.

JOSEPH M. MATHEWS, M. D.,   -   -   -   -   -   EDITOR AND PROPRIETOR.
HENRY E. TULEY, M. D.,   -   -   -   -   -   ASSOCIATE EDITOR AND MANAGER.

A Journal devoted to Diseases of the Rectum, Gastro-Intestinal Disease, and Rectal and Gastro-Intestinal Surgery.

Articles and letters for publication, books and articles for review, communications to the editors, and advertisements and subscriptions should be addressed to
Editors Mathews' Medical Quarterly, Box 434, Louisville, Ky.

BOGUS MEDICAL JOURNALISM.

The timely editorial in the *Medical Record* of recent date on *Bogus Medical Journalism* is of such value that its sentiments should be indorsed by the whole profession. Whenever a medical journal becomes so prostituted as to be turned into an advertising medium " by men who have neither interest nor sympathy with the profession save as they can make it a means of livelihood for themselves," it should be discountenanced by all honest practitioners. It is simply degrading a profession which should be honored for its integrity, and by a patronage of such the subscriber becomes *particeps criminis* to a fraud.

It is a shame, as the editor of the *Medical Record* points out, that a medical man can be " hired " to edit such a journal. It is also true that sometimes a journal is launched upon the medical public " by some doctor who is desirous of exploiting his own abilities and skill, and at the same time add to his library such books as he may be able to induce publishers to send for " review." Fortunately, however, the publishers soon discover the nature of such frauds, and cease to send any books to them. It can also be said that cheap medical journals are sometimes started apparently for the sole purpose of venting the spleen of its editor, and of quarreling with all mankind upon the slightest pretext. No such spirit should be received kindly by doctors, and a rebuke should be intimated by a refusal to take any such periodical.

In this day of good medical journals the bad can be easily assorted out, and the profession should act as the grist mill. It is but fair that an editor and publisher of a medical journal should give full compensation for its advertised rate of subscription, and such journals as do not will meet a deserved fate. It is unfair, as the *Medical Record* remarks, that journals which are put to great expense in order that the reader should have the best should be made "to compete with journals such as have been mentioned." We give our full indorsement to the well-written and appropriate editorial of the *Medical Record*.

THE AMERICAN MEDICAL ASSOCIATION.

The next meeting of this Association will be held in Atlanta, Ga., May 5th to 8th, and it bids fair to be one of the most successful in its history, held as it will be in a city which has made such a reputation for hospitality to strangers in her gates during the recent exposition. It is to be hoped that all the readers of the QUARTERLY will be present and take an active interest in the proceedings.

The choice of a route is necessarily a serious question, but with great wisdom the official line has been designated as the Queen & Crescent and Southern Railways from this section and the "bluegrass" region. From any point in the State connections can be made with the *Journal Train* by communicating in regard to time with Mr. A. Whedon, Louisville, Ky., Mr. C. A. Beuscoter, Chattanooga, Tenn., representing the Queen & Crescent and Southern Railways, and Dr. John B. Hamilton, Chicago, editor of the *Journal of the American Medical Association*.

Correspondence.

ONE HUNDRED AND THREE CASES OF INFANTILE INTUS-
SUSCEPTION; INTESTINAL DISTENSION VS. LAPAROTOMY; IN-
DIGESTION OF STARCHY FOODS, TREATMENT; CARCINOMA
RECTI; COCAINE IN MAJOR OPERATIONS; POST-MORTEM
BASSINI; TRAUMATIC APPENDICITIS; FOREIGN BODIES IN
APPENDICITIS; ANALYSIS OF ONE HUNDRED CASES OF AP-
PENDICITIS; "THE ABDOMINAL TONSIL;" APPENDICITIS DUE
TO RHEUMATISM; THE WOUND TO FIT THE CASE; PEROXIDE
OF HYDROGEN; GASTRO-ENTEROSTOMY; MILK SUPPLY OF
NEW YORK, SOURCE, INSPECTION, COST OF TRANSPORTATION,
PREVENTION OF THE SALE OF ADULTERATED MILK, MILK
ANALYSIS, "STANDARD MILK," MICRO-ORGANISMS IN MILK,
CERTIFIED MILK, ASSOCIATION OF HONEST DEALERS; EXCIS-
ION TRANSVERSE COLON; PIN IN THE APPENDIX.

NEW YORK ACADEMY OF MEDICINE, STATED MEETING, JAN-
UARY 2, 1896.

Infantile Intussusception. A study of one hundred and three
cases treated either by intestinal distension or laparotomy, formed
the subject of a paper by Dr. F. H. Wiggin.

Mr. Howard Marsh, of London, reported the first successful
abdominal section on a child under twelve months. Mr. Edmund
Owen operated on a child for intussusception occurring on the
day of its birth. The bowel was found pervious. Operation on
the next day; enemata under chloroform giving no result, an
artificial anus made by opening the first piece of small intestine
presenting in the wound. The child survived six days, dying
when nine days old.

Of the one hundred and three cases nearly fifty per cent.
occurred in about the same proportions at the fourth, fifth, and
six months. Eighty-nine per cent. of the ileo-cecal variety.

Trauma. Violent jumping up and down is considered by A. Jacobi as producing serious results.

Where the gut protruded from the anus, or the tumor (five per cent. of cases reported) was found low down in the rectum, tenesmus was present in a marked degree.

The greater the constriction the more pain and frequency of passage of blood. In two cases cure by sloughing occurred. Sixteen, or forty-one per cent., of thirty-nine cases treated by inflation or enemata recovered. Chloroform was used in three cases while distension was being practiced. Twenty-three cases died, or a mortality of fifty-nine per cent. One case, inverted while under chloroform, vomited, inspired the vomited material, and died. The average age in the fatal cases was five months. If inflation is deemed advisable, one and one half pints of tepid saline solution, at an elevation of not more than three feet, should be injected, and if the desired effect is not brought about, abdominal section should be resorted to after a few hours' rest.

Laparotomy was resorted to in sixty-four cases of infantile intussusception, twenty-one, or thirty-two and eighth tenths per cent., were successful, and a fatal result in sixty-seven and two tenths per cent. The average age in the successful cases was six and one half months, and the average hour of operating was the forty-fourth from the onset. Average age in the fatal cases, five months; average hour, one hundred and second. Incision, resection, or anastomosis was practiced in forty-five cases; twenty-four were fatal, a mortality of fifty-three and four tenths per cent. Excluding those cases operated upon previous to 1889, we have eighteen operations with the present technique, and only four deaths; a mortality of twenty-two and two tenths per cent., giving a fair idea of what may be accomplished if operated upon within the first forty-eight hours.

Dr. A. Jacobi believes that infantile intussusception is a preventable disease. The mesentery of infants has a very loose attachment, and the intestine likely to be invaginated from violence or disease, whooping cough, diarrhea, or constipation. Attention, early, to diarrhea or constipation will avoid localized peritonitis, and thus prevent many cases of infantile intussusception.

If enemata correctly applied have no effect within a few hours, call in a surgeon. With moderate pressure, one and one

half feet, more fluid can be introduced and danger of injuring
the intestine avoided.   Anesthetize, raise the hips, and while the
enema is being given knead the abdomen.   Give opium to quiet
the child after reduction; invagination will be less likely to
recur.

Dr. B. F. Curtis, in statistics collected up to 1891, found a
mortality in children and adults of fifty-eight per cent., and
questioned whether there is now so low a mortality as twenty-
two per cent.   The improvement is due to an earlier resort to
operation.   Consultation with a surgeon should be sought early,
and even the mildest cases considered as surgical.   Acute cases
demand surgical interference without wasting time with enemata.
Senn has suggested that relapses might be prevented by stitching .
together the mesentery at the time of operation.   Relapses do
occur within a few hours or even after the lapse of three months.
Injections are of diagnostic value, but if necessary to repeat
operation should be resorted to at once.   Inject a pint at a time,
from an elevation of not more than three feet, and allow this to
flow out before introducing the second.   Anesthesia is necessary.
Examine the tumor between each injection.   Make a small
opening, reduce quickly, and keep up the body heat.   Enterectomy in these young infants is very fatal.

Dr. J. Lewis Smith: We should be able to make a diagnosis
between dysentery and intussusception within thirty-six or
forty-eight hours.   In intussusception after the first few hours
the stools consist of blood and mucus, free from fecal matter.
Fluid enemata are superior to inflation.

Dr. R. Van Santvoord: *Post-mortem* examinations of infants
frequently show numerous small invaginations.   One writer
suggests that colic in infants is due to transient intussusception.
Opium may be used with care in preventing a more serious
invagination as the result of excessive action of the intestine.

Dr. Wiggin, in closing, called attention to the fact that the
successful cases treated by enemata were in the forty-first hour,
and the unsuccessful cases on an average the sixty-ninth hour.
In operative cases success was achieved in the forty-fourth hour,
in the unsuccessful cases operation was not done until the one
hundred and fourth hour.   *Delay is dangerous.*   Twenty-two
per cent. mortality is a fair estimate when laparotomy is done

during the first forty-eight hours. Make a one and one half or two inch incision and reduce the invagination within the abdomen.

Section on General Medicine; Indigestion of Starchy Foods; Treatment.

Dr. Reynold W. Wilcox, continuing his report on the Starchy Foods (see report New York State Medical Society, 1895). Treatment must take into consideration causation. Dr. Bulkley has called attention to the fact that a glass of hot water taken a half hour before meals will prevent a desire for liquids with the food, and aid in preventing indigestion of starchy foods. Excessive acidity of the stomach interferes with the conversion of starch. Sodium bicarbonate may be given in continuous large doses, without harm, solely to neutralize the hydrochloric, and will not affect the organic acids. Salines diminish the acidity of the urine and increase the alkalinity of the intestinal contents. The normal sodium phosphate, to which is added fifteen to forty grains of sodium bicarbonate, should be administered in hot water on rising. When diarrhea alternates with constipation, neither opium nor astringents should be given. Naphthol is locally irritant, and, by increasing the chlorine, interferes with starch digestion; sodium phosphate reduces the quantity of mucus.

Malt extracts have given good results, but the liquid preparations contain too much alcohol to admit of much power to convert starch. The best results have been obtained from the use of Taka Diastase.

Dr. C. M. Quinby maintained that the adjustment of false teeth and the addition of saliva by the use of chewing-gum after meals has been of decided benefit. Pancreatin has also given good results as an aid to digestion.

Dr. A. H. Smith prefers the milk of magnesia, not necessarily the proprietary preparation. A much smaller dose will neutralize a given amount of acid than will sodium bicarbonate. It does not set free gas, and acts as a gentle laxative. The addition of saliva is of service.

Dr. F. A. Burrall: A German physician recommends the use of slippery-elm, etc., between meals to increase the flow of saliva.

Indigestion of a nervous origin, due to worry, often ceases when the cause is removed. Strychnine with ipecac are of service. Flatulence, due to torpidity of the liver, may be relieved by calomel and ipecac, half grain each at night, with a mild laxative.

Dr. Wilcox: The increased flow of saliva is of undoubted benefit. Pancreatin gives good results in those cases where the stomach acidity is low.

SECTION ON SURGERY.

In order to demonstrate that " carcinoma of the rectum is not necessarily a very malignant form of disease," Dr. H. Lilienthal presented a man of fifty years of age, who when first seen was looked upon as inoperable. The rectum, bowel, and adjacent tissues were greatly involved. Colotomy was refused; alcholic. Severe hemorrhage necessitated haste, so that some of the infiltrated tissue was left behind. Saline infusion was deemed necessary and gave a good result. Twisting the lower end of the gut was attempted at a later date, but the tissues were so friable as to make this impracticable. Operation June 23, 1895. The man can now attend to his duties as a plasterer. A slight recurrence can be felt in the perineum. Nothing felt as high up in the rectum as the finger will pass.

Answering the question, Is it cancer? by Dr. Wyeth, Dr. Lilienthal said that the patient had been put on antisyphilitic treatment for one week before operation, without any effect.

Inguinal colotomy Dr. Dawbarn considers more favorable in every case.

The use of cocaine as an anesthetic in major operations was brought out in discussing a case of Dr. Wyeth's, in which he had used a four-per-cent. (25 minims) in amputating the shoulder for osteo-sarcoma. Dr. Curtis (chairman), said that he had used cocaine in several major operations; gastrostomy, exploratory laparotomy for intestinal obstruction, and found it very useful. Dr. Meyer objected to the four-per-cent. solution unless combined with nitrogylcerine. He prefers the solution of Schleich of Berlin; one tenth per cent. of cocaine, one fifth per cent. of saline solution, with the addition of one third grain morphia, acts admirably. Large quantities can be used without danger.

Dr. Wyeth has performed many hundred operations under cocaine, and never seen any toxic effect. Twenty minims of a four-per-cent. solution is as much as he uses for an operation, never puts in more than ten minims at a time. He has removed the appendix in this way.

Dr. Dawbarn prefers to give large doses of stimulants before using cocaine. Patient half numb, less sensitive to pain and depressing effects of the cocaine.

A Brief Report of a Post-mortem Examination of the Parts Involved in a Bassini Operation for Inguinal Hernia, six weeks after its performance, by Dr. George E. Brewer. Right oblique inguinal hernia, about the size of an orange, silk-worm buried sutures. Dressing removed on the fourth day, primary union. At next dressing the greater part of the wound had separated, and later granulated up. Suddenly at fifth week the patient developed a right hemiphlegia, due to cerebral thrombus, and died. Examination of the wound showed a slight thickening. When open the parts showed slight puckering in the region of the internal ring, firm pressure developed no weakening of the abdominal wall, but a small drop of whitish fluid was expressed and a minute fragment of catgut used in ligating the sac. The skin was slightly adherent to the external aponeurosis; the divided aponeurosis had firmly united; at the upper portion of the wound some of the silk-worm gut was deeply embedded in the tissues, and aseptic; the floor of the inguinal canal was firm. Between the aponeurosis of the external oblique and the floor of the canal the vas deferens and spermatic artery and a number of small veins could be seen lying separately, and healthy. None of the kangaroo-tendon sutures could be found.

Dr. W. B. Coley presented a boy thirteen years old, who had been operated on by Dr. Curtis for an appendicitis of traumatic origin. Three days before applying to the hospital he had received a blow from a push cart. Application was made for a truss to support a supposed hernia. Temperature was 100° Fahrenheit, and a tender enlargement was found in the right iliac fossa. In the hospital his symptoms became so rapidly worse that he was operated upon at midnight of the same day. The appendix was found gangrenous at the end, and an abscess the size of a walnut.

Dr. J. P. Tuttle exhibited a calculus weighing 3½ grams, removed from the appendix, which had sloughed off from the intestine. Recurrent colic was the only special symptom.

Dr. A. E. Gallant stated that he had examined nearly two hundred fresh appendices in search of foreign bodies. In one of these he found seven berry-seeds situated in the distal end of the appendix. In most of the other appendices there was apparently a foreign body, but all proved, on close examination, to be only fecal concretion.

A Series of One Hundred Operations for Appendicitis. By Dr. Robert T. Morris. In this paper classification shows that there were thirty-four acute with abscess, and four chronic with abscess. All were operated upon, except two patients, who died before his arrival. One from suppurative nephritis, the other, septic peritonitis. Several were septic at time of operation. In two the appendices had already sloughed. Where the appendix was in a gangrenous condition fecal fistula followed; all but one have closed. Saphenous phlebitis occurred in four cases, three of these were on the left side. The smallest useful incision, peroxide of hydrogen, and saline solutions for irrigation of the cavity, a wick for drainage, with accurate coaptation of the various layers. No gauze packing; no opium.

Post-operative hernias have been nil, due to accurate coaptation of the abdominal layers and suturing the cecum to the margins of the abdominal opening in suitable cases.

In thirty-eight cases of acute or chronic appendicitis without abscess, the appendices were removed through an incision of one and one half inches. No adhesions were met with in thirteen. Various degrees of adhesion in the remaining cases. No deaths occurred in those without abscess. Two cases with adhesion, the appendix was found to have atrophied to mere fibrous cord.

One case showed cancer of the appendix and six tuberculosis. All recovered.

Torsion of the appendix was met with in two cases, causing obstruction and distension with mucus.

Two concretions were found present in one appendix, causing persistent nausea and pain. The organ was in a healthy condition, the concretions expressed into the cecum, and the appendix left *in situ.* All the symptoms subsided.

Make the diagnosis, then operate, regardless of the stage to which the disease has progressed.

Dr. Beverley Robinson, in discussing this subject, advanced the theory of Bland Sutton that the appendix is the "abdominal tonsil," and inflammation in that organ arises from an underlying rheumatic or gouty diathesis, and can be cured by antirheumatic medication by the use of salol, salicylic acid, the salicylates or salicin. When abscess is present the knife must be used. The cold coil is bad; warmth, better. Avoid active purgatives and opium. Give frequent large doses of the salicylates.

Dr. C. L. Gibson considered the mortality rate as the best on record, and the absence of hernia wonderful. The method of suturing seems to be that employed by many surgeons.

Dr. B. F. Curtis, chairman, called attention to the fact that there were but thirty-eight cases in which an abscess was present and in sixty-two no infection of the peritoneum. Deaths result from peritonitis, freedom from hernia, from clean wounds.

Dr. H. Lilienthal spoke of a case of abscess in a boy followed by sinus. When excising the sinus it was found to contain a concretion in which was buried a piece of wood about one third inch long. Granting that the rheumatic theory may be a factor in the production of appendicitis, acute progressive cases, which do not respond to other treatment in twenty-four hours, unquestionably demand operation.

Dr. A. L. Fish: When operating for an appendiceal abscess, saving life is the most important factor. To accomplish this purpose it is often necessary to make an incision large enough to see as well as feel. It is not good surgery to close all wounds after operation for appendicitis.

The development of the intestinal tract is from a tube, followed by a dilatation for the stomach; then the tube folds upon itself and the lower portion folds over the upper portion, producing the descending and transverse portions of the colon. The caput coli is at first under the liver, but subsequently it grows downward, forming the ascending colon. The appendix is the same formation as the rest of the intestinal tract, and can not be the seat of an accumulation of lymphoid tissue, or properly called the "abdominal tonsil."

Dr. Curtis indorsed the fact that the size of the incision must depend on the case.

The use of peroxide seems dangerous, as the expansion of the gas might break down adhesions.

Dr. G. E. Brewer exhibited a specimen from a case of gastro-enterostomy for sarcoma of the pylorus. The jejunum had been united to the greater curvature of the stomach by three rows of sutures. Six days after operation the patient died suddenly. Autopsy showed the wound union perfect, and nothing to account for the sudden death but chronic heart disease.

SECTION ON PUBLIC HEALTH.

Some very interesting figures relative to the milk supply of New York City were presented by Dr. George B. Fowler, showing that the total daily average is 19,164 forty-quart cans, equal to 766,560 quarts. Seventy-eight and eight tenths per cent. comes from New York State alone. These significant figures show that legislation which the State Board of Health desires, looking to the supervision of the supply of milk, can by control of the home supply eventually force other States to fall in line.

The above named quantity represents the yield from about one hundred thousand cows, ninety thousand of which are in this State. To pay salaries and compensate owners of condemned cattle the State Board of Health is allowed $20,000. There are five inspectors for the whole State. The service needs $400,000, so that every can of milk can be certified.

The work of the milk inspectors for 1895 includes 72,036 milk inspections; 99,080 specimens examined; 2,677 quarts of milk destroyed; 408 arrests; 398 trials; 364 held on bail; 78 days in jail; and $12,260 fines.

In order to encourage the shipping of milk from remote points, the railroads charge a uniform rate of thirty-two cents per forty-quart can from any point on their lines.

" *The Methods Employed for the Prevention of the Sale of Adulterated Milk in New York City,*" by Mr. E. W. Martin, Ph. D., Chemist to the Health Board.

In order to systematically keep watch on the adulteration of milk, the city is divided into six districts corresponding to the judicial courts districts.

The inspectors are constantly making rounds through these districts. The registration of the thermometer and lactometer, the number of the instrument, and name of the dealer are noted. A number is given to the sample, the policeman's number who accompanies the inspector is placed on the sample for ready identification. Two clean bottles are then filled with the milk and sealed in the presence of the officer; one is sent to the department chemist of the Health Board, who gives the inspector a receipt. The second bottle is placed in the hands of the dealer, so that he can have the sample analyzed by any chemist for his own satisfaction.

A record is kept by the Health Board of each milk dealer for reference.

One dealer has been arrested six times and compelled to pay an aggregate of $1,000, from 1887 to 1893, and still continues business.

About ninety-nine per cent. of the adulteration of milk consists in the addition of water and the removal of the cream.

Milk Analysis, with Special Reference to the Detection of Adulterations, by E. J. Lederle, Ph. D.

The most frequent adulterations of milk consist of the addition of water, skimming, and both watering and skimming. Foreign substances are added in the form of antiseptics or coloring matter, for the purpose of increasing the keeping qualities or making up the loss of color from the removal of cream. No injurious aniline colors have been found.

The addition of water reduces the nutritive value, and many epidemics of typhoid fever are due to the use of water from a polluted well.

The sample of commercial milk for determining the addition of water must be well stirred with a metal disc, and when the specimen is in a bottle shaking will not mix it thoroughly, but by pouring from one bottle to another this can be easily accomplished.

The quanity of water present can be determined by evaporation; the weight of the solid being deducted the water is found by difference. For the determination of fat by chemical analysis for court purposes, Adams' process is the official method. " Fat-free " paper in strips two and one half inches wide, twenty-two inches long, are rolled into a coil and held by a loose clamp, five grams of milk are poured onto this coil, thus distributing the milk over the coil of paper, and the fat can be readily dissolved by the ether. Dry the coil in an air bath (two and one half hours), then place in a Knoffler's extraction apparatus for two and one half hours. First weigh the flask of the apparatus after distilling the ether from the flask, dry one and one half hours, then weigh, and the additional weight will be the amount of butter fat in five grams of milk. The average at this season is four to four and one half per cent. of fat.

The standard in New York State is: Water not over eighty-eight per cent., total solids not less than twelve per cent., fat not less than three per cent.

The author demonstrated the methods of determining the amount of salts, which in a fair average commercial milk is 0.72 per cent., and quite constant.

The addition of all antiseptics to milk and milk products is prohibited.

The *intelligent* use of the lactometer has resulted in securing $12,000 in fines, and is considered the most useful, simple, and rapid test for examining large number of samples milk.

For rapidly testing cream and fat the cream gauge, a test-tube twelve inches long, graduated so that each division represents one per cent. by volume, is used. Fill the tube to the 0 mark, place in the refrigerator or cold water for twelve hours, and read off the cream. Good milk will show from twelve to sixteen per cent. of cream, corresponding to three and one half to four and one half per cent. of butter fat. To ascertain the fat in one or two hours, divide the cream gauge by a mark into two equal parts. Pour milk into the tube to this mark and fill to the 0 mark with water at 150° F., containing a pinch of sodium carbonate. Shake well and place in ice-water for say an hour.

Feser's lactoscope: A pipette of milk is placed in a tube and water is added, a little at a time, shaking well, until, when held at arm's length, the black lines can be distinctly seen. The figure at the level of the mixture in the tube indicates the percentage of fat.

Of the centrifugal machine the author will speak in a future communication.

" *The Significance of Micro-organisms in Milk.*" Dr. R. G. Freeman: Sterile milk can be obtained from the udder of a healthy cow by the use of a sterile catheter. The ordinary milk of cities contains from 50,000 to 450,000,000 in one c. c. or 15,-000,000 in one drop.

The non-pathogenic bacteria are derived from the dust in buildings, the dirt on the hide of the cow, the hands of the milkman, and the pails. Mold and yeast are usually present. City milk can not be sterile so long as it is obtained from a dirty cow, milked in a dirty barn by an equally dirty man. Clean milk, rapidly cooled and quickly delivered does not contain one thousandth part as many bacteria.

Pathogenic micro-organisms enter milk by the same careless means as admit the non-pathogenic, and are contained in the dried fecal matter, sputum, on the hands of the milkman, the water, and dust. They may also enter the milk while in the udder of a diseased cow.

The most frequent pathogenic organism present in milk is the tubercle bacilli. The use of tuberculin has demonstrated that about seven per cent. of all the cattle in New York State are tubercular.

Owing to the small number present in a specimen of mixed milk, the determination of the presence of tubercle bacilli by staining is a tedious task. Positive results can only be obtained after repeated cultures for at least a month.

The harmlessness of tuberculin to non-tubercular cows, and the fact that it is so delicate a test when that disease is present, compels us to look to that substance as a protection against this form of milk contamination.

Encourage clean dairy methods and quick delivery, increase and enforce dairy inspection. Proper legislation to stamp out tuberculosis.

" *The Results of Certifying Milk.*" After several years' ineffectual effort to secure favorable legislation to control the quality of the milk supply of New Jersey, Dr. H. L. Coit, of Newark, organized a committee of seven physicians to enter into a contract with a responsible dairyman. Three or four experts were engaged by the committee and paid by the dairyman to supervise the dairy and see that the conditions of the contract were carried out. Certificates were then issued to the dairyman. The work of inspection has been conducted by Professor Leeds, of

12

Hoboken; Professor Liantard, the veterinarian; and Dr. F. M. Prudden and Dr. R. G. Freeman.

Results: Financial success for the dairyman, his sales having increased during the past two years from 800 to 2,000 quarts per day, despite the increased price, viz., twelve cents per quart.

The general standard of milk in Essex County has also been improved.

At one time in the summer the number of micro-organisms were reduced to 3,400 per c. c.

The health officer of Newark has expressed the opinion that the death-rate among infants has been lowered as a result of the improved milk supply. Dr. Snow, of Buffalo, N. Y., has already adopted the plan in that city.

"An Association of Dealers for the Distribution of Honest Milk in the Whiteman Standard Indicating Jar" has been formed and has sent out the following circular:

"The above named reputable milk dealers are making an earnest effort to improve the quality of milk served in New York City and vicinity, and maintain it at the highest standard. At considerable expense they have adopted the Whiteman Standard Indicating Jars. These jars indicate plainly and squarely the quality of milk they contain, which should show $12\frac{1}{2}$ per cent. of cream. They believe in letting their customers see at any time the quality of their milk as a guarantee of good faith.

"Do you know the kind of milk you are buying?

"Is it up to the standard required by law?

"Do you know what the standard is?

"Whiteman's Standard Indicating Milk Jars show the quality of milk in every jar, and whether it is up to the standard or not.

"That is the kind of jar we put our milk in.

"On receipt of accompanying slip properly filled out, any of the above dealers will be glad to send to your residence free a sample bottle of the milk they are furnishing in this way.

"By having your milk delivered in the Whiteman Standard Indicating Jars you can any day examine it and see if it comes up to the required standard.

"Any dealer who is selling a pure article should be willing to subject his milk to the test of these jars if required. He is like the honest conductor, not afraid of the bell punch."

That this effort has been appreciated by physicians in New York is demonstrated by the fact that one dealer has on his list forty doctors, whom he supplies with milk at twelve cents per quart, butter at seventy-five cents per pound, and cream at sixty cents per quart.

The Health Board has issued orders to all milk dealers that no milk shall be offered for sale or delivered in this city without a permit in writing from the Board. Milk shall not be stored in any place used for sleeping or domestic purposes, nor shall any be opened therein. It shall not be transferred from cans to bottles in the street, or on ferries or depots. Bottles in which it is put shall be cleansed after a prescribed fashion. Permits from the Board to sell milk must be posted in a conspicuous place in every store.

SURGICAL SOCIETY, REGULAR MEETING, FEBRUARY 12, 1896.

Dr. H. Lilienthal presented a patient from whom he had removed six inches of the transverse colon, and approximated the ends with Murphy button 1¾ inches in diameter. (MATHEWS' MEDICAL QUARTERLY, Vol. I, page 616, and Vol. II, page 130.) Despite the fact that many nodules were left in the mesentery and surrounding tissue, the patient has enjoyed excellent health, having gained thirty-five pounds. Operation June, 1893.

Dr. Parker Syms said that he had assisted in the removal of the whole transverse colon, and the patient was in good health at the end of a year. [This case was reported by the writer, MATHEWS' MEDICAL QUARTERLY, Vol. I, page 463.]

Dr. Briddon reported a case from which he had this afternoon removed a mass, lying in the middle of the omentum, which was adherent to the ascending transverse and descending colon and to the small intestines. Embraced by the hepatic flexure of the colon, in a sort of pit, was found another mass which he will remove later from behind. The growth was believed to be malignant.

Dr. Syms exhibited a toilet-pin removed from a sinus, the result of an appendiceal abscess which had been drained in 1893. The sinus persisted, and while endeavoring to close it the pin was found, thickly covered by fecal concretion.

A. E. GALLANT, M. D.

DISEASES OF THE RECTUM.

MANLEY, THOS. H., NEW YORK: FECAL IMPACTION IN THE
ANO-RECTAL SPACE. (*The Medical Brief.*)

Fecal Impaction. The colon is the great reservoir or receptacle for all the residual by-products of alimentation. At the
ileo-cecal isthmus we will note, in the herbivora, the carnivora,
the mammalia, batrachia, and all alike, the precise point at which
digestion abruptly and absolutely ends and the work of elimination commences. In the herbivora, though, we will note that a
vast pouch occupies the site of the rudimentary relic, the appendix, in man ; nevertheless, as peristalsis presses its current onward
at this precise site, the pulpy fluid mass, dehydrated partly, and
in the sheep and rodents, cylindrical fecal masses cross the valvular barrier to enter the colon.

The colon, strengthened in its longitudinal axis by broad,
strong, fibrous bands, is greatly deficient in muscular tissue for
its caliber as compared with the ileum, and hence its peristaltic
and propulsive power is correspondingly less. This is freely
movable only in its transverse segment, and hence the movement
onward of its contents, and their expulsion, is rather dependent
on the motion of the diaphragm and the contraction of the
abdominal muscles than on the vermicular tide so characteristic
of the ileum.

The Location of Fecal Impaction. Fecal impaction may occupy,
(*a*) the entire large intestine ; (*b*) the caput coli ; (*c*) the sigmoid
flexure ; (*d*) the rectum.

The *sigmoid flexure* of the colon, in the adult female, has a
lengthy, free mesenteric ligament ; probably in order to adapt its
situation for the rising gravid uterus in pregnancy. In parturition, unless previously unloaded, it may impede fecal descent,
and in any event suffer severe compression. In consequence,
after labor, when the feces reach this point, the weakened walls
expand until distension is very great. This condition presents
so many characteristics in common with suppurative formations,

that, unless one is on his guard and is cautious in examination, he may mistake it for a state requiring a serious operation. Rectal impaction of feces is the most common, and though undoubtedly the least dangerous, yet, no one who has had much experience in the practice of his profession can deny that it is a fruitful source of very serious local lesions, grave disturbances of the whole system, or even the loss of life.

It may develop *de novo*, or be secondary as one of the manifestations of certain pathological conditions of the nervous system.

It is undoubtedly one of the most potent causes of hemorrhoids, simple ulcer, or malignant disease of proctitis with resulting stricture—nay, more—inflammation of the prostate, cystitis or neuralgia, and in women pain in the ovaries, uterine displacement, and vesical irritation. The symptoms of rectal impaction are not so definite and easy of recognition as one might expect.

A patient may tell us that he is disposed to be costive, but keeps the bowels open by pills or by injections; another may tell us that the motions are rather small, but frequent; while in other occasional instances there may be a positive looseness of the bowels.

The Diagnosis of Fecal Impaction of the Lower Rectum in the Female Sex. In the woman we immediately notice that the uterus hangs suspended between the rectum and bladder.

We may affirm at the outset here, that morbid conditions of the rectum or uterus reciprocally influence the other; thus, the gravid uterus impinges on the rectum and sigmoid and leads to an impediment of the fecal current; and those who have much experience in rectal surgery know, too, that the relaxed ligaments of the womb in elderly women may allow it to sag backward and tumble over against the rectum, producing so much pressure, and giving rise to such a state of things as may lead the inexperienced to suppose a new growth is in the way. We may also assume that, of all the active causes which lead to a crowding forward of the uterus, so-called anteflexion, a loaded rectum stands at the head of the list.

Diagnosis of rectal impaction in women should not be difficult, for the reason that, particularly in the multipara, we may easily determine it through the vagina. Perhaps this very facility may lead us all the more readily to overlook it, and ignore symptoms which will seldom mislead, if we will only strive to interpret them.

To summarize, it may be stated :

1. In all cases of rectal inertia, either resulting from a low grade of inflammation or ulceration occupying any site of the large bowel, tubercular, specific or malignant, attended with persistent constipation, the perfunctory administration of an enema or cathartic is inadequate, and nothing less than a painstaking physical examination will reveal the real condition.

2. In many males past middle life, in whom we have reason to suspect inhibition of the sensory reflexes of the rectum, either from local lesion in the neural periphery, the spinal or central ganglia, particularly when vesical symptoms are present, with obstructed or forced urination, before directing any special therapy to the genito-urinary organs, we should be entirely assured, by a digital examination, that there are no fecal concretions lodged in the lower rectum.

3. A relaxed, partially prolapsed uterus may roll over and interfere with complete defecation, to give rise to symptoms of fecal obstruction ; and *per contra*, a rectal bag of hard, dehydrated or calcified fecal mass may, in turn, give rise to nearly every symptom of ovarian, tubular, or uterine disease. A thorough exploration of the rectum as a primary measure is of prime importance in all such cases.

OTIS, WALTER J., BOSTON: CLINICAL FEATURES AND TREATMENT OF EXTERNAL PILES. (*Boston Medical and Surgical Journal.*)

As each form of external piles is the expression of a definite abnormal process, which is either taking place or has taken place, and so indicates its own history that it is possible for the experienced eye to read it almost at a glance. The cutaneous variety of external pile is classified as, (1) *Redundant*, which indicates that its distinctive feature is a redundance, or a superabundance of the circumanal integument, brought about by the stretching it receives from the subjacent varicose external hemorrhoidal veins when fully distended, as during defecation. (2) *Hyperplastic*, indicating that the pendulous cutaneous tag, associated with an abrasion, fissure, or ulceration at the anal verge, is the result of an inflammatory hyperplasia; also (3) *Hypertrophic*, indicating that the swollen, thickened, radiating anal folds, associated with an eczematous inflammation, are the results of an inflammatory or irritative hypertrophy.

The site of external piles is confined to the region of radiating folds of the cirumanal integument, passing from this central point over the lower portion of the internal surface of the external sphincter and its lower border to disappear in the contiguous skin. It must be remembered that the lower border of the internal sphincter is subcutaneous, and that the external sphincter partially overlaps the internal.

External piles are not to be regarded as internal piles permanently crowded down and forced outside the internal sphincter. The external pile, covered with skin, will have the color of skin, will be dry, with no tendency to bleed; the internal pile will have the appearance of mucous membrane, moist, deep red color, with a history of bleeding, soft to the touch, with velvety or granulating surface.

External venous piles are readily distinguished from the cutaneous by their unmistakable sanguineous nature. While the existence of a varicose condition of the external hemorrhoidal vein is not pathological in adults, it becomes so when permanent enlargement has occurred by repeated and excessive venous engorgement. This variety of external pile is called the *varicose*. As a varicosity of the external hemorrhoidal plexus is always associated with and secondary to a similar condition of the internal plexus, all treatment of the external varicose pile must be directed to the removal of the cause of the varicosity of the internal plexus. The tonic effect to be derived from the daily application of cold water appears to have this double effect.

The *thrombotic* pile is produced by a thrombosis in one or more of the dilated veins of the external plexus. The rapidity of its formation is one of its chief characteristics, another its livid and bluish color. The greatest amount of relief will be experienced from incision of the vein and removal of the thrombus.

The treatment of the *hyperplastic* pile should be (1) to render the stools semi-liquid, (2) the protection of the ulcer during defecation, (3) the local treatment of the ulcer, (4) the removal of the pile. (1) Regulate the diet and administer a laxative; (2) vaseline, one ounce, acid boracic, two drachms, in ointment, applied before stool; (3) by means of a small speculum, exposure of the ulcer and touching with nitrate of silver for a few days; then application of dry powder.

The treatment of the *hypertrophic* pile should be directed solely to treatment of the anal eczema, and the most efficacious is the application of black wax and oxide of zinc ointment.

GUYON, PARIS : FORCEPS.   (*Journal de Médecine de Paris.*)

The accompanying cuts represent forceps after the model of Dr. Guyon, of Paris, which are used for the ablation of soft and vascular tumors, such as hemorrhoids.   They are constructed in three different forms, bent at a right angle, slightly curved, or more so, forming a semicircle.

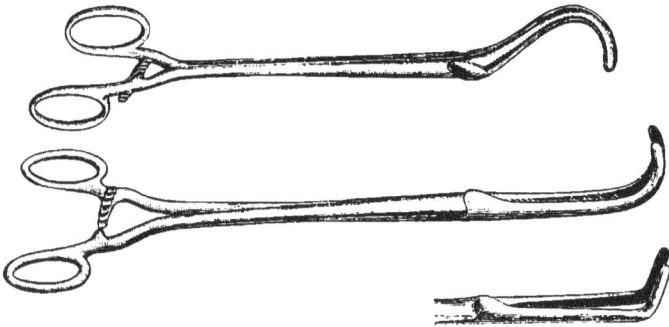

These forceps close with a spring, into which fits a little hood, so that they may be closed as tightly as necessary; by this means the base of a large tumor can be grasped, and by pressing gradually the mass will be removed.   The handles are long, so as to make it possible to work up in the rectum, the vagina, or even the back of the throat.   This instrument, made by M. Collin, has been praised by Prof. Farabeuf.

GASTRO-INTESTINAL DISEASE.

HEMETER, JOHN C., BALTIMORE: AN APPARATUS FOR WASHING OUT THE STOMACH AND SIGMOID WITH A RETURN STOMACH OR RECTAL TUBE AND A CONTINUOUS CURRENT. (*New York Medical Journal.*)

. The apparatus consists of, first, a large glass jar (Fig. 1) *a*, having a stopcock, *f*, near the bottom, graduated into divisions, each indicating one hundred cubic centimeters, and placed on an elevated platform, shelf or the like. Secondly, of a double or return stomach tube, which differs from the ordinary stomach or lavage tube of Ewald by the presence of a partition of soft

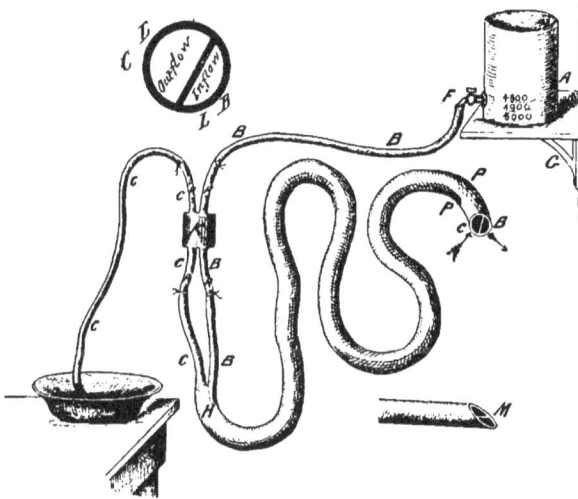

FIG. 1.

rubber running through its entire length and seen in cross section at 1, really dividing the instrument into canals, *b* and *c*. The canal for the inflow, *b*, connected at *f* with the faucet of the reservoir bottle, is only half of the caliber of the outflow canal, *c* (this is not very apparent in the accompanying drawing), a most important provision to insure at all times a greater facility to the outflow than to the inflow, otherwise the stomach may become overloaded with water, which, owing to the elevated position of the reservoir, runs into the stomach very readily and in greater quantities than could be carried away by an outflow tube

the same size as the inflow tube. The connecting piece, k, with
its four branches, $r\ r\ r\ r$, is made of hard rubber to prevent any
stretching of the two tubes, b (inflow) and c (outflow), at h, the
junction. The end of the return tube has side openings, $p\ p$, in
addition to the openings at the lower end, which does not terminate abruptly, as in the sketch, but is beveled off smoothly
and tapers, as shown at m.

In case the apparatus is to be used for washing out the rectum and sigmoid, the patient must be placed in the position suggested by Dr. Howard Kelly for atmospheric distension of the
female bladder and rectum. In this position the double rectal
tube depicted in Fig. 2, is slowly introduced, consisting simply

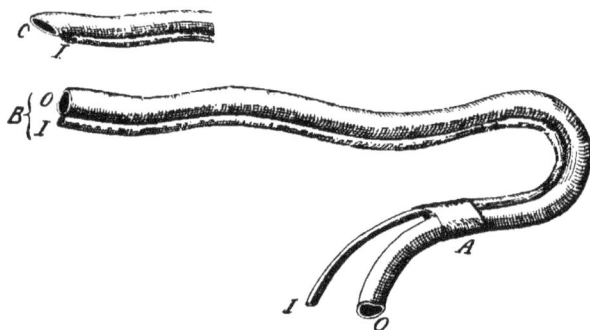

FIG. 2.

of the smaller soft inflow tube, i, united with the larger outflow
tube, o, and separating at a, not ending bluntly, as shown in
cross section at b, but tapering as shown in Fig. 2 at c. The gastric tube is useful in cases of considerable gastric decomposition
or accompanied with great masses of mucus and requiring much
time with the single tube. The inflow and outflow can be controlled by the pressure of the fingers of the patient if necessary.
The same precautions are necessary with this as with the single
tube in its employment. The double rectal tube is very useful
in membranous colitis. With Dr. H. A. Kelly's method I have
frequently succeeded in exposing the entire rectum and first turn
of the sigmoid flexure to view in direct sunlight. It is possible
to pass a soft rubber rectal tube of my pattern entirely through
the sigmoid flexure into the descending colon in the cadaver.
From these results in the cadaver it is conceivable we may yet be

able to pass a tube into the transverse colon in the living subject.
The return stomach tube is 1.3 centimeter in diameter and can
be ordered of any desired length ; the double rectal or sigmoid
tube is 2.3 centimeters broad one way, and 1.0 centimeter in the
direction at right angles to the measurement first mentioned.

PEARSE, HERMAN: GASTRIC NEUROSES; THEIR DIAGNOSIS
AND TREATMENT. (*Kansas City Medical Index*.)

The functions of the stomach are three : secretion, furnishing
the fluids and ferments; motion, mixing them with the food, and
absorption, by which the prepared portions are taken up.

The great questions are : Upon what do the functions depend ?
Why are the ferments secreted? What causes the various movements ? Why and how are the functions altered ? Secretion is
excited to activity by the presence of food in the stomach, the
nerve centers receiving impressions and originating impulses
which are sent to the muscular fiber and cells. The true diseases
of the stomach, independent of outside influences, are cancer,
inflammation, and ulcer. Aside from these, the most important
deviations from the normal gastric phenomena occur: (1) As a
symptom of chronic nephritis. Here urinology furnishes the key
to its diagnosis and treatment. (2) As a symptom of tabes dorsalis, the so-called " crisis " of ataxia. The accompanying symptoms should fix its diagnosis. (3) Hepatic insufficiency, so-called
" bilious state," in which, the hepatic cells failing to perform
their metabolistic function as rapidly as the occasion demands,
the blood current is obstructed and a general engorgement takes
place. This condition may be caused by constipation. *This is
always transient*, a few days' fasting and a mild purge followed by
a bitter tonic clearing up the trouble. (4) Dilatation from obstruction at the pylorus. (5) General neurasthenia, in which the gastric functions, like the cerebral ones, are inhibited or prevented
by an altered and unnatural condition of the central nervous
system, sympathetic as well as cerebro-spinal ; this is frequently
associated with lithemia. (6) The neuroses, hyperacidity, gas-
tralgia, colic, hypersecretion, anacidity, eructation pyrosis, cardiac and pyloric insufficiency, and the many reflexes from the
sexual organs, spinal cord, eyes, etc.

Having determined the presence of one or more of the neuroses, the treatment must be considered. Three things are

essential: (1) That the exact condition be ascertained. (2) That
the condition, if abnormal, be relieved. (3) That the exciting
cause of the condition, whether a mentality, a defective eyesight,
a diseased uterus, a cerebral edema, a secret worry, or whatever
may be causing the disturbance, be removed.

The first is to be accomplished by the stomach tube and the
chemical tests of the gastric contents, the third by shrewd and
careful judgment and examination; for the second we consider:

1. Lavage: This is useful in (1) dilation, brilliantly so; (2)
in acute gastric catarrh—engorgement—the so-called "bilious
attack;" (3) in chronic gastric catarrh, after the hepatic condition has been cured; here a little hydrastis should be added to
the water, and even a very weak solution of nitrate of silver,
well washed away, may be used; (4) in hiccough and the
vomiting of chloroform, both neuroses, the trouble is relieved
in most cases by a single cool lavage; (5) where anacidity
allows fermentation the fermentation may be stopped by lavage
after six to seven hours of digestion.

2. Hydrochloric acid should be used only where its absence
interferes with digestion, and should be supplemented by such
constitutional and local treatment as will remove the anacidity;
such measures are exercise, massage, and the stomach douche in
atony, warm general baths and the electrization of the spine in
neurasthenic anacidity, etc., regulating the diet, of course, to
leave as little fermentation to occur as possible, that is excluding
the starch and sugar from the diet.

3. Of pepsin the same may be said; we should use it only
temporarily, while general and local treatment seeks to call back
the suspended secretion of the peptogenetic cells, and depend
upon general treatment to cure the apeptic condition.

4. Miscellaneous drugs and medicines. The vomiting of the
neuroses is quite well controlled by chloroform-water, and occasionally by the old formula of equal parts of iodine and creosote,
given in drop doses.

SMITH, GREIG, BRISTOL, ENG.: EXTRA PERITONEAL CLOS-
URE OF FECAL FISTULA. (*Bristol Medical Journal.*)

Here the bowel does not protrude through the parietal opening, and there is no spur, or only a slight one. A simple fistula
lined with granulations leads from skin to bowel.

The granulations are first scraped from the sinus by means of a small, sharp Volkmann's spoon, and the parts around are purified. If there is any discharge from the intestine, a small sponge with string attached is pushed through the fistula so as to block it.

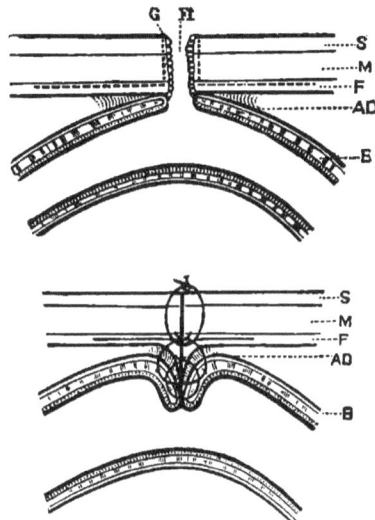

DIAGRAMS TO SHOW METHOD OF CLOSING FECAL FISTULA.

Fi. Fistula in abdominal wall communicating with bowel. G. Granulations lining fecal fistula. S. Skin. M. Muscular layer. F. Subperitoneal fascia. AD. Adhesions between bowel and peritoneum surrounding fistula. B. Bowel. Broken line in upper diagram shows incisions around fistula and in subperitoneal areolar tissue. Lower diagram shows operation finished and sutures placed.

Two incisions are now made in the parietes, with the fistula as center, down to the subperitoneal areolar tissue. Their direction is to be guided by that of the principal muscular fibers in the parietes, so as to avoid their division and thus minimize weakening of the parieties. A fistula in the middle line would have vertical incisions above and below, in the loin it would be vertical or oblique, as we desire to preserve the fibers of the internal oblique muscle and aponeurosis. The incision comes up to but does not pass through the fistula; it is carried around the fistula; the fistula with the cicatricial tissue around it is bodily removed. The parietal incision goes down to the subperitoneal areolar and fatty tissue, but does not go through it. Then with finger and scissors the parietal peritoneum with its fat is detached from the muscle all around the fistula for a distance of from one to two inches. When the separation is complete, the fistulous tract is removed down to the gut. The bowel remains attached to the parietal peritoneum by adhesions around the fistulous opening. By means of forceps placed close to the opening, the bowel, with its attached peritoneum, may now be lifted out through the incision in the parietes. If there is any difficulty in doing this, a little more detachment of the peritoneum will make it easy. The opening in the gut is now closed by infolding of the raw areolar surfaces around the fistula and sutured by Lembert's method, as if smooth peritoneal surfaces only were involved. The line of closure may be vertical or transverse as seems best. Two layers of closely placed sutures, one continuous (Dupuytren), suffice for closure. The outer row will engage some of the subperitoneal areolar tissue, and should have considerable grip of material. The sutured gut and peritoneum is pushed inside, and the parietal wound closed over it by silk-worm gut sutures in the ordinary way. A small drainage-tube laid over the line of gut suture adds to the security by preventing burrowing of fluids in case of leakage.

CABOT, A. T., BOSTON : OPERATIONS FOR THE CURE OF INGUINAL AND FEMORAL HERNIA. (*Boston Medical and Surgical Journal.*)

Inguinal Hernia. The operation which I propose, and which I have carried out in one instance, is to slit the aponeurosis of the external oblique muscle well up toward the anterior superior spine of the ilium, exactly as is done in Bassini's operation, to tie or suture and cut off the sac on a level with the peritoneal surface; then to suture the internal oblique and transversalis muscles and transversalis fascia on the inner side to Poupart's ligament on the outer side. These stitches may include the edges of the slit in the external oblique aponeurosis, thus closing the old inguinal canal.

Finally, to close the upper remaining part of the slit in the external oblique aponeurosis with a continuous buried suture, which shall also include the upper edge of Poupart's ligament.

This closure is to be done from below upward, and the spermatic cord is to be brought through at the highest point that it can be made to reach with moderate traction. Usually it emerges at about the point where it would perforate the abdominal wall by Halsted's method.

By this operation the muscular wall of the abdomen is not weakened by any incision except that through the aponeurosis.

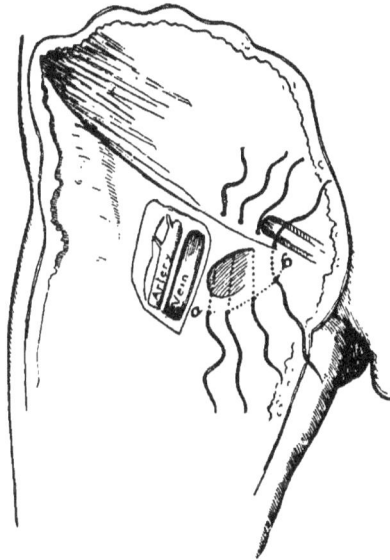

FIG. 1.

The curved dotted line *ab* represents the line of incision through the fascia. The dotted parts of the sutures run beneath the fascia, and, when knotted, they lift the flap of this fascia and attach it firmly to Poupart's ligament in the manner seen in Fig. 2.

The cord is entirely diverted from the old canal, which is obliterated by stitches. The new canal pursues an upward and outward course through the abdominal wall, which, as I have pointed out above, seems to have decided advantages over the canal running downward and inward; and it would seem possible, too, that this method of closure would afford a canal of greater strength than the short, straight one afforded by Halsted's operation.

Femoral Hernia. The difficulties attending the radical cure of femoral hernia arise from the shortness of the canal, from the proximity of the femoral vein, which interferes with the extensive placing of stitches about the femoral ring, and, lastly, from the fact that it is made up of tendinous structures that do not readily adhere by a permanent union when sutured.

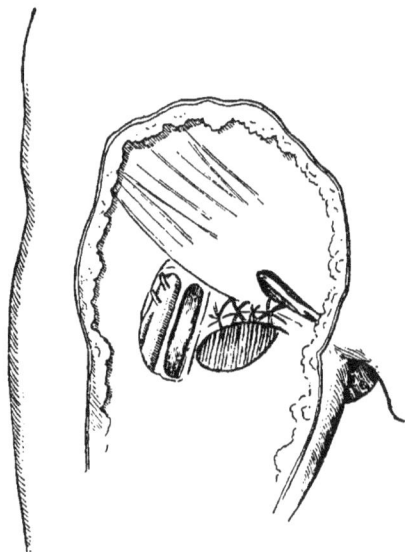

FIG. 2.

This diagram shows the manner in which the flap of fascia is pulled up against Poupart's ligament. Through the opening left below, where the sutures are knotted, the fibers of the pectineus and the tendon of the abductor longus come in view. The surface of the muscle uncovered is so small and is so near to its attachment that there would seem to be little danger of hernia of the muscles through this opening in its sheath.

The healing together of the fibrous walls of the canal is made more difficult by the necessary tension of the stitches pulling upon rigid portions of the fascia and ligaments. This tension is not great enough to prevent the approximation of the walls of the canal, as is shown by the very complete closure which Dr. Cushing obtained by his double method of suturing; but when rigid tissues are pulled together in this way, it is well known that the sutures are apt to cut through quickly.

The suggestion I have to offer is, that previous to applying sutures a semicircular incision shall be made through fascia lata just beneath the saphenous opening, the saphenous vein having been previously tied and cut away. The fascia can now be separated beneath, so that the lower wall of the canal can be readily drawn upward against the unyielding portions of Poupart's ligament, where it can be held with buried sutures without the least tension, and the whole canal is thereby tightly closed.

DEANSLEY, EDWARD, WOLVERHAMPTON, ENG. COLOTOMY. (*Birmingham Medical Review.*)

Colotomy first proposed by Littre, in 1710, as palliative operation for relief of obstruction of colon.

The first colotomy actually carried out was done, not on the sigmoid flexure, but on the cecum, by Pillore (of Rouen) in 1776 ; and the second, on the transverse colon, by Fine (of Geneva) in 1797. In pre-antiseptic times, however, either of these operations exposed the patient to the grave risks of opening the peritoneal cavity. Consequently colotomy was little practiced until the retro-peritoneal opening in the left loin, proposed by Callisen (of Copenhagen) in 1776, was practically established by Amussat in 1839. The advantages of this procedure were so great that the earlier operations of Littre and Pillore were almost abandoned till the introduction of antiseptic surgery in our own times. The safety of opening the abdomen having then been established, and the risk of intra-peritoneal leakage of feces having been obviated by operating in two stages, Littre's iliac operation again came into favor.

The form of cancerous disease of the colon which most frequently leads to obstruction is the tight annular stricture in which the presence of a new growth can often only be demonstrated by the microscope. Consequently it is rarely palpable through the abdominal wall, and is sometimes hard to find even when the abdomen is open. For these reasons it is obvious that left lumbar colotomy is only applicable in cases in which the exact situation of the stricture is previously known. In all other cases a preliminary laparotomy is essential. In cases in which the nature and position of the obstruction are entirely obscure, the ordinary median hypogastric incision is of course to be preferred.   13

In the male subject it is not always easy to or even possible
to examine every part of the colon through a median incision,
for even by passing the whole hand in the abdomen with the distended small intestines, it is not possible to decide whether every
part of the colon is empty or full.

On the other hand, the oblique lateral incision used for iliac
colotomy, not only allows the rectum, sigmoid flexure, and descending colon to be thoroughly examined, but suffices also for
making the colotomy should the obstruction prove to be in or
below the sigmoid. Should the latter, however, be found empty,
and therefore obviously below the obstruction, a second incision
can be made over the cecum and the colotomy be made there.
And whereas a preliminary median incision followed by colotomy almost always requires a second incision directly over the
part of the bowel which is opened, the right iliac incision will
in many cases not require a second.

The number of incisions is in reality a matter of small consequence if they are not too large.

Two courses are open, either resect the stricture and then
stitch the divided ends into the wound so as to form an artificial
anus, which after relief of the distension can be closed by a second operation ; or to place a temporary artificial anus in the
cecum, and to resect the stricture itself at a later time when the
conditions will allow of immediate reunion of the divided gut.
The restoration of the continuity of the bowel after the divided
ends have been implanted in the abdominal wall, presents great
difficulties owing to the adhesions and to the septic condition of
the skin around the artificial anus which lies within the immediate field of operation. In this case it is true that the same operation which restores the continuity of the bowel also cures the
artificial anus. On the other hand, when the temporary colotomy is placed in the cecum, a resection of any part of the colon
below it can be carried out during complete rest of the newly
united bowel as regards the passage of feces, and this is certainly
a great additional security.

If colotomy is intended merely as a temporary or preliminary
measure, it is important to avoid two things which are necessary for making a satisfactory permanent colotomy. In the
first place, no attempt should be made to form a mesenteric

spur by fixing a complete knuckle of intestine outside the abdomen; in the second place, the bowel should be sutured to the peritoneum only, and not to the skin or raw surface of the wound, for, as Mr. Greig Smith has shown, this leads to very dense adhesions. Both these conditions, but especially the formation of a spur, render the subsequent closure of the artificial anus a matter of difficulty. When a colotomy is intended to be temporary, it should be made as nearly as possible a simple fecal fistula in the cecum; and for this purpose the simplest method is to tie one of the smaller tubes introduced by Paul (of Liverpool), and then to return the cecum completely within the abdomen, merely keeping it in contact with the abdominal wall by the traction of the tube, aided, perhaps, by one or two peritoneal sutures. I have found the use of Paul's tubes a very simple and efficient means of completing a colotomy or enterostomy, and one which enables the abdominal wound to be treated aseptically, while permitting the immediate evacuation of the bowel. The introduction of this simple device must be regarded as one of the greatest improvements in the technique not merely of colotomy but of all operations requiring evacuation or drainage of any part of the alimentary tract.

HOUCHON: AN INTESTINAL NEEDLE WITH A SPRING. (*Paris Medical Journal.*)

This needle shows an eye which is simply closed by such a spring that the assistant who threads the needle has but to make very light pressure.

This very light needle is especially suited for sutures of the intestine, for operations for vesical and vaginal fistulæ. The curve is to be adjusted according to the use to which the needle is put.

Book Reviews.

Pediatrics. By Thomas Morgan Rotch, M. D., Professor of Diseases of Children, Harvard University, Boston. Illustrated. Philadelphia: J. B. Lippincott Company. 1896.

Therapeutics of Infancy and Childhood. By A. Jacobi, M. D., Clinical Professor of the Diseases of Children in the College of Physicians and Surgeons (Columbia University), New York. Philadelphia: J. B. Lippincott Company. 1896.

It is rare that in a single year two such valuable books from two such distinguished authors should be offered the medical profession upon common subjects, at least in part. It has occurred to the reviewer to write first of their common points, following with comments more in detail. Dr. Jacobi has been identified with the teaching of Pediatrics for so many years, besides being the pioneer in the establishing of systematic clinical instruction in that branch, and his writings being limited to the medical periodicals and brochures, that a thorough expression of his opinions upon a subject upon which he has so long been an authority, under one cover, can not help but be greatly appreciated. The book from the very start has the stamp of *originality*, some times akin to dogmatism, so emphatic are his statements, yet as one reads he knows that the statements are not mere theorizing, but the recording of the opinions of a very close observer. As one reads, many paragraphs are found which are like the aphorisms seen in the productions of that versatile writer, Mark Twain: "The choice between a badly-tasting medicine and a fine-looking funeral ought not to be difficult," "Children are what they appear, and appear what they are," "No printed rule ever supplies or substitutes brains," are paragraphs full of meaning, terse, and to the point; many similar ones being found through the text.

The work of Dr. Rotch covers a much wider field, being a thorough study of the child, normal and in disease; but one is struck with a paragraph found on page 326, especially when read in connection with one in the preface of Dr. Jacobi's book. Dr. Rotch says: " An important fact to remember in the treatment of infants and young children is that drugs *play a very insignifi-*

cant part in the actual cure of their diseases," and Dr. Jacobi says in his preface, " I believe in medicines. Advancing years and experience during a period of increasing exactness in medical methods have rather strengthened my belief than otherwise. What the knife is to the surgeon drugs are to the physician. The knife does not make the surgeon, nor do medicines make the physician ; both, however, are indispensable." One hardly knows how to account for two such different statements from such authorities.

In Pediatrics the infant is studied from birth, much space being devoted to normal development to puberty. After this follow chapters on the Hygiene of the Nursery, Feeding, the Blood in Infancy and Childhood, ¶Diseases of the Newborn, the Exanthemata, Diseases of the Nervous System, Mouth, Nose, etc., Esophagus, Stomach, and Intestines, Larynx, Trachea, Lungs, and Pleura, and a chapter on unclassified diseases, among which have been left Rheumatism, Scorbutus, and other diseases to which we think too little space has been given.

The author, we take it, has made a great mistake in arranging the work in the form of a series of lectures. By this means many expressions have been used which are common only to the lecture room, and out of place in a work like this. The illustrations throughout are excellent, and add greatly to the value of the text. They are almost without exception original photographs, taken to illustrate the special subject under discussion. The scheme followed of describing a disease or special form of it, and relating a case illustrating the subject under discussion, makes the book not only very instructive but highly entertaining.

More in detail the following points have been noted as the reviewer proceeded :

Very inadequate instructions are given as to the care of the cord of the newborn. The importance of great care in its dressing should have been emphasized. The paragraph stating that there is no need of an abdominal band after the first weeks is to be commended.

We think the term "*slight digestive attack,*" page 185, not very well chosen. The most important section of the work, in our opinion, is that upon feeding, and it is here that Dr. Rotch, to whom so much is due in developing artificial feeding, putting it on a scientific basis, and the working out and establishing milk laboratories, is seen at his best. This section is worth the price of the book alone. The only criticism one can make is that in showing the sugar-measure, on page 277, only its capacity by weight is shown, its linear measurements are not given. That the author believes in milk laboratories is shown by his preferring milk prepared there to the mother's milk in premature infants.

The word "brooder" is recommended to be used, instead of incubator, which we think is altogether a wise suggestion.

We doubt the possibility of sonorous râles causing sufficient jarring of the chest wall to make preceptible fremitus, coarse mucous râles are frequently felt.

We have had occasion before in these columns to express an opinion in regard to more emphatic expression about "maternal impressions" being essential in books of reference such as the one before us. If it is mentioned at all it should only be to condemn any such theories.

The portion devoted to surgical diseases of the newborn is very incomplete. In the treatment of hydrocele no mention is made of acid carbolic injection into the sac, and nothing is said of the use of antitoxin in tetanus. Antitoxin is recommended very highly in the treatment of diphtheria, though the chapter is slightly conservative. The chapter on Diseases of the Lung is a very complete one, and we are pleased to see that "capillary bronchitis" is not used.

The book is a highly entertaining and very instructive one, and is bound to have a well-merited sale.

Those who are familiar with Dr. Jacobi's writings find in this volume of his many familiar expressions. Often he differs from recognized authorities, but it is a pleasure to thus read one's individual opinion. He differs with Dr. Rotch in regard to the efficacy of lime-water as an antacid. Dr. Rotch states that one sixteenth part of lime-water will prove effectual in neutralizing an acid cow's milk, while Dr. Jacobi states: "The effect of lime-water is in part imaginary. If given for the purpose of neutralizing acids it is a failure." Frequently such conflict of opinion is found, but that is not objectionable.

Unfortunately "Capillary Bronchitis" is used as a distinct pathological lesion. Two distinct chapters are devoted to

Scrofulosis and Tuberculosis, which is hardly to be expected from Dr. Jacobi.

We are surprised not to find any mention of stomach washing in the treatment of gastro-intestinal troubles.

The last chapter, "Addenda," is largely a review of the latest therapeutic measures suggested in the various periodicals. Diphtheria antitoxin is here considered and recommended.

The book shows evidence of careful proof reading, except on page 56, the foot-note being of no service, referring to the March Archives of Pediatrics, the year not being given.

This book will at once take its place in the library of the student and busy practitioner, and the profession is greatly indebted to Dr. Jacobi for his careful work.

Annual of the Universal Medical Sciences: A Yearly Report of the Progress of the General Sanitary Sessions throughout the World. Edited by CHARLES E. SAJOUS, M. D., assisted by seventy associate editors, assisted by over two hundred corresponding editors, collaborators, and correspondents. Illustrated with chromo-lithographs, engravings, and maps. 1895. The F. A. Davis Company, Publishers, Philadelphia, New York, Chicago. London: F. J. Robman.

It is like the visit of an old and cherished friend to see the appearance each year of the Annual of the Universal Medical Sciences, edited by Sajous. Indeed a library without this manual is not complete at all. Rich in contents, elegant in arrangement, and beautiful in make-up, it is both pleasing and instructive to its owner. It is really wonderful the amount of useful knowledge contained in the five volumes. Every subject known to medicine and surgery is treated of, and intelligently so. It is a library within itself, at least so far as the latest advances are concerned. Whereas objections might be urged to positions taken by some of the writers, yet this could be said of any work which treats of scientific subjects.

For instance : The author of the chapter devoted to diseases of the rectum and anus in one place says : " Although nothing is found in the history of syphilis that at all corresponds with stricture of the rectum, there are lesions of other varieties which closely resemble it anatomically and clinically ; and the author is led to the conclusion that the so-called syphilitic stricture is the result of a simple proctitis arising from any one of many irritating causes." Such a conclusion, in the light of all modern investigation, is unworthy the author who writes it, or the book in which it is published. Mr. Alfred Cooper, F. R. C. S., England, in the second edition of his masterly work on syphilis, lately published, says on page 199: "Gummatous deposits and infiltration of the submucous tissues are the most common causes of syphilitic ulceration and stricture of the rectum." Mr. Cooper is not only one of the most learned syphilographers of the age, but has devoted much of his time to rectal diseases, and in the sentence above speaks that which is the consensus of opinion of all modern thinkers.

The Annual is published by the F. A. Davis Company in most excellent style, both in regard to paper, illustrations, and binding.

Pregnancy, Labor, and the Puerperal State. By EGBERT H. GRANDIN, M. D., Consulting Surgeon to the New York Maternity Hospital; Consulting Gynecologist to the French Hospital, New York, etc.; and GEORGE W. JARMAN, M. D., Obstetric Surgeon to the New York Maternity Hospital; Gynecologist to the Cancer Hospital, New York, etc. Illustrated with forty-one original full-page photographic plates from nature. Royal octavo. Pages viii, 261. Cloth, $2.50 net. Philadelphia: The F. A. Davis Company, Publishers, 1914 and 1916 Cherry Street.

This is a companion to the volume previously written by these authors, and reviewed in a former number of the QUARTERLY. As a *practical* work, a book for the practitioner and advanced student, it will answer admirably. Nothing new is offered in the chapter on Pernicious Vomiting of Pregnancy, though the résumé of the methods of treatment now in vogue is well written. This whole chapter upon the Pathology of Pregnancy is perhaps the best in the book.

The illustrations, which are intended to be the chief attraction in the book, are, upon the whole, disappointing. Plate I is very indistinct, and does not show the fine shading which is intended.

Plate II is the *exact reproduction* of three of the cuts, much enlarged, on a plate in the *American System of Gynecology and Obstetrics*, illustrating an article by Spiegelberg, the plate bearing the title of "Presence of Areola in Pregnancy." The cuts, if not copied, must have been executed by the same artist, who had forgotten the plate above referred to had been made. Even the folds of the sheets covering the breast lie in exactly the same position.

Plate X is perhaps a good photograph, but not well chosen. A photograph of the hands applied on either side of the median line, placed symmetrically, to learn the position of the dorsum, would be much more instructive.

The plates illustrating the mechanism of labor, position, and attitude of the fetus are excellent.

Plate XXVII is well chosen to emphasize the importance of catheterization by sight.

The various plates depicting the mechanism of labor are very good with the exception of the step showing the index finger in the rectum. We can not see how asepsis and this method of delivery can be reconciled. It is hardly possible to do nothing else with the finger so used, and the injury done the rectum by pressure more than counterbalances the assistance rendered.

Plate XXXV would have been much better left out, as it represents nothing intelligible. The book has much of value in it, and should have a permanent place in medical libraries.

The American Year-Book of Medicine and Surgery. By J. M. Baldy, M. D.; C. H. Burnett, M. D.; Archibald Church, M. D.; C. F. Clarke, M. D.; J. Chalmers Da Costa, M. D.; W. A. N. Dorland, M. D.; V. P. Gibney, M. D.; Homer W. Gibney, M. D.; Henry A. Griffin, M. D.; John Guitéras, M. D.; C. A. Hermann, M. D.; H. F. Hansell, M. D.; W. A. Hardaway, M. D.; T. M. Hardie, B. A., M. B.; C. F. Hersman, M. D.; B. C. Hirst, M. D.; E. Fletcher Ingals, M. D.; W. W. Keen, M. D.; H. Leffman, M. D.; V. H. Norrie, M. D.; H. J. Patrick, M. D.; Wm. Pepper, M. D.; D. Riesman, M. D.; Louis Starr, M. D.; Alfred Stengel, M. D.; G. N. Stewart, M. D.; Thompson S. Westcott, M. D. Under the general editorial charge of GEORGE M. GOULD. M. D. Profusely illustrated with numerous wood-cuts in text and thirty-three handsome half-tone and colored plates. Philadelphia: W. B. Saunders, publisher.

This book is a yearly digest of scientific progress and authoritative opinion in all branches of medicine and surgery, drawn from journals, monographs, and text-books of the leading American and foreign authors and investigators. In addition there are critical editorial comments by the compilers, and the whole is under the editorial charge of that systematic writer, Dr. George M. Gould. With such an array of authors, which contains names of some of the best-known men in the medical profession, it can be readily seen that the work is first class in every particular. It is really an epitome of most every thing of importance that has been written on medical subjects during the

year, or, what might be more properly said, a digest of monographs, medical journals, etc., issued or published during that
time. Of course it is to be expected that in copying the views
of so many there is much that is not really *authoritative*. But
the reader can sift the good from the bad and have much to content himself with. To those who are not in the habit of subscrib-
ing for a number of good medical journals this book will prove
a rare treat. Indeed, every physician would be much benefited
by adding this volume annually to his library. The book is
issued in the usual handsome style that is characteristic of the
publisher.

A Comedy of Sentiment. By MAX NORDAU, M. D. New York: F.
Thomson Neely. 1896.

There is very little plot in the story, which is one descriptive
of the woes, owing to his somewhat oversusceptible nature, of a
bright college professor, who becomes acquainted with a bright,
fascinating, but deeply-designing widow, who most artfully
weaves an illusionary web around the professor's affections,
and thus causes him to drift almost helplessly under her control,
and almost impoverish himself in gratifying her numerous
appeals for material aid. Fortunately the friends of the professor, who had been the means of introducing the two to each
other, learned of the desperate situation in which the professor
had become involved, and just in the nick of time saved him
from marrying the adventuress, who, it was subsequently learned,
had a liaison with another man during the same period she was
impaling the professor upon the point of the knife of her artful
dissimulations. Of course there was nothing in the simulated
grief of this artful adventuress to prevent her soon after marrying the unsuspecting victim, which her great powers of fascina-
tion had already prepared and kept on hand ready for just such an
emergency as followed the professor's disillusion and consequent
breaking off of the engagement of marriage. The characters in
the story are few, and the story is confined almost entirely to the
doings of the two persons in question. As the author has lately
become somewhat famous by reason of the radical ideas he has
promulgated in his publication styled Degeneration, his latest
book is likely to attract many readers.

Medical and Surgical Report of the Children's Hospital—1869-1894.
Boston. Edited by T. M. ROTCH, M. D., and HERBERT L. BURRELL, M. D.
1895. Published by the Board of Managers.

This book is an extremely attractive one, and shows that great labor has been expended upon it. Its contents are divided into three portions: The Administrative, the Medical, and the Surgical. In the first division is considered the history of the Hospital, a description of the Hospital, the Convalescent Home, and the Outdoor Relief Department.

The Medical division embraces a series of articles on the prevalent troubles in the hospital by the visiting staff, as follows: Typhoid Fever, by Gordon Morrill, M. D.; The Value of Milk Laboratories for Hospitals, by T. M. Rotch, M. D.; The Relation of an Aural Service to the Needs of a General Hospital for Children, by Clarence J. Blake, M. D.; Malaria in Children, by E. M. Buckingham, M. D.; An Epidemic of Scarlet Fever, by Thomas F. Sherman, M. D.; The Etiology of Chorea, by Charles W. Townsend, M. D.; An Endemic of Diphtheria Apparently Stopped by the Use of Antitoxin, by F. Gordon Morrill, M. D.

The Surgical division occupies two hundred and forty-four pages, and it is about as complete a treatise upon orthopedic surgery as we have seen. It contains thorough descriptions of diseases of the joints and less common surgical affections met with in children, and articles by the following authorities: E. H. Bradford and R. W. Lovett, Herbert L. Burrell, W. N. Bullard, H. W. Cushing, and E. G. Brackett.

There are numerous illustrations throughout the book, which are mostly half-tones from photographs. They are remarkably clear and distinct.

There is much to commend in this book, which, though apparently written as a history of a noble work, would serve admirably as a text-book in medical schools, especially for the surgical troubles of infancy and childhood.

Miskel, a Novel. By L. M. PHILLIPS, M. D., of Penn Yan, N. Y. Advance copy No. 2 of the Doctor's Story Series, 266 pages. New York: Bailey & Fairchild Company. 1896. Price, paper, 50 cents.

This is a story which treats largely of hypnotism and other occult arts of the far East. It is after the Rider Haggard order, full of inconsistencies and most improbable scenes. Two of the principal characters are represented as male servants from India; dumb, from having their tongues cut out, yet they are made to say most elaborate speeches in the sign language, such as the following: "Oh, my sweet one! my dove! my sunbeam! Kill thy slave that he may become thy guardian mentor!" and many others as flowery. The chief female character, endowed of hypnotic power, is made to use it to the downfall and ruin of a married man; and this part will have the effect, if widely read, of prejudicing people against hypnotism and its legitimate use by the portrayal of such disgusting scenes.

To those who relish this character of work, descriptions of weird seances, where the hero is made to shoot elephants who shed "*ghouls* of blood," etc., Miskel will while away a pleasant hour, leaving naught but confusion in its wake.

A Manual of the Practice of Medicine. By GEORGE ROE LOCKWOOD, M. D., New York, Professor of Practice in the Woman's Medical College of the New York Infirmary. With seventy-five illustrations in the text and twenty-two full-page colored plates. Philadelphia: W. B. Saunders. 1896. Price, $2.50 net.

The reviewer has taken occasion to state in previous issues when reviewing "manuals" that they tend to make a student, for whom they are for the most part written, superficial in his studies. Yet they do occupy a place in his library, giving a fulsome résumé of the subject. The one before us, by Dr. Lockwood, is one of the best manuals we have seen, being tersely written and well illustrated. If it does nothing more than impress upon the reader the importance of a more liberal use of temperature charts in the clinical history of a case, it will have fulfilled one mission. The liberal use of temperature charts through the text is to be commended.

Transactions of the New York Academy of Medicine. Instituted in 1847. Second Series. Volume X, for 1893. Printed for the Academy. 1894.

This volume contains papers which from time to time have appeared in the various medical journals, but a collection of them in one volume is valuable. Among the contributors to this volume are Charles McBurney and M. Allen Starr, Alfred L. Loomis, D. B. St. John Roosa, Charles E. Quimby, Beverly Robinson, Robert Abbe, Hermann M. Biggs, Simon Baruch, J. West Roosevelt, H. G. Piffard, and Thomas R. Pooley.

A Manual of Medical Jurisprudence and Toxicology. By HENRY C.
CHAPMAN, M. D., Professor of Institutes of Medicine and Medical Jurisprudence in the Jefferson Medical College, etc., Philadelphia. Second
edition. Revised. With fifty-five illustrations and three plates in colors.
Philadelphia: W. B. Saunders. 1896. Price, $1.50 net.

Medical Jurisprudence is a subject upon which but little
stress is made in most medical schools, and it takes such complete works upon the subject as the one before us to assist the
student in his studies. Though called a manual, it is remarkably clear, and will answer as a text-book. With the exception
of foot-notes referring to the bibliography of the subject this volume is but little altered from the first edition.

**A Text-Book upon the Pathogenic Bacteria, for Students of Medicine
and Physicians.** By JOSEPH McFARLAND, M. D., Demonstrator and
Lecturer on Bacteriology in the Medical Department of the University of
Pennsylvania. With one hundred and thirteen illustrations. Philadelphia:
1896. Price, $2.50.

This is one of the most useful books that we have seen, being a
treatise upon *bacteria*, the discussion of malaria plasmodium, etc.,
being omitted, but thoroughly describing " such bacteria as can
be proven pathogenic by the lesions or toxines which they
engender." The illustrations are, without exception, excellent,
the photo-micrographs being for the most part perfect.

The Principles and Practice of Bandaging. By GWILYM G. DAVIS, M. D.,
Assistant Demonstrator of Surgery, University of Pennsylvania. 1891.
George S. Davis, Detroit, Mich.

This is a neat little work of sixty-one pages, with splendid
illustrations of the various bandages complete and in the process
of application to the various parts of the body. This attention
to details makes the work a valuable one, such a work being
indispensable to the student.

The Principles and Practice of Medicine. By WILLIAM OSLER, M. D.,
Professor of Clinical Medicine in the University of Pennsylvania, Philadelphia. Second edition. New York: D. Appleton & Co. 1895.

This book, which had such a phenomenal sale in its first edition, needs no introduction to the profession of America.

So many important changes and additions have been made in
this edition that it is really a new work. It is a book which the

more one reads the more he becomes attached to it as a work of reference and as a text-book. The classifications are excellent, the arrangement of the text is admirable.

But one serious criticism can be made to the whole work, and that is that the author has seen best to leave out from the chapter on Diseases of the Stomach the methods of clinical examination. If this does not belong to a work of this character, we fail to see where it does. It is inconsistent to include tests for abnormal urine and not those above referred to. We trust that this omission will not be made in subsequent editions.

The consideration of all the latest theories and therapeutic measures is thoroughly up to date. We commend the work most heartily.

Transactions of the Medical Society of the State of North Carolina. Forty-second Annual Meeting. Held at Goldsboro, N. C., May 15 and 16, 1895.

This is a carefully edited volume, and contains, besides the papers read at the annual meeting, a list of the places of meeting, and the officers of each annual meeting since April, 1849, a point which it would be well for other societies to note in their annual volume of proceedings.

Stories of a Country Doctor. By WILLIS P. KING, M. E., Kansas City, Mo. 1896. Bailey & Fairchild Co., New York.

This is the first of the Doctor's Story Series, issued in paper form, and needs no introduction to the medical profession. It is a decidedly original and humorous story, told in an inimitable manner.

Books and Pamphlets Received.

WHAT MAY BE EXPECTED OF THE X RAYS.—As if to throw into the shade entirely such wonders as photographing invisible planets and researches into the ultra-violet regions of the solar spectrum, the new process may claim the unique position of delineator of the invisible in general. We see very acute pictures of bullets lodged between the bones of the lower leg, or of the bony structure of living hands, or even of whole skeletons of small animals, taken through the flesh. It is not to be wondered at that the majority of investigators are hard at work developing the capabilities of the new art in surgery and medicine. From metallurgists we do not hear a great deal, and it must be admitted that we have yet much to look forward to from them. Glass, the type of transparent substances, resists the new rays like so much wood. A photographic plateholder, with glass doors, would be a curiosity, and yet not wholly useless for the new process. The homogeneous structure of such metals as are penetrable by the X rays, as they are called, is well detailed. All the ornaments and lettering, as well as concealed flaws, cracks, etc., in the coins hitherto experimented upon have been quite distinctly rendered. Eminent authorities in gunnery have expressed their belief in the powers of the new art to reveal hidden defects in expensive gun-forging. With more powerful sources of actinism, and with photographic chemicals and plates prepared for this express purpose, we may yet be able to realize such triumphs of useful picture-making as the revealing of dangerous flaws in bridge beams or in the shafts of sea-going steam vessels.— *From " Recent Photographic Invention," by Ellerslie Wallace, M. D., in North American Review for March.*

The February number of The New Bohemian contains all the sparkle and spice which we have learned to expect in this sprightly magazine. The opening story, " The Last Man," by Elizabeth Cherry Haire, is distinguished by its *fin de siecle* qualities; and, while it is a rather rollicking love story, it contains much genuine sentiment and some splendid character sketching. " On Mount Marcy " is a thoughtful and eloquent essay by William F. Seward. It shows in a striking manner how the same environments may variously influence persons of widely different habits of thought and mental characteristics.

In a most invitingly dainty cover, seasonably typifying midwinter—its frost and snow—the February Ladies' Home Journal, resplendent with illustrations by the best artists, and attractively

varied in its literary features, is unique.   James Whitcomb Riley
sings in his sweetest, purest strain of "Little Maid-O'-Dreams,"
a fanciful little poem, the spirit of which Rosina Emmet Sherwood has winsomely presented in a full-page picture.

An American Text-Book of Surgery for Practitioners and
Students.   By Charles H. Burnett, M. D., Phineas S. Connor,
M. D., Frederick S. Dennis, M. D., William W. Keen, M. D.,
Charles B. Nancrede, M. D., Roswell Park, M. D., Lewis S.
Pilcher, M. D., Nicholas Senn, M. D., Francis Shepherd, M. D.,
Lewis A. Stemson, M. D., William Thompson, M. D., J. Collins Warren, M. D., and J. William White, M. D.   Edited by
William W. Keen, M. D., LL. D., and J. William White, M. D.,
Ph. D.   Second edition, carefully revised.   Philadelphia.   W. B.
Saunders, 925 Walnut Street.   1895.   Price, $7 cloth, $8 sheep,
$9 half russia.   For sale by subscription only.

The Treatment of Malignant Tumors by the Toxines of the
Streptococcus of Erysipelas; Conservative Surgery on the Battlefield and First Aid to the Wounded; Early History of Vaginal
Hysterectomy; Etiology, Pathology, and Treatment of Intestinal
Fistula and Artificial Anus.   By N. Senn, M. D., Chicago.   All
but last reprint from Journal American Medical Association.

Electricity in the Treatment of Exophthalmic Goitre, Supplementary Report on the Success of Electrolysis in the Treatment
of Strictures of the Urethra.   By Robert Newman, M. D., New
York.   Reprint from the Journal of the American Medical Association.

Various Fractures; Simple and Compound.   A Clinical Report of Fifteen Cases.   By Thomas H. Manley, M. D., New
York.   Reprint from American Medico-Surgical Bulletin.

A Combined Uterine Irrigator and Drainage-Tube, with Indications for Its Use.   By O. Grothan, M. D., St. Paul, Neb.
Reprint from Journal of American Medical Association.

Granular Lids; with Cases in· Practice.   By A. Britton
Deynard, M. D., New York.   Reprint from the Medical and
Surgical Reporter.

Deformities Following Fractures of the Shafts of Bones, with
Observations on Treatment.   By Thomas H. Manley, M. D.
Reprint from the American Medico-Surgical Bulletin.

The Sensory Nervous System in Diagnosis; The Reflexes.   A
Contribution for Medical Students.   By Charles H. Hughes,
M. D.   Reprinted from the Alienist and Neurologist.

On Some Difficulties in Reference to the Early Surgical
Treatment of Appendicitis.   By Carl Beck, M. D., New York.
Reprint from Journal American Medical Association.

The Limitations of Surgical Operation as a Means of Relief or Cure in Epilepsy. By Thomas H. Manley, M. D., New York. Reprint from the International Journal of Surgery.

Some of the Newer Problems in Abdominal and Pelvic Surgery in Women. By Charles P. Noble, M. D., Philadelphia. Reprint from the American Journal of Obstetrics.

Report of One Hundred and Eleven Additional Cases Operated upon with the Anastomosis Button. By J. B. Murphy, M. D., Chicago. Reprint from Medical News.

Tendon Graftings; a New Operation for Deformities following Infantile Paralysis. By Samuel E. Milliken, M. D., New York. Reprinted from the Medical Record.

A Consideration of Certain Doubtful Points in the Management of Abortion. By Charles P. Noble, M. D., Philadelphia. Reprint from Therapeutic Gazette.

The Necessities of a Modern Medical College. By E. Fletcher Ingalls, M. D., Chicago. Reprint from Bulletin of American Academy of Medicine.

Notes on An Epidemic of Acute Anterior Poliomyelitis. By C. S. Caverly, A. B., M. D., Rutland, Vt. Reprint from Journal American Medical Association.

The Etiology and Symptomatology of Pruritus Ani. By Lewis H. Adler, jr., M. D., Philadelphia. Reprint from the Philadelphia Polyclinic.

Radical Cure of Inguinal and Femoral Hernia. By William L. Rodman, A. M., M. D., Louisville, Ky. From the American Practitioner and News.

The Techniques of Maunsell's Method of Intestinal Anastomosis. By F. Holme Wiggin, M. D. Reprint from New York Medical Journal.

A New Operation for Congenital Ptosis, with Report of Two Cases. By T. C. Evans, M. D., Louisville. Reprint from New York Medical Journal.

Tumors of the Mammary Gland. By W. L. Rodman, M. D., Louisville, Ky. Reprint from the Journal of the American Medical Association.

A Case of Dermoid Tumor of Both Ovaries, etc. By Charles P. Noble, Philadelphia. Reprint from American Journal of Medical Sciences.

Syphilitic Ulceration of the Rectum. By James P. Tuttle, M. D., New York. Reprint from Journal of American Medical Association.

14

Address on State Medicine. By Henry D. Holton, A. M., M. D., Brattleboro, Vt. Reprint from Journal American Medical Association.

Technique of Emptying the Uterus in Inevitable Abortion. By Charles P. Noble, M. D., Philadelphia. Reprint from Codex Medicus.

Treatment of Puerperal Sepsis. By Louis Frank, M. D., Louisville, Ky. Reprint from the American Journal of Obstetrics.

Rational Drug Therapy in Typhoid Fever. By O. Grothan, M. D., St. Paul, Neb. Reprint from Wilkinson's Omaha Clinic.

The Rectum and Urethra as Related in Disease. By James P. Tuttle, M. D., New York. Reprint from New York Polyclinic.

The Treatment of Ano-rectal Fistula. By James P. Tuttle, M. D., New York. Reprint from New York Medical Journal.

Modern Treatment of Diseases of the Stomach. By John Ford Barbour, M. D. Reprint from the American Therapist.

Modern Surgery of Serous Cavities. By Merrill Ricketts, M. D., Cincinnati, O Reprint from The Railway Surgeon.

The Treatment of Epilepsy. By Allison Maxwell, M. D., Indianapolis, Ind. Reprint from Indiana Medical Journal.

Hypertrophic Rhinitis. By Edward J. Bermingham, M. D., New York. Reprint from the New York Medical Times.

Sleep in its Relation to Diseases of the Skin. By L. Duncan Bulkley, M. D., New York. From the Medical Record.

Brain Surgery for Epilepsy. By B. Merrill Ricketts, M. D., Cincinnati, O. Reprint from Cincinnati Lancet-Clinic.

Rhinological Don'ts. By Edward J. Bermingham, M. D., New York. Reprint from the Texas Medical Journal.

Biographical Sketch of Dr. Charles H. Hughes, M. D. Reprint from Physicians and Surgeons of America.

Nephritis of the Newly Born. By A. Jacobi, New York. Reprinted from the New York Medical Journal.

Movable Kidney. By Charles P. Noble, M. D., Philadelphia. Reprint from Gaillard's Medical Journal.

Subphrenic Abscess. By Carl Beck, M. D., New York. Reprint from the Medical Record.

Septic Peritonitis. By Louis Frank, M. D., Louisville, Ky. Reprint from the Medical News,

Notes and Queries.

"DON'T'S FOR CONSUMPTIVES, or the Scientific Management of Pulmonary Tuberculosis," is the title of a book which, under the authorship of Dr. Charles Wilson Ingraham, will soon (about February 10th) be issued by the Medical Reporter Publishing Co., of Rochester, N. Y.

The complete work of thirty-five chapters is devoted exclusively to the general management of pulmonary invalids, no reference whatever being made to drug treatment.

The object of the author is to supply the physician with a practical work, and at the same time, by eliminating technical terms, reduce the text within the easy comprehension of the intelligent patient.

The book will be printed on 72-pound antique book paper, bound in cloth (imitation morocco), with title in gold leaf. Price, $1.75.

DR. JOSEPH B. BACON, of Chicago, an occasional contributor to the QUARTERLY, Professor of Diseases of the Rectum in the Post Graduate Medical College, has recently been appointed Clinical Instructor in Gynecology in the Northwestern University and Surgeon to the Woman's Hospital, both positions of honor and well merited. Dr. Bacon will contribute an article to the next QUARTERLY, describing an original and important operation with reports of cases cured by the procedure.

"THE HAPPY MEDIUM."—We are in receipt of a handsome illustrated brochure, issued by the Medical Fortnightly, of St. Louis, which reflects great credit on that enterprising journal. The book contains thirty-two pages, incased in a unique embossed cover, and twenty-five half-tone portraits of its staff, including Dr. Frank Parsons Norbury, Managing Editor; Drs. Hubert Work and T. A. Hopkins, Associates; Charles Wood Fasset, Secretary, and twenty-one department editors.

THE New York State Medical Reporter is now edited by Charles Wilson Ingraham, M. D., vice H. Bronson Gee, M. D., resigned. The following is the staff of the Reporter: Charles Wilson Ingraham, Editor, Binghamton, N. Y.; George R.

Fuller, Manager Publication Department, 15–27 South St. Paul Street, Rochester, N. Y.; Ferdinand S. Randall, Manager Advertising Department, 60 Liberty Street, New York City. Publishers, Medical Reporter Publishing Company.

THE *Clinical Recorder*, edited by Wm. S. Gottheil, M. D. The first number appeared February 1, 1896. It is published by G. A. Sykes, 66 World Building, New York, and is to be a quarterly publication. Number 1 of volume 1 contains articles by Dr. H. J. Garrigues, Carl Beck, S. Henry Dessan, Louis Fischer, William S. Gottheil, Augustus Gollet, Fred B. Valentine. Subscription price, $1 per year.

AT the time of the next meeting of the American Medical Association in Atlanta the National Confederation of State Medical Examining and Licensing Boards will convene in Room 1, The Aragon, Monday, May 4th, at ten o'clock A. M. This meeting will undoubtedly be productive of great good.

THE *Medical Council*, edited and published by J. J. Taylor, M. D., "devoting special attention to obstetrics, diseases of women, diseases of children, and *stirpiculture*." Subscription price, $1 per year.

THE *Canada Medical Record* is now owned and edited by the Faculty of Medicine, University of Bishop's College. Its present address is P. O. Box 2174.

NOTICE TO CONTRIBUTORS.

Articles and letters for publication, books and articles for review, and communications to the editors, advertisements, or subscriptions, should be addressed to EDITORS OF MATHEWS' MEDICAL QUARTERLY, BOX 434, LOUISVILLE, KY.

All necessary illustrations will be furnished free of expense to authors where they send black and white drawings—or negatives—with their MSS.

All articles for publication in the QUARTERLY will be considered only with the distinct understanding that they are contributed to it exclusively.

Alterations in proof-sheets are charged at the rate of 60 cents per hour, which is the printer's rate to the journal.

Reprints may be had at printer's rate if request is made upon proof when it is returned.

All manuscript must be received by the first of the month preceding its publication.

Remittances may be made by check, money order, draft, or registered letter.

The Editors are not responsible for the views of contributors.

PLATE I.—The First Step in Colonoscopy.

MATHEWS'

MEDICAL QUARTERLY.

"ALIS VOLAT PROPRIIS."

| Vol. III. | JULY, 1896. | No. 3. |

Original Contributions.

PROCTO-COLONOSCOPY AND ITS POSSIBILITIES: BY A NEW METHOD.*

(*Illustrated.*)

BY THOMAS CHARLES MARTIN, M. D.,

CLEVELAND, OHIO.

It is almost universally true that surgeons who devote their energies to any field of anatomy embracing a tubular or hollow organ have devised some sort of mechanical contrivance, hoping to facilitate the obtaining of a better view of the part. The multiplicity of these attempts, the complexity of the devices projected, and the paucity of advantage resulting have been such that the great body of general medical practitioners have come to regard the speculum-making tendency of the specialist as an indication of *fin de siecle* degeneration, though not a particularly harmful form perhaps.

However, the very fact of the great number of rectal specula is sufficient evidence that all that could be desired in the way of inspection of this organ has not yet been achieved, and each additional device proposed is indicative of a restlessness and very wholesome desire for further progress.

*For the excellent drawings which illustrate this article I am indebted to Dr. Hubert Spence. He has very satisfactorily shown the extent to which the parts are exposed by this method of examination, and has accurately represented the relation of each viscus to all the other organs of its community.

To my assistant, Dr. Snearer, I am under obligation for the perfect negatives from which Messrs. Endean and Enwright have made the photographs which indicate the several steps of the procedure.

I am also very greatly indebted to Dr. C. M. Thurston for assistance in the preparation of the specimens shown by photograph.

15

The cylindrical tube and the beveled blade retractor are the essential elements of all these more or less elaborate rectal specula. Sims' simple vaginal speculum has been modified by Hegar, Van Buren, Helmuth, and Kelsey without practically increasing its usefulness. Single and double pairs of blades have been hinged together, with blades parallel or at radiating angles, with very creditable ingenuity and in almost as endless variety as in vaginal specula. Blades and cylinders have been fenestrated until there remained but little of them when in use but their pain provoking margins and their windows filled with pouting membrane. The results of all this expenditure of inventive and constructive power have until recently been most meager, the fenestrated tube giving but an imperfect view of one small longitudinal portion of the mucous membrane, while the bi-valve or tri-valve speculum permits at most an exposure of but two or three narrow areas, and that, too, to a depth not exceeding 7.62 cm. (three inches), the lateral exposure seldom exceeding more than half the anal circle.

Many years ago Bodenhamer endeavored to obtain a view of the sigmoid flexure† of the colon by joining two cylindrical tubes at an obtuse angle, and placing in their angle a mirror for the reflection of the image of that portion of the mucous membrane which prolapsed on the tube's distal end. The diagnostic value of this method, as well as of any specular examination of the rectum in the semi-prone or dorsal posture which exposes to view only that part which prolapses into the speculum, is really but little greater than are the subjective symptoms of the patient.

Very recently Dr. Howard Kelly made known‡ a method of inspection of the rectum which, when compared to the methods commonly practiced, is of very much greater practical value. His speculum consists of a cylindrical tube provided with an obturator and is similar to his cystoscope save that it is larger, and it does not essentially differ from the rectal specula of Andrews. With his characteristic genius Dr. Kelly inverts his patient to the knee-chest posture and applies to the rectum manipulations similar to those which distinguish his cystoscopy, thus employing a posture and a law of atmospheric distension which

† Ball, Surgery of the Rectum.
‡ Annals of Surgery, April, 1895.

were first employed by Otis years ago: a posture and principle which, by the way, rectal surgeons have been rather slow in borrowing from the gynecologist. A little practice in Kelly's method teaches the operator how to carry the end of the procto-scope away from the prostate, the uterus, or an obstructing valve, in order to obtain the atmospheric distension necessary to command a view of the rectal ampulla. In a few cases and under some conditions even this manipulation fails to secure the desired result. This method exposes to view a very much larger area of the rectum than is shown by the methods commonly practiced, but the diameter of the cylinder used is so small that the surgeon is perplexed because there is so limited a portion of the part accessible for surgical manipulation.

In the sigmoid flexure the surface exposed to view by Kelly's method is usually limited to that portion of the gut which falls across the end of the colonoscope. The method is simple, and when applied to the rectum alone is free from the danger of accidental injury to the patient.

The credit of priority in the use of the hollow cylinder and of the prone posture with elevated hips for rectal inspection properly belongs to that distinguished master of rectal surgery, the elder William Allingham. However, the technique by which he secured the posture and by which he introduced the tube§ were so elaborate and inconvenient that his method has been but little employed.

There are certain instruments which are necessary to a complete visual inspection of the lower gut.

I shall now proceed to describe the method which I have formulated for ocular examination of the rectum and sigmoid flexure. The instruments employed are named in the time order of their use:

To support the patient in the genu-acromial posture a pair of firm uprights is necessary—those which I have devised may be clamped to the edge of the usual operating-table, or if that table be not at hand may be screwed into the surface of an ordinary kitchen or dining-table. The uprights telescope for portability.

My rectal speculum is aseptic, may be manipulated with ease, and is designed so as to obstruct the field of vision as little as possible. (Fig. 1.)

§ Allingham's Lectures on the Rectum and Anus.

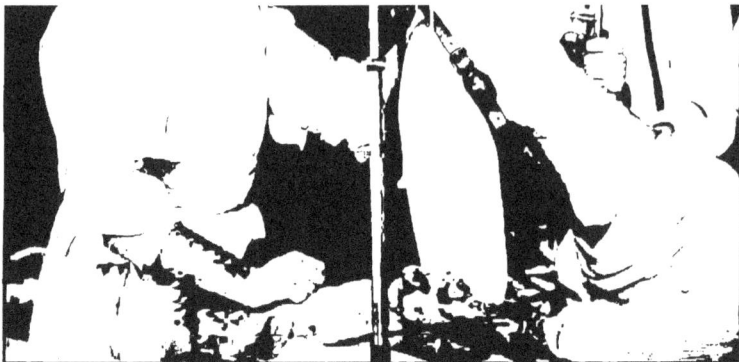

PLATE II.—The first step in Proctoscopy, showing the correct position of the patient, assistant, and operator. The speculum introduced and the parts divulsed. The operator commanding under reflected light a view of the rectum as high as the promontory of the sacrum.

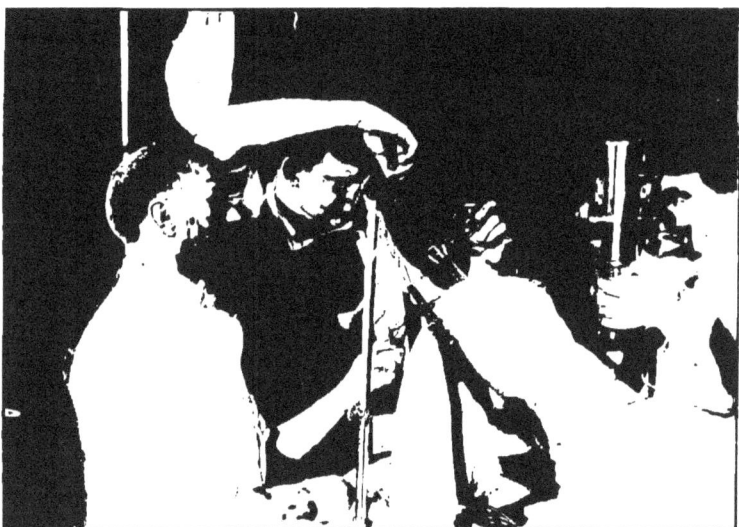

PLATE III.—Showing the Colonoscope introduced and the operator manipulating the colonoscopic mirror.

PLATE IV.—Correct position of patient, assistant, and operator, showing sound in third position.

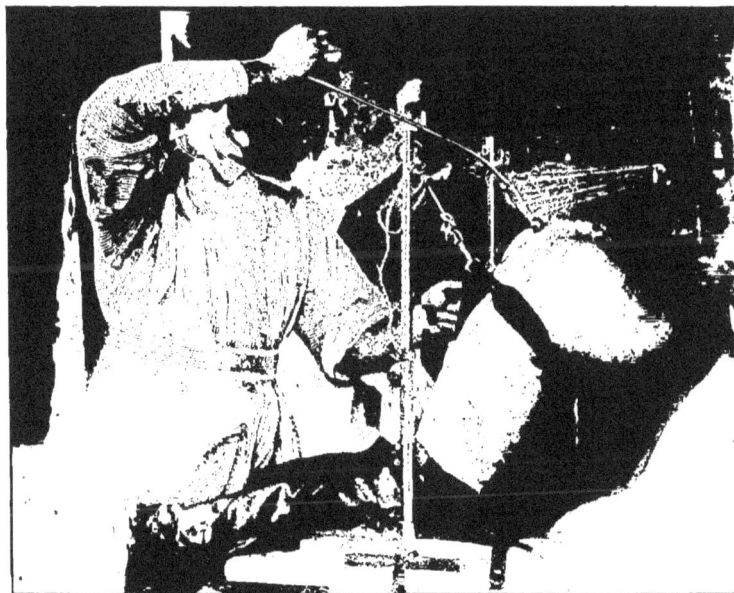

PLATE V.— Illustrates an incorrect position of the first assistant, lamp, and patient. and the correct position of the operator.

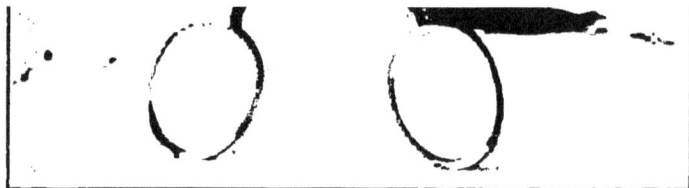

PLATE VI.—A photograph of a posterior view of the rectum and lower half of the sigmoid flexure. The specimen was taken from an emaciated male cadaver still in a state of *rigor mortis*. Preliminary laparotomy was performed and the bowels found in normal position, the sigmoid occupying the left iliac fossa. After emptying the bladder a small opening was made in the upper sigmoid, a hose pipe introduced, the anus opened and the gut thoroughly flushed; the residue of water was then emptied, the part replaced to its normal situation and the abdomen closed. The subject was then fixed in the knee-chest posture and a finger passed through the abdominal wound; the anus divulsed and the length of the gut affected by the atmospheric inflation digitally ascertained. At this limit it was ligated. By degrees the gut was filled with melted paraffin, which was poured in at the anus. So soon as the substance had hardened the coccyx was removed, symphysotomy performed, the mesenteric and other attachments divided and the firm inflexible cast lifted out. Before removal its angles with reference to the perpendicular had been carefully observed; the photographs show the position of the rectum and lower sigmoid under atmospheric distension in a subject fixed in the genu-acromial posture.

It will be noticed that as the sigmoid filled it rode out of the left iliac fossa across the middle line to the right—a fact not infrequently observed in proctoscopy.

PLATE VII.—A photograph of the same specimen pictured in Plate VI. Side view genu-acromial posture.

The hook-mirror faces away from the operator and is provided with a shaft 20.32 cm. (eight inches) in length. This mirror may be used as a hook with which to pull aside obstructing valves. (Fig. II.)

The proctoscopic mirror faces the operator and is 3.81 cm. (an inch and a half) longer than its fellow. (Fig. III.)

In determining the direction of the upper end of the rectum and the position of the colon I employ as a sound a strong metal rod 45.73 cm. (eighteen inches) in length, having a limited range of flexibility and provided with a series of interchangeable bulbs. This instrument may be converted into a convenient applicator by substituting a piece of absorbent cotton for the bulb. (Fig. IV.)

Artificial illumination is required for inspection of the sigmoid flexure. This I have easily accomplished by equipping the ordinary light condenser with an arm bracket and drop-light gas-tube connection; an assistant directs this light into the head-mirror worn by the operator, from which it is reflected into the colon.

The two instruments essential to sigmoidoscopy are a long handled mirror (Fig. V) and the colonoscope (Fig VII). My colonoscope is 38.10 cm. (fifteen inches) in length and about 2.54 cm. (an inch) in diameter when in the bowel. It consists of a proximal handle ring and a distal steel bulb, connected by six strong steel spokes screwed safely into the bulb and secured firmly in the handle; two of these spokes may be removed to facilitate approach to a diseased area. For convenience in packing the handle of the instrument has been made detachable. (Fig. VI.)

These instruments constitute a working set. After a little practice the searching sound may be dispensed with and the colonoscope immediately made to find *its own way* into the colon.

Three other instruments, however, I have found of use in the treatment of disease located high up in the rectum, the cup-curette, a very long insufflator, and the elevator. (Fig. VIII.)

The preparation of the patient for colonoscopy consists in a thorough evacuation of the bowels by means of a three liter (about three quarts) flush of the colon eighteen hours prior to the time of inspection, the administration of a cathartic twelve hours before, and of a rectal enema of not more than 124 grams (four ounces) of a ten-per-cent. watery solution of glycerine one hour before the seance.

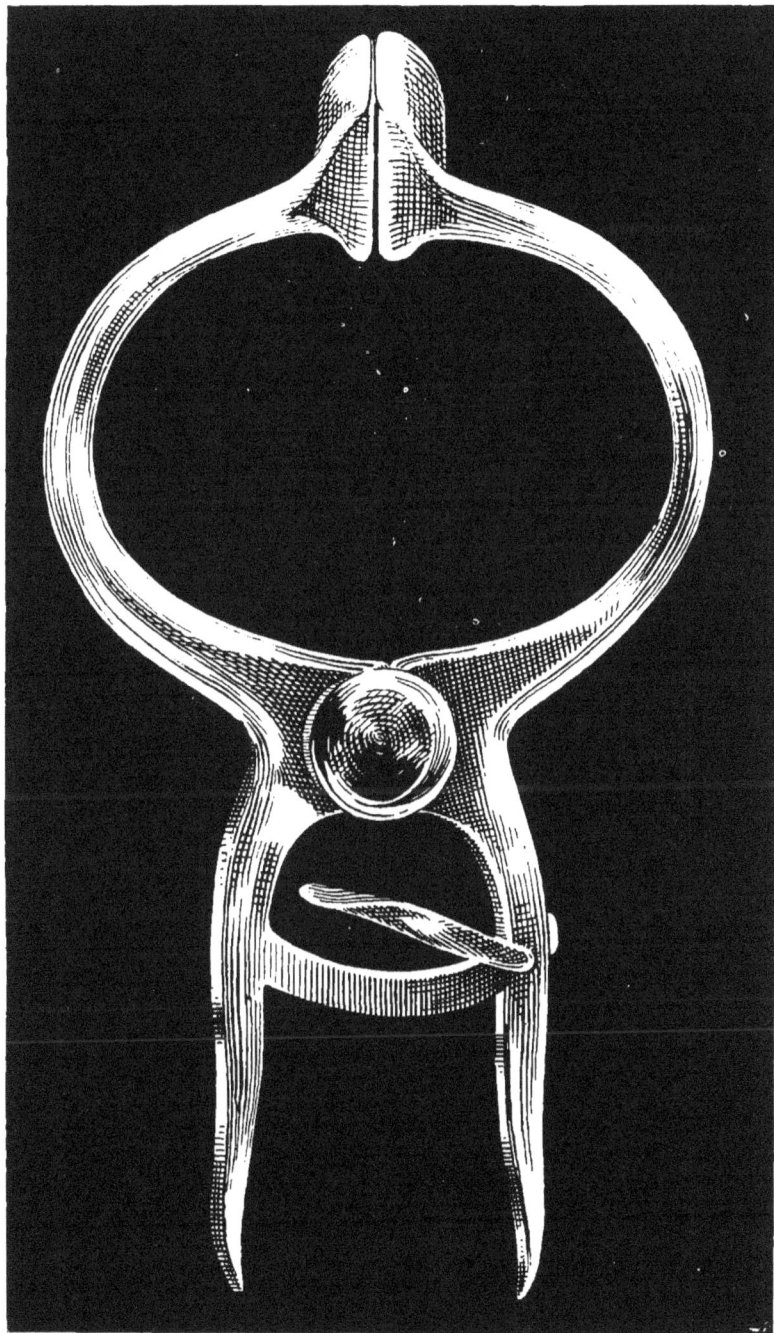

FIG. I.—The Rectal Speculum.
(Full size.)

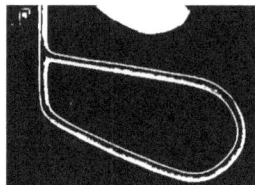

Fig. V.—The Colonoscopic Mirror.
(Half size.)

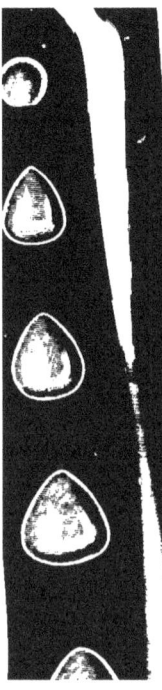

FIG. VI.—The colonoscope with a spoke, removed and the handle detached. (Half size.)

FIG. VII.—The Colonoscope. (Half size.)

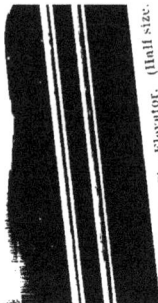

FIG. VIII.—The Elevator. (Half size.)

The posture of the patient is secured by the following steps:
The patient is first anesthetized in the dorsal decubitus and
drawn down until the nates present at the end edge of the table,
then the thighs are flexed to a right angle and the patient turned
to Sims' posture; the buttocks are now carried to the side edge
of the table, and finally the trunk is elevated by raising the hips
till the patient forms a tripod resting on knees and shoulder. The
subject may be quickly secured in this position. (Plates I and II.)

My assistant has given the anesthetic to patients in this position about a hundred times, and it has proven not only a possible
posture,|| but many times the best one for the patient. Inversion
for anesthesia is universally conceded to be most desirable in
case of infants and very young children, but I should be as hesi-
tant of placing the very old or subjects of diseased vessel walls
in this position as I am, under the same conditions, of placing
them in Trendelenburg's posture.

The manipulation of the instruments necessary to inspection:
It is important that the instruments be kept immersed in water of
38° C. (100° F.), so that, when introduced, fog will not gather to
obstruct the view. On account of the very different anatomical
formation of the two parts, preliminary sphincter dilatation‡ is
not so necessary to proctoscopy as is dilatation of the female
urethra to Kelly's cystoscopy, for in the performance of its function the anus often expands to as great a degree as is required for
proctoscopy. Immediately, therefore, the nates may be separated
by the finger and thumb of the left hand and the oiled speculum
blades inserted to their arms; a quick divulsion of the anus will
be followed by an immediate expansion of the rectum and a part
of the sigmoid under the pressure of the inrushing air. The
lower of Houston's semi-lunar valves may now be pushed aside,
the hook-mirror hooked over that valve at the junction of the
middle and upper portions of the rectum, and the proctoscopic
mirror placed opposite and in advance; now, if the two mirrors
be intelligently manipulated, the gut may be visually examined
to a depth of 25.4 cm. (ten inches). The transverse diameter of
the rectum is variable; while in some places not more than 2.54
cm. (an inch), in others it is sometimes more than four times that

| Vide Andrews' Diseases of the Rectum, page 22.
‡ Kelly. Annals of Surgery, April, 1895.

diameter, so that the rectum may present to the eye of the imaginative student the appearance of a chain of urinary bladders communicating one with the other by means of irregularly elliptical openings set at varying axes, and bounded by the non-parallel borders of Houston's valves. In the normal rectum the air pressure smooths the mucous membrane evenly over the entire surface of the gut. The normal mucous membrane of the ampulla appears at first shiny, wet, and bluish gray. As it dries, under the influence of gravitation the blue, venous tint fades out of the gray and the wall becomes pink tinged; presently it assumes the appearance of parchment, painted at rare intervals with ramifying little arteries crowded and overlapped by the larger companion veins which are more ramifying and which more suddenly dive into the bowel wall. In time over all there comes a sheen as of collodion varnish, and the vascular pictures fade, save that one beneath the promontory of the sacrum. The high yellow light on the free border of the semi-lunar valves takes no part in this panorama of color. These phenomena appear exactly as described only in the healthy rectum; in disease the color varies much.

The flexible bulb-tipped sound may now be bent 5.08 cm. (two inches) out of line, oiled and carried into the open sigmoid, the handle then rotated 180° while at the same time the instrument is carried forward; the valve-like obstruction which is now met may be overcome by quickly directing the sound's bulb-end first at one angle and then at another until it is passed, and by elevating the handle the searcher glides on *almost of its own accord* to the diaphragm. The instrument may now be withdrawn, straightened one half and the performance repeated; thus it may be determined if the gut may be straightened out in the middle line parallel with the vertebral column to receive the colonoscope. This instrument at a temperature of 38° C. (100° F.) should be oiled, and with a regard for the sigmoid deviations determined by the sound, it may be introduced, and gently, almost of *its own weight* it will sink to the depth already reached by the sound. The long mirror if well warmed, spread with warm glycerine and placed within the colonoscope, will reveal to the eye of the operator those concave areas whose convexities are presented toward him. And thus, with no more applied mechanical

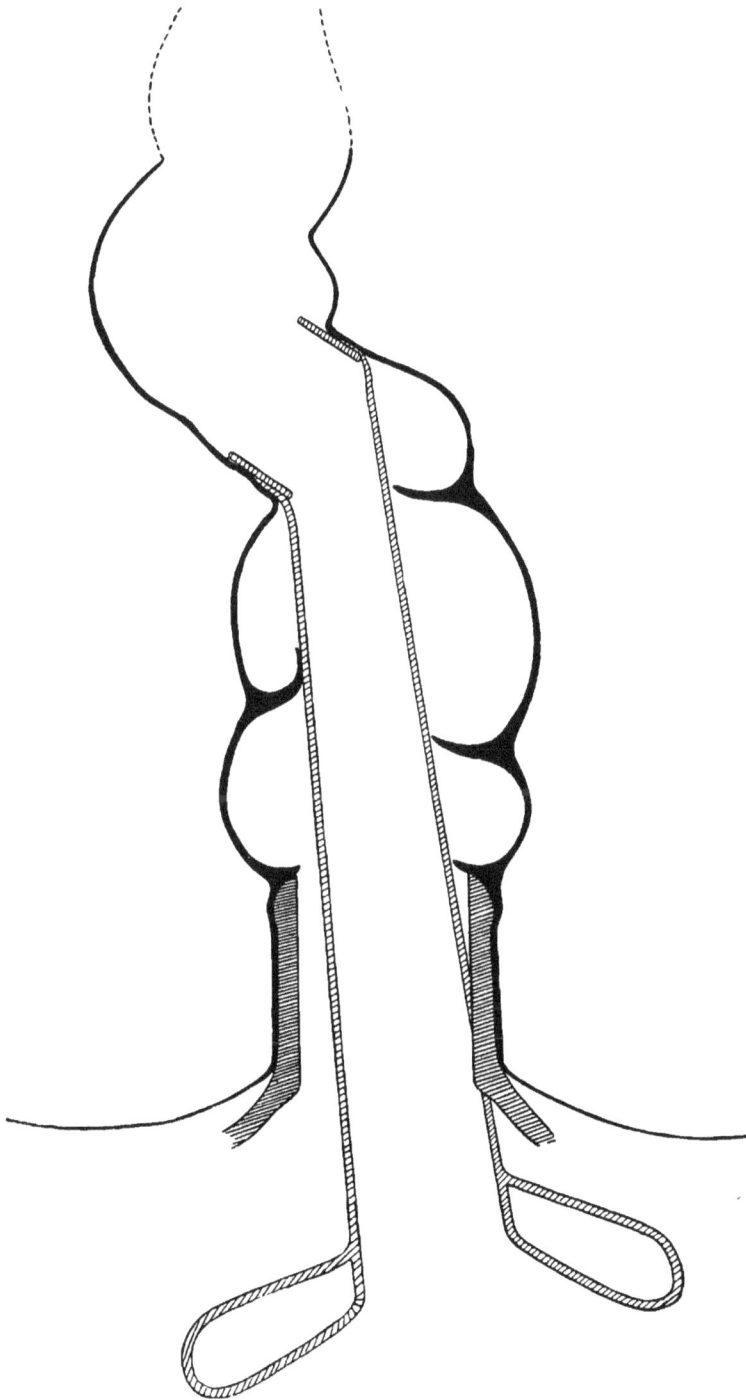

FIG. IX.—Diagramatic Drawing of a Tranverse Vertical Section showing the method of using the Mirrors. (Half size.)

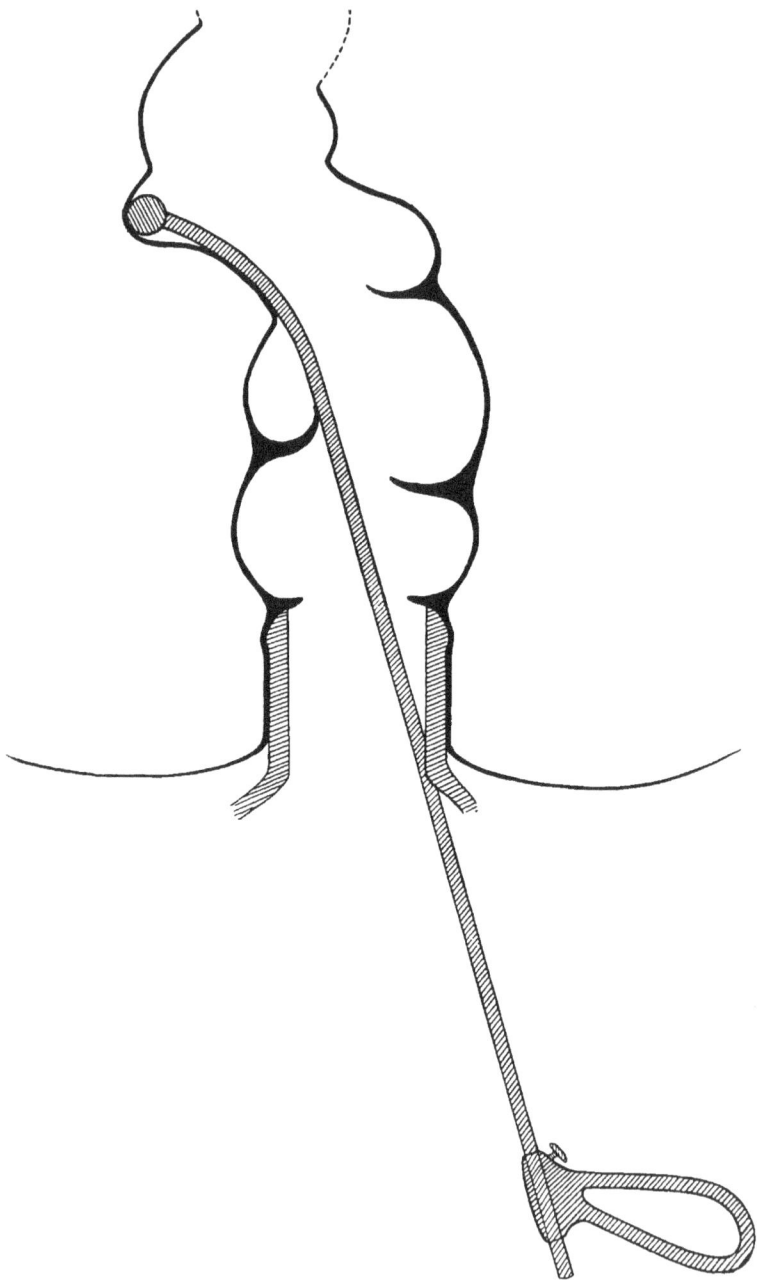

Fig. X.—Diagramatic Drawing of a Transverse Vertical Section showing the sound in the first position. (Half size.)

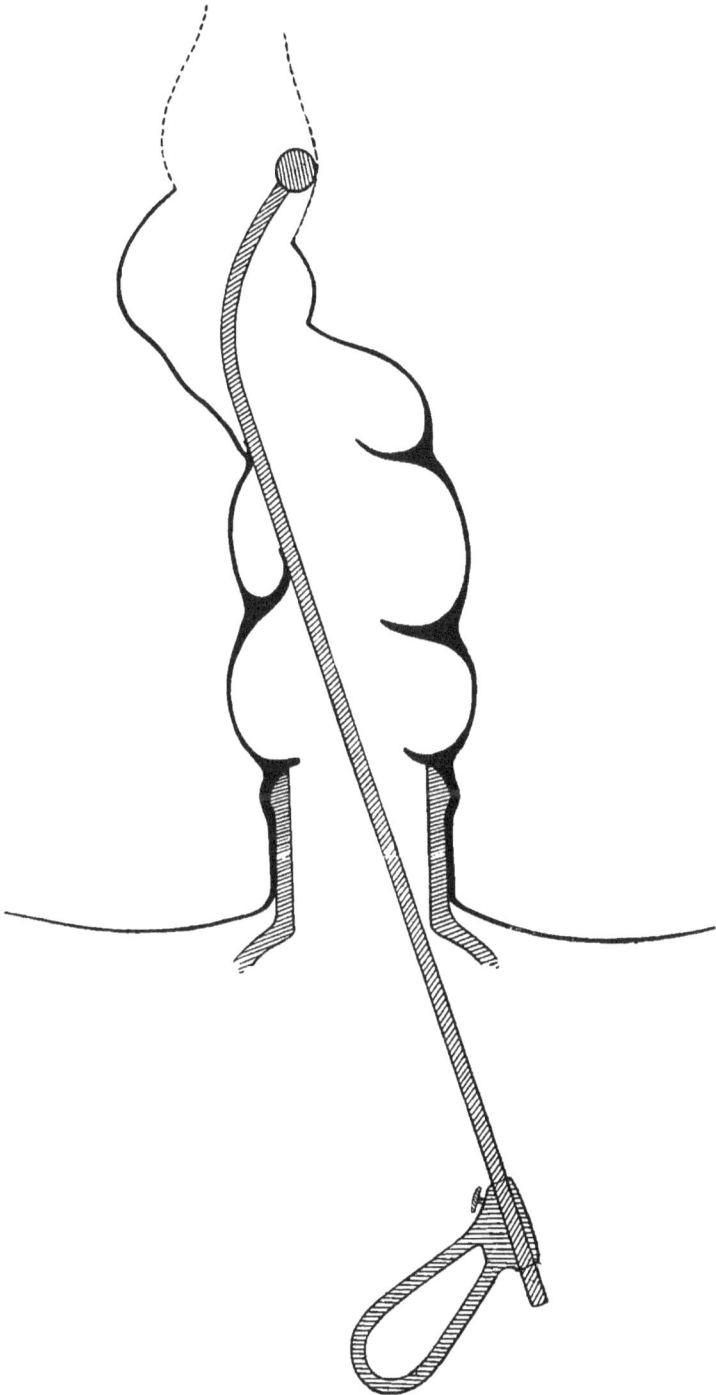

FIG. XI.—Diagramatic drawing of a transverse vertical section, showing the sound in the second position. (Half size.)

violence than is necessary for laryngoscopy, colonoscopy may
be achieved.

The great advantage which this method of visual inspection
affords the surgeon is the ability to declare to his patient with an
unerring certainty that there is or is not disease of the ampulla
or the lower half of the sigmoid flexure.

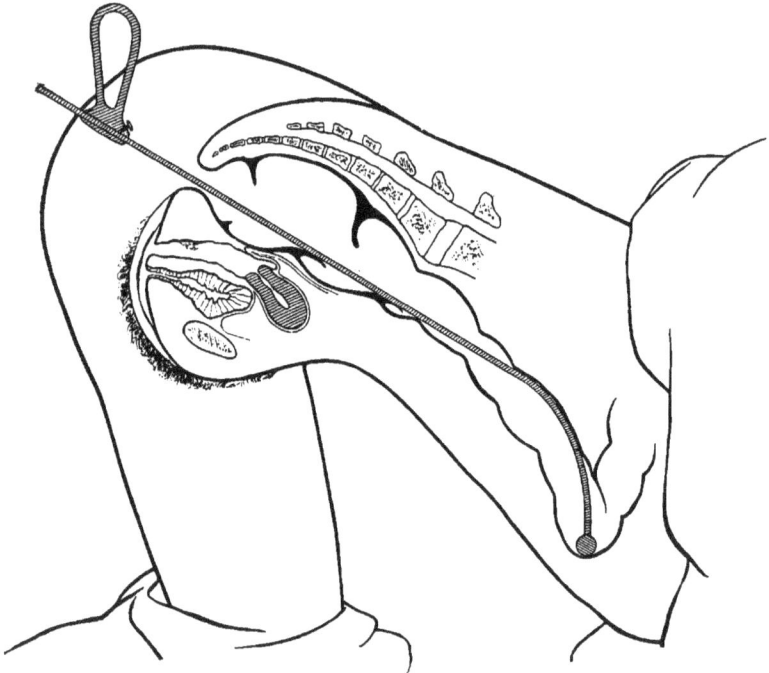

Fig. XII.—Diagramatic drawing of a longitudinal vertical section, showing the
sound in the third position.

Observation by this method has taught me that in nearly all
cases of disease at the anus there is congestion of the rectal ·
mucous membrane at the ampulla, presenting the usual appearance of congestion; sometimes a diffused proctitis of a mild
character attends hemorrhoids and fissure.

Those cases in which there is no apparent lesion at the anus, and which are in a general way diagnosed as catarrh of the rectum, will at once have their real cause, such as a high-up rectal polypus, congenital or organic stricture or ulceration, positively diagnosed, and will be made accessible for an intelligent treatment.

Inflammation of the sigmoid flexure of the colon will be readily recognized also. New growths or ulceration may be seen, and by means of a long-handled curette scrapings made in order that the microscopist may determine their exact character.

Stricture of the rectum and sigmoid need no longer be regarded as of only doubtful presence, and this method proves positively, even to the casual observer, how fallacious is the rectal sound as usually employed in the diagnosis of stricture ; I have repeatedly proven how easy it is for an entering or returning bulb-sound to be caught and held by Houston's valves§ and to elicit those signs which are generally considered diagnostic of organic stricture of the rectum and sigmoid flexure.

Prolapsus of the rectum and intussusception may be determined and in many cases, I believe, corrected almost instantly by this process, and the treatment of these conditions, because of the greatly increased accessibility afforded by this method, is made almost as easy as inspection.

Lesions of the sigmoid are accessible to surgical manipulation, and the gut for the depth of 35.56 cm. or 40.64 cm. (fourteen or sixteen inches) is readily seen and reached for the purpose of topical application.

Vesico-rectal and vagino-rectal fistulæ are apparent at a glance, and, I believe, may be operated upon through the atmospherically distended rectum with more than the usual success.

If this method of ocular examination be practiced, I am convinced there need be no longer any excuse for calling an undiagnosed disease of the rectum or sigmoid flexure obscure disease.

The Parsons Company, of Cleveland, make the instruments which are necessary to the practice of this method.

791 Prospect Street.

§ A paper demonstrating the existence of Houston's valves and pleading the surgical importance of their universal recognition will be published subsequently.

HEMORRHOIDS AND THEIR TREATMENT.

BY WILLIAM V. LAWS, M. D.,

Assistant to Chair of Clinical Surgery, Kentucky School of Medicine.

LOUISVILLE.

[Written for MATHEWS' MEDICAL QUARTERLY.]

With the possible exception of bad colds (coryza) the hemor-rhoidal disease is the most frequent with which the civilized race is afflicted. It appears, from the earliest historical accounts and from the most antiquated medical writings, that hemorrhoids, or piles, were recognized as constituting a common disease at a very remote period, having been spoken of by Hippocrates and other old writers.

The word piles (from *pila,* a ball or swelling,) is, as implying no theory or symptom, perhaps, on the whole, a more suitable term.

Piles are peculiarly a disease of middle life, very rarely met with either in the young or very old. There are few people who live to be fifty years of age without at some time experiencing this trouble in a greater or less degree. It is no respecter of persons, attacking rich and poor, active and sedentary, male and female, alike.

Hemorrhoids, to my mind, are something more than mere varicosities of the blood-vessels, as they are sometimes defined. They are real vascular tumors, originating in an abnormal con-dition of the blood-vessels of the lower end of the rectum and anus. They can be seen, defined, handled, and grow by plastic infiltration.

It seems rather useless to seek for the causes of a malady which is so universal beyond a few which are well recognized and manifest. Among these are straining at stool, pregnancy, affec-tions of the internal organs which interfere with the return of venous blood, and constipation. The erect position of man, to-gether with absence of valves in the veins of the hemorrhoidal plexus, might be added.

The time-honored classification of hemorrhoids into internal and external, while not strictly accurate, is probably the best division for clinical purposes. Those below the external sphinc-ter muscle are considered external; those above, internal.

Of external hemorrhoids we have the venous or thrombotic, and the cutaneous, and the combined or compound external pile. The first variety consists of an inflammatory enlargement of a piece of skin near the anus, generally of a rounded form, of a soft feel, and a livid or blue color, containing a blood-clot in the center. The second variety is nothing more nor less than a hypertrophied condition of the superfluous tags of skin sometimes found around the anus. The combined or compound external pile seems to be a combination of the two preceding varieties, hence partakes of the nature of both.

The symptoms of hemorrhoids generally come on suddenly after a costive stool. At first it is merely a sense of fullness and uneasiness at the anus; but as inflammation progresses the pain becomes extremely severe, so that the strongest man may be thereby quite incapacitated for business. Sometimes marked constitutional symptoms occur, such as furred tongue, headache, loss of appetite, feverishness, etc.

There are two modes of treatment, the palliative and the radical. If the first plan is to be adopted, about the first thing to tell the patient is to quit manipulating the tumor by trying to put it up within the sphincter, which they will invariably have attempted, only with the result of having bruised and inflamed the parts that much more.

An ointment of equal parts of extract of opium and extract of belladonna may be applied to the tumor, followed by a warm poultice; at the same time the bowels should be freely cleared and a slight unstimulating diet, with rest in bed, prescribed.

Instead of the ointment, the absorption of which I do not have much faith in, we may use frequent ablutions of cold water to an advantage. The inflammation will usually subside in a few days, but it leaves behind a thickened projection of skin, ready at any time to again inflame at the slightest provocation.

If the radical treatment is to be pursued, I do not believe that incising the tumor and turning out the clot is the proper way, for it generally takes as long to reduce the inflamed tags of skin left as it would for nature to absorb the clot.

Excision is undoubtedly the only way to radically cure this form of trouble. If several tumors are to be dealt with, it will probably be better to give the patient a general anesthetic, as the pain is considerable. If there is only one tumor the injection of cocaine into the skin around the base of the tumor will answer the purpose. Catch up the tumor with a pair of pile-forceps, carry your knife completely around its base, thereby excising it.

Some authors warn us about cutting away the fold of skin about the anus too freely. In my experience I have always had to regret not having cut enough. I think the dangers of contraction are very slight if any judgment is used at all in operating. Occasionally a small arterial branch will be cut, and a ligature will have to be applied. Dress the parts with iodoform gauze, cotton, T-bandage, etc., put the patient to bed, control the pain with morphia hypodermatically, allow bowels to move on third day, and patient to get up on fourth or fifth day.

Internal Hemorrhoids. It is only necessary to divide internal hemorrhoids into two classes, the capillary and venous. The capillary hemorrhoid, or strawberry pile, as it is sometimes called from its appearance, is in reality an erectile tumor, composed of the terminal branches of the arteries and veins and the capillaries which join them.

This tumor is never of large size and never projects very far into the cavity of the rectum. They strongly resemble an arterial nevus with a granular surface, and the membrane covering them is always of extreme thickness. This accounts for the chief symptom which distiguishes them clinically from the other variety, the free hemorrhage which follows the slightest bruising of their surface, even in the act of defecation. Some of the standard authors claim that the capillary pile is simply the incipient stage of the venous variety.

The one symptom of a capillary hemorrhoid is the daily hemorrhage, and as this occurs at the time of defecation without pain, the patient may be entirely ignorant of the daily loss of blood until the loss is so great that the constitutional symptoms begin to appear. The anemia produced by this constant drain on the system is sometimes very great. This form of piles can be radically cured by applying a caustic, such as nitric acid or Monsel's solution, to its surface, or, as Dr. Mathews prefers to do, by catching up the bleeding surface with a pair of forceps and ligating it with a silk ligature.

The second variety of internal hemorrhoids, or venous piles, consists of a varicosed condition of the submucous venous plexus, situated just above the external sphincter muscle, which, by inflammatory action and plastic infiltration, has resulted in the formation of tumors.   These tumors resemble very much the venous external pile, with this important difference, that the latter is covered with skin and does not bleed; whereas the former, being covered only by a mucous membrane and more exposed to injury by the passage of feces, is exceedingly prone to hemorrhage.   After they have existed for quite a length of time bleeding from them may entirely cease.   According to my conception of what constitutes an internal hemorrhoid, it is very plain to see that there is no non-operative curative treatment of the disease.   Of course a mere congestion or varicosity of the hemorrhoidal plexus, due to some obstruction in the return circulation, can be relieved by removing the cause by appropriate treatment.   The two cardinal symptoms in internal hemorrhoids are protrusion and bleeding; in fact, by the time the patient presents himself to you for treatment you can make out a diagnosis by simply listening to his description of the symptoms.   It is never wise to neglect a thorough examination, to make sure the trouble is not complicated by or depending on something of a more serious nature.   The palliative treatment of this condition consists in perfect daily regularity of the movement of the bowels and the free use of cold-water applications several times a day.   A cold-water enema before each action of the bowels will also be found very effective.   Ointments and suppositories are not of much use in the treatment of internal hemorrhoids.   Injection of carbolic acid, which can not be considered any thing more than a palliative method, is too dangerous and unsatisfactory to bother with.

We have several reliable methods for the radical cure of internal hemorrhoids.   I will place first in the list, ligature, applied according to the method of Allingham, Smith's clamp and cautery method, and the clamp and excision method which I will describe.   Since 1888 Dr. Outerbridge, of New York, abandoned the use of the ligature and clamp and cautery, and treats all cases of hemorrhoids by excision, and unites the cut edges by continuous catgut suture, using the ordinary pile clamp to hold the edges together while being sutured.* Dr. Earle, of Baltimore, considering this the most rational method of treating hemorrhoids, devised a clamp which facilitates the introduction of the suture, and reported same in the last issue of MATHEWS' MEDICAL QUARTERLY. He describes the mode of application as follows:

"The pile is caught with catch-forceps at its most prominent point, pulled out and down, and then the clamp forceps are applied as near the base of the tumor as may be thought proper. After being closed as tightly as possible, the part of the tumor above the beak of the forceps is cut off close; then the suturing is begun at the distal end of the clamp—the end of the suture is caught with a pair of catch-forceps instead of being tied, in order that the running suture, when complete, may be drawn from both ends—and is continued over and under the clamp until the whole cut surface is included, but is not drawn tight. The clamp is now loosened, when it can be easily slipped out between the suture, and the two ends of the suture drawn sufficiently tight to bring the cut edges in nice apposition and control all hemorrhage. The two ends are now made fast by a knot in each, which should be made close down to the mucous surface.

" While this clamp is best adapted to the removal of individual hemorrhoids, either external or internal, yet I have recently used it in a typical case for Whitehead's operation with great satisfaction, where I removed internal hemorrhoids from the entire circumference of the rectum, and external varicosities from two sides and on the anterior surface. In this case I used the clamp at right angles to the long axis of the rectum, whereas it is generally used parallel to it.

" What is claimed for the instrument? It will very much facilitate the time needed for the operation, generally; it adapts the cut surfaces more evenly than can be done without them; it exposes the least possible amount of raw surface to inspection, and it produces very little contusion of the compressed surfaces, which is rendered still less injurious by its occurring just at the edge of the cut surfaces."

923 Fourth Avenue.

*Gallant: MATHEWS' MEDICAL QUARTERLY, Vol. 1, No. 4, p. 518.

TREATMENT OF HEMORRHOIDS.*

BY H. R. COSTON, M. D.,

FAYETTEVILLE, TENN.

Hemorrhoids is a disease of such universal prevalence that its treatment is always full of interest to medical men when assembled. I shall not attempt a critical review of the many modes of treatment which have, from the earliest history of medicine, been brought forward as curative agents in this disease, but shall consider briefly and conservatively the palliative, injection, ligature, and clamp and cautery methods of treatment. Of these, I will say in the beginning, that my preference is decidedly for the clamp and cautery operation, and as we proceed I shall attempt to show you why it is so.

Palliative treatment is never to be used in a well-established case of hemorrhoids, when the patient's consent to a radical operation can be obtained, unless there are grave reasons why an operation should not be performed, such as serious heart lesions, advanced phthisis, etc. In acute hemorrhoidal disease, palliative or medical treatment sometimes allows nature an opportunity to heal her own ills. In such a case the bowels should be kept in a soluble condition, the patient should abstain from animal food, stimulants, and tobacco.

He should take a full enema before going to stool, and after the bowels have been emptied he should bathe the parts thoroughly in cold water. The tumors should be carefully reduced after each evacuation. After reducing the piles a suppository or an ointment of subsulphate of iron should be introduced into the rectum, opium, cocaine, or belladonna may be added, *pro re nata;* carbolic or nitric acid may be applied to bleeding points, taking care not to touch the healthy surface with it. Ice may be used as a suppository when there is much inflammation present.

In acute cases as in all others, whatever complications exist, such as torpid liver, uterine displacements, stricture of the rectum or urethra, vesical calculus, enlarged prostate, phimosis, etc., must receive proper treatment, else the treatment directed to the piles will result in failure.

*Read before the Tennessee State Medical Association, April, 1896.

Treatment by Injection. This method is not so reliable as either the ligature or clamp and cautery operations, but occasionally cases are met with in which it is necessary to use it because of an absolute refusal of the patient to submit to more thorough methods of treatment. The substance injected—carbolic acid, creosote, nitric acid, fl. ext. ergot, kino, or any other of the host of remedies which have been lauded as curative—should be thrown into the center of the tumor, not under it. Ulceration and abscess are very prone to follow in this method of treatment, and relapse almost sure to occur. Cure, if cure is obtained, is slower in being attained, only one tumor being injected at a time. It is impossible to determine the amount of sloughing that will occur, and this leaves an open sore for the ingress of microbes.

This treatment is unreliable, attended with pain and danger, and is the ideal operation of advertising charlatans and should receive no encouragement from the regular profession.

Operation by Ligature. Before any radical operation, the bowels should be well emptied; the diet should have been light for the preceding twenty-four hours. The patient should be profoundly under the influence of an anesthetic—I prefer chloroform. The patient should be placed in the extreme lithotomy position, with the legs well drawn up by Kelly's leg-holder or by a sheet folded diagonally and passed under the knees and neck of the patient, and the opposite corners tied together; this arrangement keeps the field of operation well under the surgeon's control.

Next introduce the thumbs, back to back, into the rectum and distend the sphincter until there is no resistance. Wash out the rectum with a large stream of water. Take hold of the tumor with a tenaculum forceps and draw it well down and make an incision at the base of the tumor at the junction of the skin and mucous membrane. Dissect well up under the tumor to its upper limit. The blood-vessels enter from above, and we need not be afraid of cutting them if we keep close to the muscular coat of the bowel. After the pile is raised well out of its bed, pass a ligature around the pedicle and draw the tumor well down and tie the ligature as high on the pedicle as possible, cut off the bulk of the tumor, being careful to leave sufficient stump to prevent the ligature slipping off. The smaller piles should be dealt with first.   After cutting the ligatures short, return the pedicle within the bowel.   An opium or belladonna suppository may be introduced to allay pain.   The bowels should be moved by enema the third or fourth day.   Keep the patient quiet for ten days or two weeks.

Clamp and Cautery Operation.   The preparatory treatment is the same for the clamp and cautery operation as for that by the ligature.   The instruments necessary are few and simple : a good thermo-cautery apparatus, a Kelly or Smith's pile clamp, a double tenaculum, and a pair of scissors.   These are all that are needed. Anesthetize the patient thoroughly and tie up the legs, distend the sphincter, and wash out the bowel as previously described. Take up one tumor at a time with the tenaculum forceps and make an incision with the scissors at the junction of the skin and mucous membrane, put on the clamp *across* the tumor—not lengthwise of the bowel—with the lower blade in the incision and screw it down tightly ; cut off the top of the tumor, leaving about one fourth of an inch of stump above the blades ; apply the cautery at a dull red heat and burn the stump well down to the clamp ; carefully unscrew the clamp and remove it, taking care that the eschar be not loosened in its removal ; follow the stump through the blades with the cautery.   Treat each pile in like manner.   If the tumors are small the top need not be cut away, but may be burned away with the cautery.   If the tumors are too large to be grasped by the clamp, split them into parts, in line of the bowel, with the scissors, treating each half as a separate tumor.   Be careful not to take up the healthy mucous membrane between the piles in grasp of the clamp and there will be no contraction of the anus or rectum.   Carefully return the stumps inside the bowel.

Open the bowels with a saline on the second day, using an enema when the desire for stool is felt.   The patient may be up from the third day, but should keep his room for a week.   I have had a patient plowing after one week indoors.

If there is pain or tenesmus after the operation, introduce an opium and belladonna suppository ; but if the sphincter is thoroughly distended, and you do not touch the cutaneous surface with the cautery, there will be but little pain following the operation.

Now, why do I prefer the clamp and cautery operation over all others? or over the ligature as the next best?

First. In the operation with the ligature you tie up the most sensitive of all nerve ends, and they are sure to resent it by intense pain, which will continue until the stumps slough and the ligatures come away. In the cautery operation you have no such to contend with; the nerve end is simply cut away and cauterized, and there is nothing left for it to do but cicatrize, and it is left in the best possible condition for this.

Second. The ligature may slip and secondary hemorrhage occur; after the clamp and cautery operation, there is no danger of *secondary*. If hemorrhage occurs it does so immediately, and the operator can only blame himself for it.

Third. There is no danger of a recurrence. Kelsey, of New York, and Smith, of King's College, London, both support me in this statement, and they have had a vast experience with this operation.

It will be admitted by all that recurrences do follow the ligature operation.

Fourth. Convalescence is much more quickly completed, for the reason that it begins at once under the eschar produced by the cautery and would be completed by the time the ligatures came away should the two operations be used on separate tumors in the same case at the same time.

Fifth. The mortality following the clamp and cautery operation is practically *nil*.

Sixth. The cautery operation requires less care from the physician after the operation.

Seventh. There are no unpleasant *sequelœ*.

GASTRO-INTESTINAL DISEASE.

THE TREATMENT OF POST-OPERATIVE INTESTINAL OBSTRUCTION.

BY A. ERNEST GALLANT, M. D.,

Instructor in Operative Surgery, New York Post-Graduate Medical School and Hospital; Assistant Surgeon, Lebanon Hospital; Gynecologist, Northern Dispensary; Assistant Gynecologist, Roosevelt Hospital, O. P. D.

NEW YORK CITY, N. Y.

During my service as interne in the New York Cancer Hospital, pain with increasing rapidity of pulse, and more or less elevation of temperature, sometimes nausea or bilious vomiting, due to intestinal distension following celiotomy, occurred in nearly every case. This led to a careful study of the recognized methods of overcoming so serious a condition. The energetic use of calomel, magnesium sulphate, etc., enemata of sweet oil, turpentine, castor oil, and soapsuds carried well up into the colon, failed in many cases to give relief before the third, fourth, and even the fifth days. In other instances the insertion of a long rectal tube would be followed by discharge of gas and temporary abeyance of the symptoms. Four cases died, fifty-four, eighty-one, ninety-two, and one hundred and five hours after operation without massage, having passed no gas or fecal matter, the fatal issue being attributed to septic peritonitis. In the writer's mind the intestinal obstruction must have been a factor in bringing about so sad a sequel.

The excellent results reported by D. Graham (Recent Developments in Massage), A. Kellgren (Technic of Ling's System of Manual Treatment as Applied to Surgery and Medicine), and R. Hirschberg (Massage of the Abdomen), in constipation, indigestion, dilatation of the stomach, etc., lead to the idea that this plan might be serviceable in post-operative obstruction.

IMPORTANT POINTS.

The following cases are introduced in order to illustrate certain points which are deemed important:

Cases 3 and 5 demonstrate the fact that by *early massage* we can give immediate relief, independent of the use of laxatives or enemata, and that in these cases the bowels move more readily and with less stimulation than when massage is delayed.

Laxatives tend to increase gas formation and intensify peristalsis, thus adding to the pain and discomfort (1 and 2).

Enemata can not reach above the ileo-cecal valve and set up colitis, rectal tenesmus (2), and proctitis (1), followed by exhaustive diarrhea (1 and 2).

The rectal tube does in many cases give partial relief.

In every case passage of gas from the bowel gave immediate relief and comfort; even in Case 4 an improvement in pulse and comfort of the patient was noticeable.

The point of greatest pain and " stoppage " (5) was felt to " give way " as massage was being practiced, and Case 6 experienced much relief as massage was carried on, and expressed it by kissing the hand of the masseur.

Obstruction to the flow of gas and resulting distension must be the main factor in producing the pain, and if we relieve this condition, the bowels can be moved at leisure, and the patient will not suffer (3, 5).

Intestinal cramps from gas or too active cathartics may be relieved by repeating the rubbing at frequent intervals.

Infection of the line of wound union or damage to structures involved in the operation by massage thirty hours after operation is not likely to occur.

ILLUSTRATIVE CASES: 1. Service, Dr. Outerbridge. Pyosalpynx and cystic ovary firmly matted down by dense adhesions. Tympany necessitated insertion of long rectal tube thirty-two hours after operation; magnesium sulphate thirty-fourth hour; oil enema thirty-eighth hour; high enema, glycerine half ounce, turpentine, two drams, magnesium sulphate, half ounce, in warm water, one quart, forty-ninth hour; calomel and soda, five and ten grains, fiftieth hour; soapsuds enema, one quart, fifty-sixth hour; calomel and soda repeated, fifty-ninth hour; rectal tube inserted, sixty-first hour; and fifteen ounces of previous enemata drained away, but no gas or fecal matter passed.

Seventy-two hours after operation the writer masséed the colon, from the cecum upward, across the transverse and down the descending colon to the sigmoid flexure. This was continued for fifteen minutes, and a free discharge of gas gave immediate relief, and was followed later by large movements from the bowels. The rubbing was repeated several hours later and gave great comfort to the patient. Respiration, temperature, and pulse fell to nearly normal. It would be difficult to imagine a greater degree of distension possible and life continue. The intestinal peristalsis and proctitis did not return to normal until ninety-six hours after operation.

2. DR. OUTERBRIDGE: BILATERAL SALPINGO-OOPHOREC-TOMY FOR LARGE FIBROID. Cathartics and enemata up to the fifty-first hour. No gas or fecal matter passed; marked disten-sion; nausea and vomiting. Massage of colon for ten minutes caused free discharge of gas and two large fecal movements within the next three hours. The laxatives given before massage set up rectal tenesmus, so that an opium suppository was given the sixty-seventh hour. Intestinal cramps were relieved by repeating the massage.

3. DR. OUTERBRIDGE: FOR BILATERAL PYOSALPYNX, RUP-TURE OF SAC WHILE SEPARATING ADHESIONS, AND INFECTION OF PELVIC CAVITY. Thirty-two hours after operation tympan-itic distension gave great discomfort. Rectal tube gave no re-lief. Massage of abdomen for fifteen to twenty minutes, and turning on the side resulted in passage of a large quantity of gas, and made the patient comfortable. Laxatives were not given until the fifty-sixth hour after operation.

4. DR. PORTER: FOR BILATERAL PYOSALPYNX AND APPEN-DICITIS. Escape of pus from rupture of the tubes during opera-tion; appendix sloughing. Intestinal distension caused much pain. Rectal tube introduced the ninth hour; no relief. Mass-age at the thirtieth hour caused passage of gas and feces. The patient died eight hours later from shock following the long and tedious operation.

5. DR. W. W. VAN ARSDALE: FOR RECURRENT APPENDI-CITIS. A private case which was placed in my care after operation. Six hours after, well-marked borborygmus. Abdo-men became more tense through the night, with attacks of pain most intense in the right iliac fossa. Thirty hours after opera-tion pain very distressing. Patient said that "it felt as if the gas got to a certain point and could get no further." This point cor-responded to the ileo-cecal region, and was the seat of greatest

pain. At this time the dressings were removed, and massage along the colon for five minutes. While doing so, the patient felt as if something gave way, and the gas moved along freely. She was then turned on the side and was very comfortable. Gas passed freely. Three hours later the patient requested to be rubbed, as "it felt so good." Laxatives were given and the bowels moved freely the next day. At the patient's request the massage was kept up three times daily.

6. DR. GALLANT, AT LEBANON HOSPITAL: Service, Dr. Parker Syms. For Inguinal Hernia—Woman, sixty-five years. Hernia began twenty-two years ago, increasing in size after birth of each of four children. Has had three attacks of very obstinate obstruction. The thin-walled sac extended half way to the knees and was twenty inches in circumference. It was found to contain nearly all the small intestines and two-thirds of the colon. Intestines returned to the abdomen, sac cut away and the edges brought together to close the ring. On the second morning, thirty-two hours after operation, Dr. Syms telephoned me that the patient had intestinal obstruction and that I must operate if necessary. The patient was found suffering from marked distension, and no benefit had been secured from the salts and turpentine enemata which had been administered through the night. Removing all the dressings, massage was practiced for fifteen minutes, gas passed freely from the bowels, and patient was left in a very happy frame of mind, very comfortable. The bowels moved freely the next day. (See plate opposite.)

Modus Operandi. Loosen abdominal binder and remove all dressing, except the gauze in actual contact with the wound. Lubricate the hand with oil or alboline (vaseline is too sticky). Begin the rubbing with the whole hand formed into a hollow, the fingers held together, palmar surface next the skin. First rub lightly from as near the cecal region as possible; the hand is carried upward over the ascending colon, across the umbilical region and down along the descending colon well into the left iliac fossa. Gradually increase the pressure of the hand. By using the closed fist one can dip down more directly upon the colon. Very often gas can be heard at once to move through the intestinal tube. As soon as gas passes from the bowels the operator may feel assured that no difficulty will be expe-

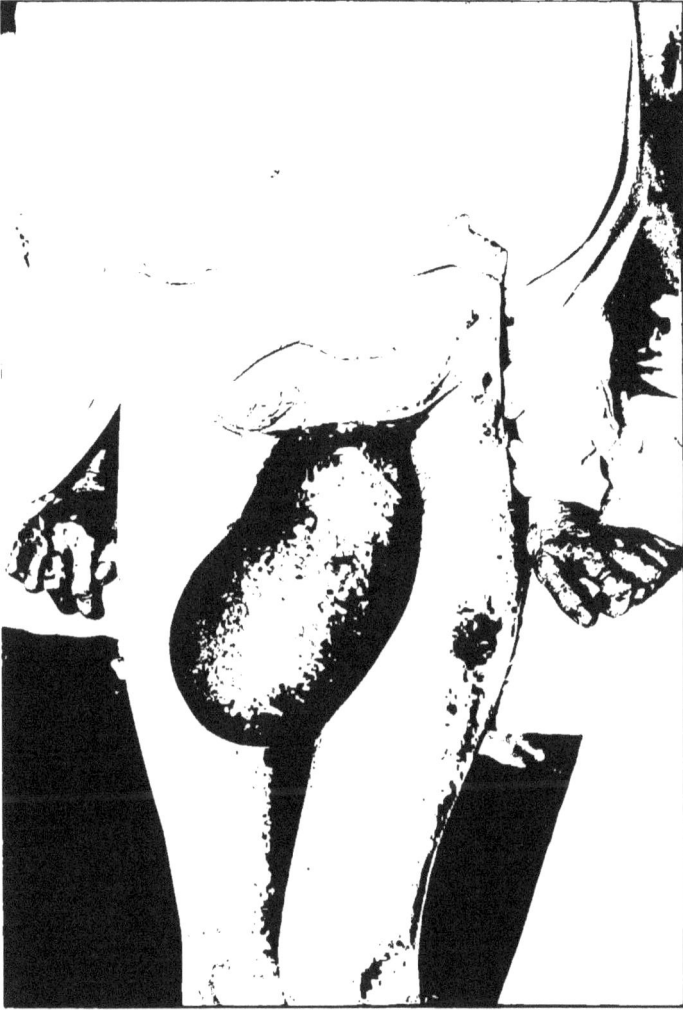

rienced in moving the bowels.   Do not give up in less than half an hour.   The patient will request that massage be repeated.

The three factors by which the result (relief of distension) is brought about are :

1. Loosening the abdominal bandage.

2. The friction of the intestines.

3. Changing the patient from the dorsal to the lateral position ; each and all tend to " take out the kinks " in the intestines, overcome the obstruction, and allow the free exit of gas.

10 West Thirty-sixth Street.

PERFORATIVE ULCER OF THE STOMACH.*

BY WILLIAM J. GILLETTE, M. D.,

Professor of Gynecology and Abdominal Surgery, Toledo Medical College ; Surgeon and Gynecologist, Toledo Hospital.

TOLEDO, OHIO.

Habershon describes three forms of ulceration of the stomach : First, a superficial ulceration affecting only the mucous membrane ; second, a follicular ulceration ; and third, perforating ulcer, acute and chronic.   This third variety constitutes the subject of this paper.

Of all rapidly fatal conditions to which the human economy is liable, but few are more deadly than perforating ulcer of the stomach.   The pathology of acute perforating ulcer is involved in much obscurity, though many theories have been advanced, notably the embolic theory of Virchow, the theory of its nervous origin, and that the perforation is due to digestion of the stomach wall where, for any reason, arrest of the circulation has taken place.

The theory that it is of nervous origin seems to have the greatest number of advocates, though the experiments of Pavy and Panum amply prove that digestion of the stomach wall always or generally follows obstructed circulation.   But to all these theories there are grave objections into which I will not enter now ; and Habershon, no doubt, says correctly that, " It is very evident there must be some perhaps specific predisposing

* Read before the Ohio State Medical Society, in Columbus, May 28, 1896.

cause for this intractable affection of whose precise character we are still ignorant."

The ages at which these perforations occur vary according to the sex of the patient: In the female between the ages of fourteen and thirty, and in the male much later in life; in the female the average period of life being twenty-seven years, and in the male forty-two years.

The point at which perforation occurs varies, but most frequently happens on the anterior wall of the stomach, as large a proportion as eighty-five per cent. occurring here, according to Brinton. Though perforation of the stomach is usually followed by fatal peritonitis, yet in a small proportion of cases when the ulceration has been sufficiently slow in forming to allow a barrier of lymph to be thrown out protecting the peritoneum, the general cavity has escaped infection. The probabilities of this occurring vary according to the position of the ulcer in the stomach wall. When the perforation is through the anterior wall adhesions are not likely to be formed. In a report by Herman Eichhorst I find that of seventy-five cases at this point there was complete perforation sixty-four times, "whereas, in thirty cases at the cardiac extremity it occurred but twelve times."

When adhesions have had time to form recovery is the rule, but in the acute perforations when the contents of the stomach are suddenly poured into the general or peritoneal cavity the case is almost of necessity fatal.

In 1883 Miculicz first operated for this condition, but unsuccessfully. Operations have been performed, however, a number of times, mostly by continental surgeons, with a twenty-per-cent. rate of recoveries; Kriege, of Berlin, is said to have had the first success.

I can see no reason why an operation should not be as successfully performed in these cases, if recognized in time, as in many other emergencies of abdominal surgery, as for instance in ruptured tubal pregnancy, etc. The stomach can be reached and sutured at all points upon its surface even posteriorly, as can be demonstrated upon the cadaver.

To be successful, however, the operation should be performed very early after the accident. Michaux, of Paris, who reports a successful case, says within fifteen hours of the accident, for as a rule the time is short after perforation until death occurs. Habershon estimates at from five to twenty-four hours, though cases may live a number of days after; one of mine lived five days after perforation.

Recognizing the great importance, then, of early diagnosis of perforation, the principal points upon which we must rely for a diagnosis should be carefully studied. It is an astonishing fact that acute perforations of the stomach often occur in those well nourished, and the patients themselves have not been aware that any thing of importance was the matter with them until suddenly seized with pain. Symptoms of perforation, as laid down by Eichhorst, are as follows: "Acute pain, signs of grave collapse, extensive tympanites and symptoms of perforative peritonitis (pain and disappearance of hepatic and splenic dullness); vomiting may be absent." Of the value of the disappearance of hepatic and splenic dullness as evidence of gas free in the abdominal cavity one can readily convince himself by experiment upon the cadaver; on forcing air through an aspirating needle and distending the abdominal cavity but moderately, dullness over these organs at once disappears, except in the rare cases where they are cemented by inflammatory adhesion to the abdominal wall.

Of course gas free in the abdominal cavity is found after perforations at any point of the alimentary tract, and even if it is determined that there has been a perforation, its location before operation is open to great uncertainty; but I have no doubt in the vast majority of cases the location of severest initial pain, or the point where the pain begins, will be found to be near the point of perforation, and I believe, unless positively contradicted, we should make our exploratory incision with reference to it. It has been my fortune to see within a short time three cases of perforation of the stomach.

The first one, in the care of Drs. R. J. Walker and J. D. Ely, of Toledo, O., a domestic, nineteen years old, who had always been well prior to what was at first considered to be a severe attack of colic. The pain was most severe in epigastric region. The girl had been about her usual household duties up to the time of the attack and feeling as well as usual. Had never complained of any disturbance of the stomach. The abdomen began soon to distend, and the diagnosis was changed to peritonitis, and the patient treated accordingly. *At no time was there vomiting.*

The case continued to do badly, and I was called in on the evening of the fifth day with view to operation. I found the patient suffering severely with greatly distended abdomen, especially in epigastric space. Tympanitic note over liver and spleen well marked.

I diagnosed perforation of the intestinal tract somewhere, but did not place sufficient stress upon the fact of the severe pain over the stomach to even think of locating it there. The patient was evidently dying, though her mind was clear, and though she urged me to operate, evidently recognizing her desperate condition, I felt compelled to decline, and she died four hours after my visit. On *post-mortem* the next day a perforation as large as a copper penny was found upon the anterior wall of the stomach near the cardiac extremity. The stomach contents were largely in the peritoneal cavity.

The mother of this girl died in a similar manner.

Had a correct diagnosis been made sufficiently early I am convinced she would have had a fair chance for recovery by operation, for the perforation could have been reached without difficulty.

The second case I saw a month since, with Drs. McVety and Minton, of Toledo, at the Toledo Hospital. The patient, a Pole, thirty-two years old, a laborer, had all the evidences of robust health, a hearty, well-nourished, fine-looking fellow; had complained for some time, however, with some stomach disturbance, but not severe enough to interfere with his labor or his food; was suddenly taken down with severe pain over the epigastric region; the abdomen began rapidly to distend. After forty hours from the onset of this attack I was called in consultation with a view to operation for peritonitis. I found the man evidently dying. *There had been no history of vomiting except at the onset of the attack.* Tympanitic note over the liver and spleen well marked, and a diagnosis of perforation at some point was made.

The man died within an hour from the time I first saw him. A *post-mortem* revealed a perforation on the anterior wall near the pyloric extremity, a point which might easily have been reached by operation.

I find another case reported in the *American Medical Compend,* for April, 1895, by Dr. Dickey, of Tiffin, Ohio, which presents almost exactly the same line of symptoms. The doctor was called in consultation to see a healthy, vigorous Irish farmer, fifty-seven years old, who up to the hour of his illness had been in exceedingly good health; said he "had never felt better in his life;" was seized with sudden pain in the abdomen. In this case there was vomiting. A diagnosis of intussusception was made by the attending physician. When Dr. Dickey reached the patient he was dying; was past all hope of good from operative interference. He died in two hours after. *Port-mortem* revealed an ulcer the size of a five-cent piece near pyloric extremity on anterior surface, " with clean cut edges, as though cut out with a saddler's punch."

I have also notes of yet another case, with almost exactly the same history, occurring in Toledo, but these three cases illustrate sufficiently the points I desire to make in this paper.

I am convinced that perforating ulcer of the stomach is much more common than is generally supposed; for here are four cases occurring within less than a year, three of them under my own observation in Toledo, and the others within but a few miles of that city. In not one of these cases was an accurate diagnosis made of the condition before death, though in the hands of very competent physicians.

It would seem from *post-mortem* appearances that, if a diagnosis could have been made early enough and an abdominal surgeon of even moderate skill operated at once, these cases would have had a fair chance for recovery.

It would further appear that the careful consideration of a few diagnostic symptoms ought to lead to correct diagnosis.

Tympanitic note over liver and spleen will indicate gas free in the peritoneal cavity, and necessarily perforation at some point. In all of the cases mentioned the severe pain was over the epigastric region; and I believe it is true that when a perforation has taken place the point of severest pain is near it. Absence of vomiting with evident peritonitis is certainly significant. These symptoms, together with rapidly rising pulse and evidence of severe abdominal illness, ought to lead without delay to exploratory incision.

THE SIGMOID FLEXURE AND ITS MESO-SIGMOID.

[SYNONYMS: S-loop, S-romanum, pelvic loop, iliac colon, sigmoid colon, flexura ilica, pelvic colon, Omega loop,
flexura sigmoidea.]

BY BYRON ROBINSON, B. S., M. D.,
Professor of Gynæcology in the Post-Graduate School.
CHICAGO, ILL.

In 200 cases it was found that the average length of the sigmoid was $18\frac{1}{3}$ inches; the longest sigmoid was 33 inches, the
shortest was 5 inches; the average length of the meso-sigmoid
was $3\frac{1}{2}$ inches; the longest meso-sigmoid was 10 inches, the
shortest was 1 inch. In 52 females the average length of the
sigmoid was $17\frac{1}{2}$ inches; the longest sigmoid was 33 inches, the
shortest was 5 inches; the average length of the meso-sigmoid
was $3\frac{3}{4}$ inches; the longest meso-sigmoid was 10 inches, the
shortest was 1 inch. In 148 males the average length of the sigmoid was 19 inches; the longest sigmoid was 32 inches, the
shortest was 6 inches; the average length of the meso-sigmoid
was $3\frac{1}{3}$ inches; the longest meso-sigmoid was 10 inches, the
shortest was 1 inch. In 200 cases adhesions were present 170
times or 85 per cent.; in the 37 females adhesions were present
32 times or 86 per cent.; in the 113 males adhesions were present
101 times or 89 per cent.; in 150 cases the inter-sigmoid fossa
was present 126 times or 85 per cent.; in 113 males the inter-
sigmoid fossa was present 97 times or 85 per cent.; in 37 females
the inter-sigmoid fossa was present 29 times or 78 per cent.; in
150 cases Gruber's fold was present 108 times or 72 per cent.; in
150 cases Gruber's fold had adhesions 24 times; in 113 males
Gruber's fold was present 80 times or 70 per cent.; in 113 males
Gruber's fold had adhesions 17 times; in 37 females Gruber's
fold was present 27 times or 73 per cent.; in 37 females Gruber's
fold had adhesions 7 times.

To compare the transverse colon with the sigmoid, the following may be noted:

In 150 cases the transverse colon averages in length $23\frac{1}{2}$
inches; the longest transverse colon was 30 inches the shortest

was 8 inches; the average length of the transverse meso-colon was 4 inches; the longest transverse meso-colon was 9 inches, the shortest was 2 inches. In 113 males the average length of the transverse colon was 23½ inches; the longest transverse colon was 30 inches, the shortest was 14 inches; the average length of the transverse meso-colon was 4⅓ inches; the longest transverse meso-colon was 9 inches, the shortest was 2 inches. In 37 females the average length of the transverse colon was 22½ inches; the longest transverse colon was 30 inches, the shortest was 8 inches; the average length of the transverse meso-colon was 4¾ inches; the longest transverse meso-colon was 7 inches, the shortest was 2 inches.

The development of the sigmoid and meso-sigmoid is interesting, as its final condition is the result of a long process of growth and adjustment. The final position of the meso-sigmoid is the result of environments. At first the meso-sigmoid inserts itself a'ong the mid-dorsal line, but finally in adult life assumes a curved line from the lower pole of the left kidney to the point on the sacrum where the posterior surface of the bowel loses its peritoneum. The human sigmoid begins to develop its adult characteristics about the tenth week of fetal life. As fetal life differs much in development, only general statements can be made as to the stage of growth in regard to time. The age and length of the fœtus should be reported. The most apparent cause of the peculiar adult insertion of the meso-sigmoid is the growth of the left kidney. In the third fetal month the left kidney gradually steals away the meso-colon descendens and appropriates it to cover itself. The meso-colon descends, at the same time moves toward the left. To say that the rapid growth of the lateral abdominal walls necessitates appropriation of the left blade of the meso-colon descendens to cover them, is, I think, not in accord with phenomenon of growth. It seems to me the growing kidney alone is the chief cause for the disappearance and appropriation of the meso-colon descendens. And as my examinations show that the existence of a meso-colon descendens is an abnormality, the long meso-sigmoid appears more striking from its size and shape.

In the evolution of the sigmoid the most noted factor is the left kidney. In the tenth fetal week it is plain to note that the meso-sigmoid reaches to the internal border of the left kidney. As the bodily growth progresses the meso-sigmoid advances to the anterior face of the left kidney and, finally, to the outer border of the kidney found in adults. The adult position of the ascending colon may be assumed by the eighth to the twelfth week of fetal life, that is, the descending colon assumes a vertical and maybe a sinuous course. It may be noted that old views considered that the sigmoid began at the iliac crest, but modern anatomy starts the sigmoid and ends the descending colon at the external border of the psoas muscle. The upper point of the meso-sigmoid often reaches the radix mesenterii and not infrequently the point where the duodenum turns into the jejunum. The whole meso-sigmoid may be distorted by inflammation on the lower or the left blade, as it lies on the psoas muscle, or by old cicatrices on its upper blade, especially on Gruber's fold. On account of the variation of the situation and course of the sigmoid and meso-sigmoid, it has induced many ingenious views from well-known anatomists. The unequal S-shape to the sigmoid is chiefly due to the length of the ligamentum Gruberii. At the end of the third month the meso-sigmoid is generally a lightly curved line reaching from the under pole of the left kidney to the middle of the third sacral vertebra. However, not infrequently the meso-sigmoid is quite concave to the left, or the upper point of the meso-sigmoid, as it begins to leave the mid-dorsal line, may be pointed to accommodate the inter-sigmoid fossa formed by the spermatic vessels. In general, at the fourth month the meso-sigmoid reaches the fourth lumbar vertebra, and then is carried to the left by the lower end of the growing left kidney. If the meso-sigmoid is elevated by Gruber's fold during the rotation of the intestinal loop, it will not allow the meso-sigmoid to leave the mid-dorsal line so far down, for Gruber's fold may be so contracted by the cecum dragging it upward as it travels across the abdomen to the under surface, that the upper point of the meso-sigmoid may reach as high as the flexura duodeno-jejunalis. Now since Gruber's fold is attached to the sigmoid at different points, it will induce different sizes to the portions of the S. I have observed that Gruber's fold is frequently attached just below the middle, thus leaving the upper portion of the S the larger. If we look at the upper loop of the

S, it will be noted that the sexual gland and the kidney are responsible for it. If we observe the lower loop in its fetal development, it may be seen that the sexual gland later changes from the lower end of the upper loop and begins to develop that. As time progresses from the fourth month the loops simply grow on slowly. From the sixth fetal month the loops are distinctly pronounced. The sexual gland descends with the kidney, grows rapidly as do the two loops of the S., meconium rapidly accumulates and irritates the mucous membrane, inducing blood flow and more active growth and elongation of the loop. The loop may grow very rapidly and so elongate that it must seek room in the right iliac fossa or among the small intestines.

In one new-born subject Dr. Lucy Waite and I found that the very long sigmoid rested in the form of a volvulus with the twist situated at the left pelvic brim at the junction of the sacrum and iliac bones. There is a portion of the descending colon, extending from the iliac crest to the external border of the psoas muscles, which is frequently more fixed than other portions of the vertical colon, but we must consider that it does not belong to the S-iliac. The unnumbered and indefinite classification given to the situation and periods of development of the sigmoid are of little use, as the situation varies indefinitely, with an organ possessing a long pedicle and no special time of growth. Yet as to growth one can note no special change in the first two months, but a very decided change in the third and fourth months. One could well say that in the first two months of fetal life the meso-sigmoid is in a primitive condition. It turns neither to left nor to right, but holds its ancient course in the mid-dorsal line, until the end of the second month or thereabouts. The meso-sigmoid is really the old original primitive mesentery of the digestive tube.

The least variation in the meso-sigmoid, of course, occurs while it is inserted in the mid-dorsal line up to the third lumbar vertebra. This period when the meso-sigmoid is inserted into the mid-dorsal line may be called the primitive period, when it has been subject to but little change as to position. From the tenth fetal week we may note many interesting changes of growth and position.

The primitive period may be succeeded by the period of transition from the child condition to that of the adult. In this period, from the beginning of the third month the portion of gut extending from the lower pole of the left kidney to the third . sacral vertebra forms two varying loops, an upper and a lower.

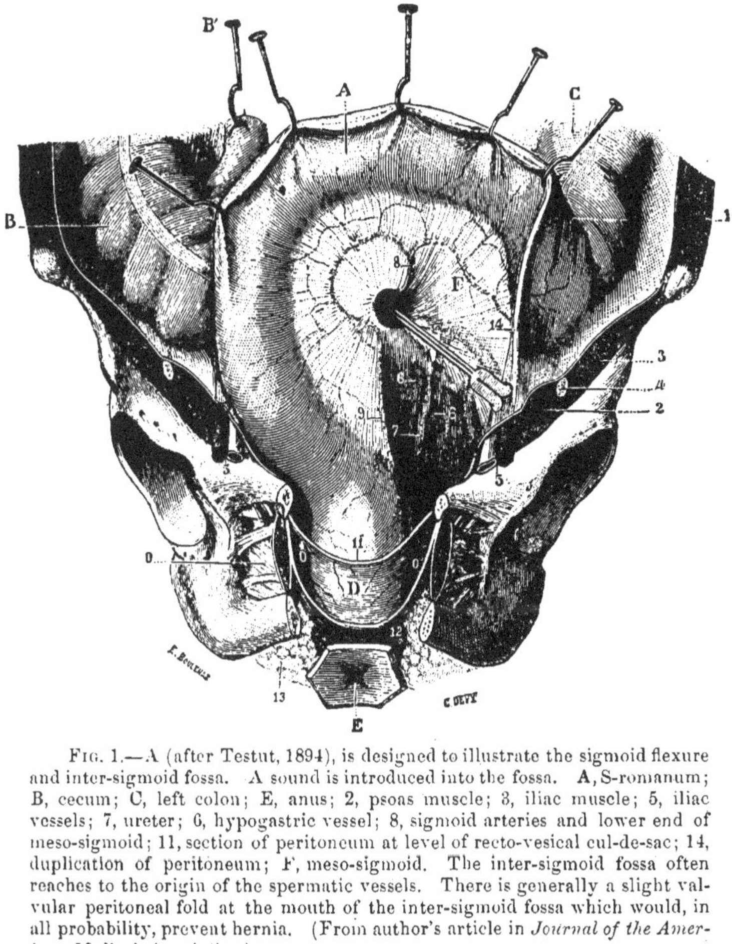

FIG. 1.—A (after Testut, 1894), is designed to illustrate the sigmoid flexure and inter-sigmoid fossa. A sound is introduced into the fossa. A, S-romanum; B, cecum; C, left colon; E, anus; 2, psoas muscle; 3, iliac muscle; 5, iliac vessels; 7, ureter; 6, hypogastric vessel; 8, sigmoid arteries and lower end of meso-sigmoid; 11, section of peritoneum at level of recto-vesical cul-de-sac; 14, duplication of peritoneum; F, meso-sigmoid. The inter-sigmoid fossa often reaches to the origin of the spermatic vessels. There is generally a slight valvular peritoneal fold at the mouth of the inter-sigmoid fossa which would, in all probability, prevent hernia. (From author's article in *Journal of the American Medical Association.*)

The two loops are divided by Gruber's fold of peritoneum at first, but subsequently the whole sigmoid may be represented by one large loop. In general, the upper loop (that is, between the

244   ROBINSON: THE SIGMOID FLEXURE.

external border of the psoas and the fold of Gruber) is at first
.the larger, but later the lower loop may become the larger loop
as the fecal matter accumulates and exercises its irritative mechanical power. The sigmoid may show a loop in the right iliac
fossa, a loop high up among the intestines, a loop in the pelvis,
and a double loop which lies partly in the right iliac fossa and
other localities. Occasionally one can find the double or single
loops entirely in the pelvis. During all the stages of transition
the line of the root or that of the insertion of the meso-sigmoid
is generally straightening out, that is, the radix meso-sigmoidea is
becoming a shorter line as growth proceeds. The acute angle
between the lower mesentery situated in the mid-dorsal line and
the mesentery extending from the lower pole of the left kidney
to it is increasing and fast becoming a curved line. In short, the
root of the meso-sigmoid is gradually descending to the adult
situation.

During the period of transition from fetal primitiveness to
that of the adult angle between that on the mid-dorsal line and
that reaching from the lower pole of the kidney to the mid-dorsal
line may be situated at any point from the middle of the sacrum
to the duodeno-jejunal bend or the second lumbar vertebra.
During this period the angle between the lower and upper segments of the meso-sigmoid is becoming larger. At first the
angle may be quite acute, but in adult life it may be almost 180
degrees or a straight line. The shifting of the root of the meso-
sigmoid toward the left is a gradual process. The chief factors
are the left kidney, the expanding packet of small intestines, the
large liver, and the rapid growth of the lateral abdominal walls.

A third period of the meso-sigmoid may be called the adult
period, when the meso-sigmoid is fixed. It begins at the external
border of the psoas muscle and ends in the pelvis at the third
sacral vertebra where the rectum loses its posterior serous covering. The loop lies partly extra and partly intra-pelvic. In the
adult it hardly ever lies in the right iliac fossa but chiefly in
front of the lumbar and sacral vertebræ. In the newborn the
sigmoid lies chiefly outside of the pelvis from physical reasons.
There is but little room in a child's pelvis for any organ except
rectum, bladder, and uterus. To resume, at first there is a double
loop in the sigmoid, a little later one loop remains in the iliac

fossa, the other ascends in the abdominal cavity, and the two loops finally coalesce into one large one as it is generally. The great rôle in forming the two parts of the S-loop is done by Gruber's fold. At first it forms the large part of the loop above the smaller below, but changes and bowel contents gradually make the lower loop of the large and, finally, the two parts of the S-loop appear as one inseparable sling. The accumulation of meconium through its irritative and mechanical power is responsible for the large loop and changes from its primitive condition.

The situation of the sigmoid at birth may be described as follows: The gut passes from the iliac crest to the psoas muscle, the loop then crosses the pelvis at a level with its brim. It then reaches more or less in the right iliac fossa (20 per cent.), but most frequently simply comes in contact with the right psoas muscle. It then extends, according to its length, more or less up among the intestines. The whole S-loop, however, may be in the pelvis in man or woman. The French contend that to do a colotomy on the newborn, incision should be made in the right iliac fossa, but this proposition will not generally hold true according to my examinations, or according to Boucart, who examined 150 infants and found the sigmoid loop in the iliac fossa forty-three times (twenty-two per cent.). There is little utility in describing a descending, transverse, and ascending loop in the sigmoid. The disagreement of authors in regard to the situation of the sigmoid is due to its changing position from emptying and filling with feces and gas, from the peristalsis and changes and situation of other viscera, especially those of the pelvis are: (a), the size of the uterus; (b), the size of the bladder; (c), the size of the rectum; (d), the size and situation of the cecum; (e), the size and situation of the sigmoid; (f), the amount of small intestines in the pelvis.

I can see no advantage in the name iliac or pelvic colons, as neither may correspond to its indicated name or situation. The long meso-sigmoid with its neuro-vascular pedicle indicates that it has a wide range of motion and that the situation of the loop will be normal wherever the mesentery allows it to range (except where it becomes fixed). Any mobile organ is dislocated when it is permanently fixed.

From years of investigations in the abdominal viscera, I would claim that one can only generalize two positions of the sigmoid flexure. The first and most frequent is when the sigmoid flexure lies suspended by Gruber's fold in the pelvis. The second is where the sigmoid flexure lies above the pelvis against the dorsal wall, chiefly to the front of the lumbar vertebræ and to their left. Most other positions of the sigmoid are uncertain, temporary, and variable.

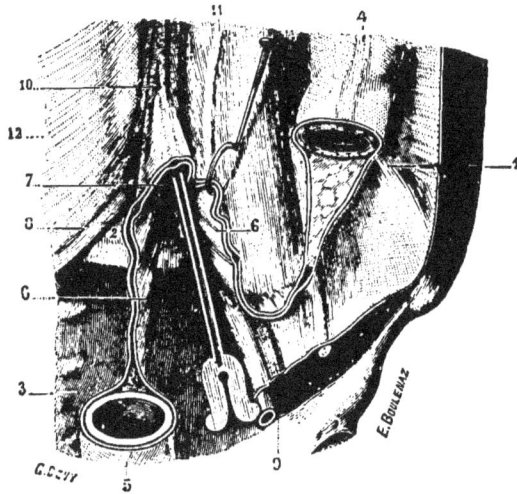

Fig. 2.—B (after Testut, 1894) is designed to represent the insertion of the meso-sigmoid. 1, iliac crest; 2, fifth lumbar vertebra; 3. sacral vertebra; 4, left colon; 5, rectum; 6, meso-sigmoid, cut edge; 7, inter-sigmoid fossa with a sound in it; 6 rests on the psoas muscle, over which the meso-sigmoid runs. Peritonitis occurs in the meso-sigmoid at the point where it crosses the psoas in seventy-five per cent of adults. The predisposing factor in the peritonitis is the contraction and relaxation of the psoas and iliac muscles. (From author's article in *Journal of the American Medical Association*.)

It may be here stated that in the shifting of the ligamentum peritonei two processes must occur, viz: Either there must be coalescence of peritoneal layers or there must be displacement.

From a long study of the peritoneum I have been led to favor displacement, that is, readjustment, but I am willing to admit that I can not satisfactorily fit displacement (or even coalescence) to all cases, and especially the formation of the inter-sigmoid fossa or Hensing's fossa, discovered by him in 1844. In the shifting of the insertion line of the meso-sigmoid (radix meso-sigmoidea) toward the left, either its left blade coalesces with the dorsal layer of peritoneum, shifts leftward, that is, readjusts itself, traveling leftward. The readjustment or displacement theory in regard to the insertion line of the meso-sigmoid is quite plainly applicable to all the meso-sigmoid except the inter-sigmoid fossa, to whose formation the coalescence theory seems to fit better. Toldt has asserted that in a shifting mesenterial insertion there is a white line to show the border of coalescence of the two layers of peritoneum. This line I have watched for scores of times, but I fail to find it as Toldt intends it. I think that the white line Toldt speaks of is a white line, not of coalescence, but which frequently arises where the peritoneum suddenly changes its direction from covering a viscus to cover parietes.

The mesentery of the sigmoid represents that which has been only partially disturbed by displacement or coalescence. The two branches of the S-iliac are a vertical primitive branch which inserts itself directly in the mid-dorsal line. This we will call the lower limb of the loop. The upper limb of the loop reaches from the lower pole of the kidney or still lower down to the point on the vertebral column where it converges with the vertical or primitive loop. The convergence of the upper and lower limbs of the sigmoid loop is on the vertebral column at a varying point according to age and individual differences. At the third fetal month the rotating umbilical loops begin to drag on the mesentery of the descending colon. The dragging falls chiefly on a fold of the mesentery of the descending colon which may be called the ligamentum mesenterico-meso-colon, but which I shall designate in honor of the Russian anatomist, Gruber's fold.

The double-bladed fold of peritoneum which stretches from the meso-sigmoid to the root of the mesenterium may be named the ligamentum mesenterico-meso-colon. The end which is inserted into the root of the mesentery varies generally in its insertion from the middle of the mesentery to its lower end. To observe this fold of peritoneum the meso-sigmoid should be put on a stretch. In some cadavers Gruber's fold is very large, but in a few bodies it is very small or perhaps may occasionally be wanting. This fold is no doubt acquired during the rotation of the navel loop. On Gruber's fold as an axis no doubt, arises volvulus of the sigmoid, yet it prevents volvulus by stretching the whole meso-sigmoid in an elevated position and also by its broad insertion. If the meso-sigmoid were allowed to fall into the pelvis like a loop of small intestine it would doubtless suffer volvulus more frequently.

Gruber's fold is what I understand as especially put on the stretch while the cecum traverses across the abdomen in front of the duodenum. It suspends the sigmoid mesentery and keeps the S-iliac out of the pelvis to some extent, also preserves the S-iliac from volvulus. At three months' fetal life one can observe the rapid displacement of the descending colon to the left. This displacement, which I believe it is, and not coalescence as Toldt advocates, so changes the primitive mesentery by dragging it from the mid-dorsal line that the meso-sigmoid demands new study. There is an angle distinct and plain at the junction of the ascending lower limb of the S-iliac and the upper limb of the sigmoid loop, at the left side of the vertebral column. Now, this angle demands attention, as from it starts the point of a cone whose base ends at the brim of the pelvis, at the sacro-iliac joint. This cone is the inter-sigmoid fossa which is due, so far as I can observe, chiefly to the spermatic vessels. Yet some may say with apparent evidence that the inter-sigmoid fossa is due to the sigmoid artery or the lower pole of the left kidney. The white line which lies to the left of the descending colon, claimed by Toldt as evidence of coalescence, is in my opinion not yet proved. I have carefully examined this line, but think the evidence is insufficient for any interpretation given. I will here note a wide variation in the position of the descending meso-colon and meso-sigmoid. The line of insertion of the descending meso-colon may be anywhere between the internal border of the left kidney and the internal border of the same. Or, it may be far out beyond the kidney against the lateral abdominal wall. Again, the meso-sigmoid may assume a varied line of insertion. In a case examined by Dr. Lucy Waite and myself, the meso-sigmoid began in the middle of the external border of the left kidney, with a relatively long mesentery. From this point the insertion of the meso-sigmoid passed in direct line to the third sacral vertebra. This was a case with an enormously long and wide meso-sigmoid, and yet the radix meso-sigmoidea was not inclined to be abnormal. The abnormality was in the lower end of the descending colon, having a long mesentery reaching from the middle of the left kidney to join with the meso-sigmoid.

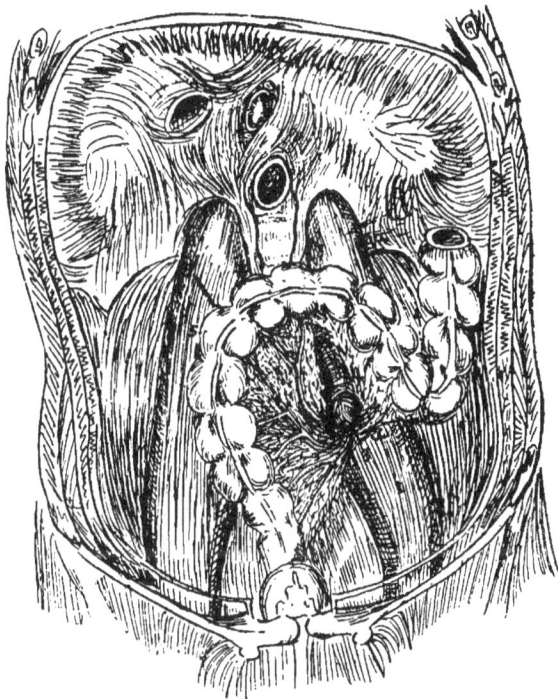

FIG. 3.—(After Byron Robinson.) This figure I sketched to illustrate the meso-sigmoid, the sigmoid loop, the psoas muscle, and the inter-sigmoid fossa. 1, psoas muscle; 2, line of insertion of meso-sigmoid; 3 points to the inter-sigmoid fossa; 4 is placed on the iliac muscle at the junction of the lower end of the colon and upper end of the sigmoid loop; 5 is near the hook which suspends the S-iliac. (From author's article in *Journal of the American Medical Association*.)

The extensive variation of the meso-colon at the lower end of the descending colon is very significant to the colostomist, for by attention to it he may avoid subsequent prolapse of the mucosa. By dragging down the colon well into the wound the meso-colon descendens is put well on the stretch and avoids any further prolapse.

In adults the line of insertion of the mesentery of the lower loop of the sigmoid varies the least, yet it is generally located to the left of the mid-dorsal. The mesenteric line of insertion of the upper limb of the sigmoid generally reaches from the external border of the psoas muscle, at about the iliac crest, to a point below the divergence of the iliac arteries. The chief displacement occurs in the upper loop. It should be distinctly remembered that much of the evolution of the sigmoid occurs after birth. The angle of the junction of the lower and upper limbs of the sigmoid loop gradually widens as the meso-sigmoid shift downward and to the left.

The left face of the meso-sigmoid is of further interest, as a cone-shaped depression is found there as well as most of the inflammatory diseases which attack it. If one pulls the sigmoid upward, especially in an infant, there will be seen, on the left blade of the meso-sigmoid an opening generally situated at the pelvic brim near the sacro-iliac joint close to the root of the meso-sigmoid. The opening is round or oval-shaped and lies at the entrance of a cone-shaped depression, which varies in depth from a simple depression on the surface all the way up to the duodeno-jejunal flexure. The entrance of the inter-sigmoid fossa on the left face of the meso-sigmoid varies considerably in position. The mouth of the fossa may be situated midway between the root of the meso-sigmoid and the sigmoid loop. It may occupy any position, having a radius occasionally of an inch and a half. The entrance may be round, oval, elongated, with a sharp border, or even a considerable of a valve, which prevents to a certain extent, retro-peritoneal hernia in this fossa. The long axis of the cone lies with its point toward the liver and base to the left iliac fossa, i. e., obliquely outward. The sigmoid artery lies above it and to its right border, while the spermatic lies below and to the right border.

The inter-sigmoid fossa scarcely occurs before the six months of fetal life, and I examined an eight months' fetus with no trace of one ; in general it is rare at birth and in adults it exists frequently, but is nearly always distorted by peritonitic adhesions due to the psoas muscle as a predisposing factor to inflammation

in the meso-sigmoid.   It generally has a peculiar valve-like fold
at the entrance of the fossa, which prevents the engaging of
omentum or bowel loop.

The fossa varies in depth from one to two inches.   Its upper
point or apex may reach to the pancreas or duodeno-jejunalis
fold.   Its lower end or base generally opens at the upper margin
of the sacro-iliac joint, but frequently lower along the border of
the same joint.   The base has varied in size in three hundred
autopsies, so as to admit the tip of the little finger to the admission of the tips of three fingers to the first joints.

The inter-sigmoid fossa generally lodges the index finger to
the first joint.   The variation of the opinion of authors in regard
to this fossa in the meso-sigmoid is remarkable and radical.
Henle and Perignon observe that the fossa ascends behind the
dorsal parietal peritoneum.   Toldt says it lies in front of the
dorsal parietal peritoneum.   Treitz and Treves say the fossa lies
between the blades of the descending meso-colon.   Waldeyer
says it is the substance of the sigmoid.

In the above views are contained three distinct and separate
anatomic situations absolutely irreconcilable.   The matter is yet
unsettled.

As to the origin and method of formation of the inter-sigmoid
fossa, I think it is the most difficult and unsettled of all retro-
peritoneal fossæ.   The first views as to origin were given by the
gifted but lamented Treitz, who attributed the formation of the
inter-sigmoid fossa to the descension of the sexual gland (ovaries
or testicles).   By the descent of these sexual glands a fold of
peritoneum was formed, the plica genito-enterica, and by a dragging on this fold of the meso-colon descendens the fossa was
formed.   This theory is only probable, yet a similar fold I can
find on the opposite side in fetuses and infants extending from
the cecum to the ovary—it is not quite so apparent in the male.

However, such folds actually exist, but to interpret them as
Treitz has is not sufficiently apparent to consider the matter
settled.

The second views as to the origin and method of formation
of the inter-sigmoid fossa arose from the celebrated Berlin anatomist, Waldeyer.   He attributed the formation of the fossa to
the internal spermatic and hemorrhoidal vessels.   This theory

appears to me to possess the best rock and base of explanation
of origin and development of the inter-sigmoid fossa. I am
thoroughly convinced that the fossa duodeno-jejunalis or some
of the cecal fossa arise from vessels projecting the peritoneum
into folds producing pockets or fossæ. Hence, according to
Waldeyer, the inter-sigmoid fossa in origin and development, is
due to the internal spermatic and superior hemorrhoidal vessels.

A third theory as to the origin and development of the inter-
sigmoid fossa is that of readjustment or displacement of the
insertion of the meso-sigmoid. The insertion line shifts to the

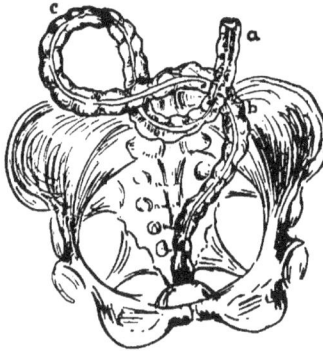

FIG. 4.—(After Byron Robinson.) This I sketched from a *post-mortem*
examination of a six-months-old child, in which Dr. Lucy Waite and I performed the autopsy. We found an enormously long S-iliac, as shown in the
figure, reaching into the right iliac fossa and existing in the state of a volvulus;
A and B represent the mesenteric axis around which the loop C has rotated.
The volvulus at the autopsy did not seem twisted enough to cause obstruction.
The figure illustrates the notable long S-iliac found in the newborn. (From
author's article in *Journal of the American Medical Association.*)

left, but for some reason the inter-sigmoid fossa was found during the shifting, perhaps due to the vessels.

A fourth theory as to the origin and development of the inter-
sigmoid fossa is that of coalescence. This is not less than one
hundred years old, and was chiefly advocated by Miller and
Meckel. Its latest and most powerful advocate is Toldt, of
Vienna. The view in this theory is that the left blade of the
meso-sigmoid coalesces with the dorsal peritoneum, especially
over the anterior face of the kidney, particularly down to its

lower pole, but between the internal border of the kidney and the median line, i. e., along the ureter, the blades of peritoneum do not coalesce and a cone-shaped slit is formed resulting in the inter-sigmoid fossa. It is given as a reason that the peritoneal blades over the ureter do not coalesce, because there exists a depression, and no resistance is made, inducing no coalescence. I would suggest that it will require all four theories and perhaps some additional facts to make the whole matter of origin and development of the inter-sigmoid fossa clear.

Jennesco notes that the fossa is almost constantly present in the newborn, with which my investigations certainly do not agree. The type of the inter-sigmoid fossa will constantly change from infant to adult life on account of seventy-five per cent. of peritoneal inflammations occurring in the meso-sigmoid of adults, which will always distort the shape of the fossa and often obliterate it. The inflammatory cicatrices often divide the entrance and cavity of the fossa into two or more apartments.

In our observation of the last one hundred and fifty cases the fossa existed in eighty-three per cent. This includes the distinct depressions which existed, though no pointed cone existed. Authors declare that the inter-sigmoid fossa exists all the way from fifty to eighty-five per cent.

It may be that the theory of coalescence so ably advocated by Toldt finds its most crucial test of existence in the formation of the cone-like inter-sigmoid fossa. But the doubtful wedge lies in the view. Why does the coalescence miss this inter-sigmoid cone? Why does the coalescence miss any portion of the surfaces and select others? In the whole investigation of the peritoneum the coalescence theory has not convinced me that it shows as much natural evidence as the theory of displacement.

Treves reports two cases of strangulated hernia in the inter-sigmoid fossa—one by Lawrence and one by Eve.

The meso-sigmoid consists of three layers, viz: tunica serosa —two layers, a right and a left, and a middle layer—the real neuro-vascular visceral pedicle. I shall designate this middle mesenterial layer as the membrana mesenterii propria.

1. The tunica serosa is the peritoneal layer which faces the right and left side of the meso-sigmoidea. It is the endothelial layer composed of flat, polygonal plates, and joined edge to edge

so as to form a continuous membrane. These endothelial cells are slightly elevated in the center where the nucleus bulges upward. The nucleus is often eccentrically located, and can be beautifully brought into view with logwood as a stain. If nitrate of silver solution, one half per cent., be applied to a fresh piece of the tunica serosa of the meso-sigmoid, the outline of the cells are marked off by dark lines. The dark lines are a precipitate of albuminate of silver. Under a high power the dark line divides into two dark lines with a white intervening space which is crossed transversely at the irregular intervals by very fine dark lines.

[TO BE CONCLUDED IN OCTOBER ISSUE.]

CHRONIC GASTRIC DYSPEPSIA.*

BY K. S. HOWLETT, M. D.,

BIGBYVILLE, TENN.

[Written for MATHEWS' MEDICAL QUARTERLY.]

The stomach is probably the organ of the greatest interest to the physician. It must be closely watched and guarded in every disease, for upon the degree to which it retains and carries on its normal functions depend in a great measure the success of the treatment and the ultimate outcome of the case. It is peculiarly susceptible to influences outside of itself, and there can scarcely be any grave pathological lesion in any portion of the economy, however remote, without its accompanying gastric symptoms. Hence it is not surprising that this organ has been the subject of thorough and enthusiastic investigation, both as to its physiological functions as well as the pathological deviations therefrom. And yet so difficult and complicated are many of the problems that arise that there is perhaps no other organ about which there is so much confusion and contradiction or so many points of vital interest which still remain in doubt.

The profession must forever remain indebted to the happy chance which brought the patient, Alexis St. Martin, with his accidental fistula, under the observation of so competent and

* Read before the Tennessee Medical Society, Chattanooga, Tenn., April 15, 1896.

practical an observer as Dr. Beaumont. From his ingenious investigations was obtained the first correct idea as to the true function of the stomach. They also gave an impulse to the further study of the natural process of digestion, and pointed out the way in which such study could be carried on, and there soon followed an extended investigation upon this line by the establishment of artificial fistulæ in the lower animals. Later years have brought into general use that most serviceable though simple instrument, the stomach tube, which not only saves the profession from the unsatisfactory waiting for rare accidental fistulæ in man, and from the wrath of the anti-vivisectionists, but has also added greatly to our diagnostic resources by enabling us to study more accurately and rationally the physiological and pathological phenomena of gastric digestion. But even this, together with all other diagnostic helps has its limitations, and it is but pleasant to indulge the imagination in the marvelous possibilities of the recently discovered Roentgen rays, and to cherish the hope that soon will be discovered some modification of their action by which the observer can look into or photograph this as well as other vital organs and behold the morbid growths and abnormal processes going on there.

By the term "chronic gastric dyspepsia" I intend to include only the inflammatory and purely functional disorders of the stomach. Except the organic change or degeneration, which is probably present in every case of long-standing dyspepsia, all the essentially organic conditions will be excluded from this paper. Although I do not desire to underrate the importance of differentiating between a dyspepsia due to an inflammation and one of neurotic origin, yet the two have so many things in common, in their etiology, clinical history, and (to a lesser extent) treatment, and they are often so closely allied, one causing or being caused by the other, that it seems to me that there is nothing improper in speaking of them together.

Etiology. Any thing which lowers the vitality and depresses the general system may predispose to dyspepsia; old age, sedentary habits, constipation, intestinal putrefaction and defective elimination, sexual excess, the commoner blood diseases, tuberculosis, mental strain or worry, exposure to extreme heat or cold, prolonged and exhausting physical labor. Neurasthenia and hysteria belong especially to the nervous form, while diseases of the heart and portal system act more directly in causing gastritis by producing changes in the circulation and venous congestion of the mucous membrane.

A mistake of cause for effect may be easily made just here. Heredity, in my opinion, plays a very doubtful part, and scarcely deserves mention at all.

Among the exciting causes common to both forms are the regular and intemperate use of condiments, tea, coffee, tobacco, alcohol; overtaxing the stomach by the excessive quantity or indigestible quality of food, irregularity of meals, rapid eating and imperfect mastication and insalivation (of which indiscretion Americans especially are said to be guilty), a too hasty return to normal diet after acute diseases, and the attempt to bring on a rapid convalescence by irritating bitters and forced feeding. On the other hand, a too rigidly restricted diet kept up too long, with the attention constantly and morbidly fixed upon the stomach and digestive process, is not an infrequent cause.

Among the things most likely to be followed by nervous dyspepsia are sudden and great grief, mental shock, excitement and disappointment, intensely fixing the mind upon business or study immediately after eating, or severe physical exercise at the same time.

Chronic gastritis may follow acute inflammation or functional indigestion, or it may be caused by indigestible, decomposed, or fermenting food in the stomach, the taking of articles too hot or too cold, the frequent use of irritating purgatives or other drugs, in fact any thing which keeps up a constant irritation of the mucous membrane.

Pathology. The pathology of functional dyspepsia consists of either an atonic condition or a hypersensitiveness and exaggerated irritability of the nerves, resulting on the one hand in diminished secretion and motion, and hence difficult and incomplete digestion, or, on the other, in an increased sensitiveness and general irritability, with hyperacidity or hypersecretion, or even normal condition of the gastric juice, and motion unimpaired. I am aware that a great many (possibly a majority) of investigators take the ground that there is no hypersecretion or hyperacidity without some inflammatory or organic lesion. Ewald takes the position that the function of what he calls a highly vegetative organ depends principally upon the activity of its own cells and very little upon the nervous system; that the purpose of the numerous nerve fibers that go to the stomach is simply to connect it with the rest of the body so that the stomach can recognize and fulfill the demands of other parts of the system, or the entire organism take cognizance of its condition; that, if we could remove from it every nervous element without injuring the other tissues, it would still secrete, contract, and absorb quite well. It seems to me, however, that a certain amount of external nerve-force is necessary to the proper performance of its function, and that, when the supply of such force is cut off or lessened by any cause, there must certainly result diminished activity, just as added irritation results in increased activity.

Ewald (who probably stands second to none in the zeal and energy with which the study of this specialty has been prosecuted) says that a perfectly normal mucous membrane is among the greatest rarities, and reports only two as having come under his observation. This mucous membrane differs materially from that of the mouth and esophagus. It is covered with an epithelium, which is supposed to possess mucinogenous properties and sends prolongations into the excretory ducts of the glands. It is so intimately connected with the glandular structure that even the mildest inflammations involve the glands themselves. Hence, "gastro-adenitis" gives a better idea of the true pathology than the commonly used "chronic gastritis." In this the mucous membrane presents a yellowish-gray color, with injected areas of scarlet or brownish-red. It is covered with a layer of firmly adherent but delicate mucus, is usually thickened with papillary projections or sometimes polypoid growths. Then there is the minute anatomy of glandular inflammation with interstitial infiltration, etc., accompanied by closure of the ducts and degeneration of the protoplasm of the cells, the cells themselves becoming granular, shriveled up, or entirely destroyed. This process of course has its corresponding interference with the normal secretion, and it seems to be the opinion that even in the earlier stages and milder cases the gastric juice is either insufficient in quantity or vitiated in quality.

If the process is not checked, we have either a progressive destruction of glandular tissue or a marked activity of interstitial connective tissue, leading to a hypertrophic proliferation. In either case we finally have complete destruction of the normal secretory function and even disappearance of the secreting parenchyma.

Symptomatology. There is a most unpleasant though familiar train of symptoms which, though coming on earlier and being perhaps more prominent in the nervous dyspepsias, yet is always present to a greater or less extent in chronic disorders of the stomach. It is often described by the sufferers as a condition of general irritability. Their ill temper is provoked by the most trivial things ; they are nervous, easily excited, low-spirited, languid, and drowsy. They have a disinclination for and are easily exhausted by either mental or physical exertion. They complain of shortness of breath, distressing dyspnea, palpitation of the heart, and sleeplessness, or more often of sleep, which, though deep and prolonged, is restless and unrefreshing ; of dull headache or fullness about the head, dizziness, and vertigo. The pulse is somewhat accelerated and sometimes irregular in the earlier stages, later becoming small and weak. There is generally loss of flesh, more marked in the inflammatory condition. In the latter also there is not infrequently a slight and irregular rise of temperature, with lips dry and cracked, and thirst as a prominent symptom. The mouth is dry, with an unpleasant, somewhat nasty taste, the tongue clean or but slightly coated in functional dyspepsia ; in gastritis more heavily coated, especially at the base, reddened, sometimes covered with swollen papillæ, or later, pale and indented by the teeth.

The local symptoms referable to the stomach are much the same in both forms. There is a sense of fullness and distension, uneasiness and discomfort, coming on immediately after eating in gastritis, later in the functional form ; pyrosis, flatulence, and frequent eructations of gas or an acid fluid more or less offensive to taste and smell, the belching giving only temporary relief. Nausea is common, vomiting rare, although the patients feel that vomiting would give relief. There is slight tenderness on pressure, seldom severe pain, but an uneasy, in gastritis a decidedly burning sensation in the region of the stomach. There is loss of appetite always present, except in certain neuroses when there is excessive hunger, the patients saying they have a dog's appetite; or in other cases there is a feeling that taking food will give relief, which it does for a time, while the opposite is true in gastritis. Constipation, more or less obstinate, is a constant symptom, though it sometimes alternates with diarrhea, the stools presenting a foamy, soapy appearance. The scarcity of urine, with abundant deposits of urates, the liver disorders, slight jaundice, etc., belong more especially to the inflammatory type.

Diagnosis. The diagnosis can only be made by excluding the organic diseases. In gastric ulcer the pain is acute, boring, localized, is rare when the stomach is empty, is increased when pressure is made over the site of the ulcer, and is usually completely relieved by an alkali; hyperacidity is the rule, and belching rare. The vomiting of fresh blood confirms the diagnosis. In cancer, beside the peculiar character of the pain, the fetid belching, frequent vomiting, and the composition of the vomited matter, there is one test that is now considered conclusive, viz., the absence of hydrochloric acid and the presence of any considerable quantity of lactic acid after the usual test meal.

Of course, when the tumor and characteristic cachexia, with the dry skin, pale and sallow complexion, make their appearance, the diagnosis is easily made.

Dilatation from pyloric obstruction, if very marked, can be determined by its history and by careful physical examination. Einhorn's electric light, introduced into the stomach to show its outlines, is of valuable aid to the other physical signs. When it is only slight and results from weakening of the muscular wall or an atonic condition the diagnosis is almost impossible. Here, however, it is only a sequel to or part of the disease under consideration anyway. For the rational and successful treatment, however, it becomes necessary to make a differential diagnosis between the different varieties of dyspepsia as well as the various stages of each variety. When this can not be done by a careful consideration of the symptoms which I have already attempted to bring out, then the test meals, stomach tube, and chemical tests must be brought into use. There is as yet no entirely satisfactory way of testing the motor power of the stomach. Leube's method of determining the length of time required by the stomach to rid itself of its contents after the test breakfast or an ordinary meal and Ewald's salol test, or Huber's modification thereof, are the ones most frequently used.   We do not find increased motility except in that condition of compensatory hypertrophy, the result of nature's efforts to force undigested food into the intestines before it ferments too much.   It is impaired in atonic dyspepsia and in the advanced stages of chronic gastritis.   In functional dyspepsia we may have nothing abnormal in the chemism, or we may have hyperacidity or hypersecretion.   In the atonic variety and in chronic gastritis the acidity is probably at all times lessened and the digestive power decreased, and when atrophy is fully established it is altogether absent.   In simple chronic gastritis the stomach contains a small quantity and in the mucus form an abundance of mucous fluid while fasting.   In atrophy it is perfectly empty.

Treatment—Dietetic and Hygienic.   As errors in diet play an important part in the causation of dyspepsia, so proper dieting plays an important part in its treatment.   And yet we may have too much of even a good thing.   Where "eternal vigilance is the price of liberty," true liberty is never obtained, and a dyspepsia only kept in abeyance by constant dieting is never cured.   I firmly believe that dieting may be so strict and kept up so long that the stomach becomes atonic from simple lack of its accustomed work.   Hence we should see that additions are made to the diet as early and as rapidly as possible.

The notable lack of uniformity in the dietary directions given by various writers is sufficient evidence that each patient must to a certain extent be a law unto himself.   I will mention, however, a few articles in common use in this country which seem to me to be especially objectionable, and which should be forbidden in all dyspeptic cases: candies and confectioneries and cakes of all kinds, desserts with rich sauces, pastry, hot biscuit, fried flour or meal cakes, tough meat of any kind, fresh pork, sausage, fried bacon, cabbage, kraut, new potatoes, turnips, cheese, fried or hard cooked eggs, pickles and strong condiments, especially vinegar, cold drinks and ices, tea altogether, and coffee only in moderation. The regular use of alcohol, even in moderate quantities, and of tobacco in any form should not be allowed.   The following foods may be allowed in ordinary dyspepsia or gradually added to the list as digestion improves: light bread, stale or toasted, plain corn-bread, soda crackers, oatmeal, rice, soft-cooked eggs, milk and butter, breakfast bacon, boiled ham, tender beef, chicken, birds, and usually fish and oysters, nearly all kinds of vegetables, except those in the preceding list, and fruits, except those that are very sweet or very sour.

Every person, especially dyspeptics, should be made to appre-ciate the fact that digestion begins in the mouth, and should be instructed to keep the mouth clean and the teeth in as good con-dition as possible; to masticate the food thoroughly and to retain it in the mouth until it is fully incorporated with the saliva. To insure this and to prevent the gastric juice from being diluted it is well to prohibit drinks of all kinds with the meals. They should have meals at regular hours, and should throw off as much as possible all business worries and mental cares at the table and avoid eating too great a quantity or too great a variety.

The hygienic measures consist of avoiding those things which have been mentioned as playing an important rôle in the causa-tion of these troubles, taking a short rest before and after meals, taking active though not excessive or fatiguing exercise, cold baths and vigorous rubbing in the morning, and following other directions as to fresh air, recreation, etc., common to any other trouble. As to special diet for each form of dyspepsia, where there is hyperacidity or hypersecretion, the patient should avoid all substances which are liable to excite or stimulate the glands, and should take all kinds of meats and albuminous foods in abundance and the starches and fats sparingly. In the atonic and inflammatory varieties the bulk of the food should be reduced as much as possible, and it should be prepared in a way to make it as tender and digestible as possible, or it may be par-tially or wholly predigested.

In the absence of much flatulence, with the motility not seriously impaired, the carbo-hydrates are admissible, and as they add much to the nutritive value of the food they should be given, though with due care. I have usually been disappointed in an exclusive milk diet. In atrophy, especially in old people, more stimulating articles of diet may be used, and wine and weak alcoholic beverages are probably beneficial.

In functional dyspepsia very little medicine is required, as we must depend chiefly upon the correction of the personal and business habits as well as the hygienic surroundings and dietetic errors of the patient. Hyperacidity and hypersecretion require alkaline remedies; the bicarbonate of soda with the addition of magnesia and rhei, if constipation is present, given two or three hours after meals. In continuous secretion the nitrate of silver wash or spray has been suggested. The administration of morphine and atropine is in my opinion not judicious in this or other forms of dyspepsia. When there is much nervousness, sleeplessness, etc., the bromides serve a good purpose.

In the earlier stages of gastritis measures to allay the irritation and reduce the inflammation are called for. Bismuth and hydrocyanic or carbolic acid make an excellent combination; nitrate of silver is said to be superior to bismuth as a sedative and astringent, and should be given in doses of one-eighth grain three times a day. Occasional doses of calomel have a good effect both by its local action and by stimulating the outflow of bile, relieving portal engorgement and duodenal congestion. Lavage with large quantities of warm water, either plain or alkalinized by the bicarb. or chloride sodium, teaspoonful to a pint, is a popular and useful remedy, and should be used once or twice a day, preferably in the morning an hour or two before breakfast. It not only removes the mucus and remnants of undigested food, but also stimulates secretion by its mechanical action as well as by the principle of Bartholow: "An alkali on one side of a secreting membrane and an acid on the other is a condition favorable for osmosis." The same result may be accomplished to a certain extent by having the patient drink large draughts of hot salt water an hour before breakfast.

Hydrochloric acid is the most universally used and highly lauded of all medicinal remedies where there is diminished secretion. The theory is that it supplies the acid deficiency, stimulates the formation of pepsin (which action it retains until complete atrophy has taken place), as well as its own secretion, and besides is an efficient antiseptic and antifermentative. It is given according to the requirements of gastric secretion in doses of five to twenty or thirty drops with or immediately before meals, and in as little water as possible. Notwithstanding the universal indorsement which this remedy has received, from my observation I am inclined to think that it is oftentimes given in too great a quantity and for too long a time. I can not agree to the proposition that it stimulates its own secretion. Bartholow's rule must work both ways, and an acid on both sides of a secreting membrane is not a condition favorable for osmosis. Nature usually responds to the need of the system and attempts to supply any ingredient that is lacking. If this acid is continuously supplied from without, it seems to me that the power of secretion would become lessened and dormant, just as the right arm would become weakened and powerless if the other arm did all of its work. Hence hydrochloric acid should be given only temporarily to supply its own deficiency, assist the digestion, and thereby promote nutrition, and should be withdrawn as the glands become more able to fulfill their normal function. The same objections apply with still greater force to pepsin. The theory upon which this remedy, together with lactopeptine, pancreatin, and that class have been so widely used is certainly a most unscientific and disappointing one, and they are fast being abandoned, and are only mentioned now to be condemned by the majority of writers.

To prevent excessive fermentation and putrefaction, medicines with more decided antiseptic properties are used; salicylate of bismuth, salol, creosote, etc. I have had better results from creosote than from any remedy of this class. A prescription first suggested to me by Dr. J. H. Wilkes, of Columbia, Tenn., containing creosote, bismuth, and bromide of sodium in camphorwater, with the addition of cascara or rhei, if constipation is present, has been most satisfactory to me in this special phase of dyspepsia. Nux vomica, in some of the bitter infusions given before meals, stimulates secretion and motion, and may be given with benefit for a long time. Belladonna and ipecac are useful stomachics, and massage and electricity are probably excellent adjuvants. To relieve the constipation cascara is indispensable, and is well suited to the condition of the stomach in these troubles.

DILATATION OF THE STOMACH.*

BY A. P. BUCHMAN, A. M., M. D.,

Professor of Diseases of the Digestive System, Fort Wayne College of Medicine.

FORT WAYNE, IND.

I conceive that the definition for dilatation of the stomach ought to include all of the factors entering into its production. To be full and concise at this time would require much more time than is at my command to-day, and I will therefore limit my work to the effects of stomach dilatation upon the general organism, feeling that, while very few physicians in daily general practice are aware of the immense importance of the subject and are constantly overlooking it in their work, they will yet readily accede to the possibility of its relative frequency once their attention is directed in that direction.

Indigestion, I am quite certain, is always either caused or accompanied by abnormal fermentation, or rather what I have called "fermentation proper," in contradistinction to digestion proper, which I look upon as a ferment action. In every case where fermentation has been present for any length of time the inevitable result is an exaggerated distension of the stomach, with no longer the power to normally retract; this is necessarily then dilatation.

As I have already stated, the fundamental cause is fermentation; yet there are other causes, acting coincidentally, which ought to be enumerated at this time. Excessive and rapid distension, brought about by eating too quickly, or by eating and drinking too often, thus keeping the stomach constantly filled and never allowing it time to assume its normal anatomical state. Again, when meals are taken too close together, never allowing the stomach time to empty itself, this, together with other bad stomach hygiene, will inevitably result in an overdistension which in time reduces muscular power and leaves the organ in a state of dilatation.

Another potent cause of distension which is of importance is the present method of stomach lavage, a method of attempting in a very crude way to apply the asepsis of surgery to medicine.

* Read before the Upper Maumee Valley Medical Society.

When two or more pints of water are forced into the stomach one or more times in twenty-four hours, the stomach soon fails in its power of retraction, and hence in the effort at cleansing the stomach may be and often is irreparably injured. Again, cicatricial constrictions, resulting from cured ulcers, or from cancer of the pylorus, or from a chronic catarrh of the mucous membrane, must be classed among the mechanical causes of dilatation.

Distensions following or accompanying general nervous debility, and especially that state of irritability which exists among hysterical people and ataxics, causing variations in the energy of the central nervous system, may result in a dilatation, but is not likely to do so. Some observers have noticed an intermittent distension in exophthalmic goitre. I have on several occasions found dilatation to exist in two or three members of the same family, which, upon close investigation, led me to conclude that there was in these cases an hereditary inadequate stomach, which had been badly cared for, or rather not cared for at all in an hygienic sense. Lastly, the study of muscular degeneration as it occurs anywhere in the body can be applied to the stomach; atrophy can be the result of fatty degeneration, to the use of alcoholic intoxicants, or to inflammation. In typhoid and other infectious fevers it often arises at the beginning and even before their commencement, and it is an open question whether it was not the previously dilated stomach that prepared the way for the introduction and elaboration of the infectious agent.

In analyzing these various causes of stomach dilatation it becomes very evident that they must be sought for in taking account of the patient's general history, that is, if its frequency in the history of diseases of the digestive tube is not fully recognized it is apt to be overlooked. The appetite often remains unaffected, or, if influenced at all, is augmented, hence in taking account of symptoms this contingent is enough to arouse suspicion. Also, ingestion is not painful or in any way distressing, but in two or three hours after taking food the stomach feels uncomfortably full, when eructations, first of inodorous gas followed by bad-tasting and fetid gases, then regurgitations of acid, which demonstrate the fact that anomalous fermentations are going on. The hydrochloric acid has no odor. The acid odor is always due to acetic or lactic acid fermentation. The feces are

19

generally doughy or sticky, stinking, acid—and although soft, are expelled with difficulty. When the stomach has been in a state of permanent dilatation for a long time, the consequent development of acid in the whole length of the digestive tube gives rise to a generally paralyzed condition of the mucous lining; to consequent thickening, which is not of inflammatory origin, but is the result of blood stasis and therefore of hypernutrition of a low grade, upon which abnormal state follows ulcerative gastritis, and the host of other phenomena which can in a general way be classed under the title, " Malignant Gastritis.'' The classification of the remote symptoms which invariably accompany dilatation of the stomach is a work which as yet has not received the attention its importance really demands. When a patient applies to his physician complaining of insomnia, vertigo, impairment of sight, weakness of one or more of the groups of muscles which control the movements of the eye, painful contractions of the muscles of the calf of the leg, or of the feet, or hands and arms, he will almost certainly regard the statement that such symptoms have their origin in an indigestion and dilated stomach with considerable hesitancy and doubt. These phenomena may all be regarded as reflex, and disappear early in the history of the treatment of cases. Their early disappearance often leads both physician and patient into the error of supposing that the agency producing them has been cured, while in fact the condition is only bettered temporarily by better stomach hygiene and diet regulation, through which fermentation has been reduced to the minimum. But there are other conditions that can not be explained upon the hypothesis of reflex phenomena, which are as important, in fact infinitely more so, for the reason that their operation is through another channel, viz., the blood and lymph circulation, and result in general and local intoxication.

The field of operation in this particular becomes almost infinitely extended, and the influence of such toxic substances upon the general organism and upon special organs, especially the emunctories, is worthy of the closest scrutiny. The skin frequently suffers, as its office in carrying out of the body effete and toxic elements is as important as that of the kidneys or lungs.

While it may seem far-fetched to link onto the dilated stomach many skin affections, I am quite satisfied that it is at least a most philosophical way of accounting for them, and the proof of the correctness of the hypothesis must lie in the fact that such abnormal skin manifestations rapidly and permanently disappear under proper treatment of the digestive system. The kidneys, standing as they do at the gateway through which an enormous quantity of matter which has served its purpose in the organism must be passed, are sensitive to the action of toxic substances upon them, whether such substances be the result of faulty metabolism or whether they arise directly from faulty digestion. The renal tissue is irritated by toxic agents, and in consequence we have an albuminuria which disappears when the condition of the stomach has been relieved.

The lungs, as emunctories, are subject to the same influences by toxic agents generated in the dilated stomach as are any of the other organs of the body.

In the foregoing I have very briefly touched upon a few of systemic conditions which are to be found as direct outgrowths of stomach dilatation, while at the same time fully recognizing the fact that the dilated stomach is not possible as a pathologic state distinct and separate, but is one link in the chain which completes the cycle of diseases of the digestive tube.

The question that now remains is: What can we do to relieve these sufferers, and is the condition amenable to curative treatment?

To the last question I am now prepared to give an affirmative answer, knowing and realizing fully the position I take in becoming responsible for the statement.

The mode of treatment can only be given in a general way, as every case is a law unto itself.

The first thing to be considered is thorough stomach cleansing, and the manner of obtaining this end is a question of prime importance. I have already pointed out the dangers of forcibly washing out the stomach by the use of the stomach tube. My experience with it has been somewhat extended, and while it is true that many of the reflex phenomena disappear with its use, I am convinced that in the great majority of cases the condition is finally only aggravated, and from this fact has arisen the generally accepted statement that dilatation of the stomach is incurable.

The use of from one half to one pint of hot water taken very slowly—from fifteen to thirty minutes should be consumed in drinking it—an hour, at least, before each meal has a decidedly better and an altogether more soothing effect than the method of lavage. This clears the walls of the stomach of its slimy, sour, yeasty coating, and at the same time encourages the normal stomach peristalsis. The use of hot water taken in this way must be continued and persisted in regularly for a long time. I am in the habit, especially in the few first months of the treatment, of directing the use of from one to one and a half table-spoonfuls of listerine in a half pint of hot water, to be taken thirty minutes before retiring. This insures cleansing and disinfection of the stomach and prevents the insomnia, which is such a frequent accompaniment of stomach dilatation.

The question of foods is one of immense importance, inasmuch as the origin of the difficulty undoubtedly lies in wrong feeding. To begin with, all fermenting and all fermentable foods must rigidly be excluded until such time as the digestive forces are toned up to a point at which they will care for them normally. At first they must be given in small quantities and must consist of the most easily digested of their class.

The albumins are admissible from the beginning, in fact must be wholly depended upon in the early part of the treatment. When the salivary fluid shows normal strength and power a few well-cooked vegetables can be taken at the morning and noon meals. The dietary requires constant and intelligent watching, especially with regard to the quantity taken at each meal.

As to the exhibition of drugs, while it is true that they are of secondary importance, yet their importance must be recognized. Tonics and mild stimulants are first in order, then such as go to increase the red corpuscles should receive consideration. However, there can not be any hard and fast lines followed in either diet or drugs for any length of time in any one case, as every such case is a law unto itself and will require the exercise of ingenuity and skill to make a satisfactory showing.

MATHEWS' MEDICAL QUARTERLY
"ALIS VOLAT PROPRIIS."

Vol. III.	LOUISVILLE, JULY, 1896.	No. 3.

JOSEPH M. MATHEWS, M. D., - - - - - - EDITOR AND PROPRIETOR.
HENRY E. TULEY, M. D., - - - - - ASSOCIATE EDITOR AND MANAGER.

A Journal devoted to Diseases of the Rectum, Gastro-Intestinal Disease, and Rectal and Gastro-Intestinal Surgery.

Articles and letters for publication, books and articles for review, communications to the editors, and advertisements and subscriptions should be addressed to

Editors Mathews' Medical Quarterly, Box 434, Louisville, Ky.

KENTUCKY STATE MEDICAL SOCIETY.

This Society held its forty-first annual meeting in Lebanon, on the 10th, 11th, and 12th of June, as the guests of the Marion County Medical Society.

The meeting was not up to the standard of previous meetings in interest or scientific attainment. This was notable especially in the absence of those who had signified their intention of taking part in the discussion of the special subjects, and were announced on the program; and the absence of all but two of the ten who were down on the program for papers on the last day. This is greatly to be deplored, and the Society can not help but suffer from it. But the social features, the entertainment of many of the members in the homes of the people, and the reception given by the County Society at the home of Mr. McElroy, assisted by the Lebanon ladies, will long be remembered.

The election of Dr. R. C. McChord, who was chairman of the Committee of Arrangements, to the presidency for the ensuing year, was a well-merited recognition of fealty to and honest work in the Society. The choice of Owensboro as the next place of meeting was also a wise one, as the western part of the State has not entertained the Society for a number of years.

The next meeting bids fair to give rise to a good deal of parliamentary sparring over the question of the revised constitution. The report of the committee, which on the President's ruling was laid over one year before it could be adopted, was, on the last day of the meeting, brought up again on motion to "reconsider the vote" disposing of it, and the constitution, save that portion relating to the mode of election of officers, was passed.

That there is a need of a change from the existing method of election of officers can hardly be denied, nor can it be said that the proposed method is greatly superior. For many reasons the election of the officers in open meeting will be more satisfactory and give less opportunity for political "wire-pulling" and "log-rolling" than any other method. The politics of the meetings are the greatest drawbacks to successful scientific work.   T.

SPECIALTIES AND THE GENERAL PRACTITIONER.

We are so heartily in accord with the following editorial, which appeared in the *American Gynecological and Obstetrical Journal*, that we print it in its entirety:

There is no question of the fact that the domain of the specialty is being more and more invaded each year by the general practitioner. Of no specialty may this be more truly said than of gynecology. It is well to consider the causes of this incursion and its effects, first upon science, then upon the specialist, and finally upon the general practitioner himself. The relation of the patient naturally presents itself in the consideration of the effect upon science.

A great causative agent is the fact that the specialties are overrun with very young men. So soon as an interne has graduated from a special hospital, he acts upon the assumption that he is entirely fitted to become a specialist in that branch. But, having no practice, and being generally poor, he can not afford to wait for special work only ; hence he calls himself a specialist for the extra fee there is in it, and gladly accepts anything else which comes along at the same time. Of course this is a parody upon specialism, for no medical man believes—and indeed not even the intelligent interne himself—that a short eighteen months in a special hospital could possibly create him an expert; the seed there planted has hardly had time to germinate, and time and experience are necessary to prove if the soil even was worth the planting.

The cause next in importance is produced by the short-sighted effort on the part of some specialists to attract the general practitioner, in the matter of consultation, by a disloyal and more generally ignorant effort to belittle their own specialty and to show how easy it is for every one to become his own specialist. We read through an article recently, by a supposed gynecologist, in one of our contemporaries, entitled "The General Practitioner his own Gynecologist." This is an example of what we mean. The author, in our opinion, did not show any evidence in his article of being a gynecologist, but this made the article in question none the less vicious in its intended effect upon the general physician.

A third cause is the accepted temptation to the general practitioner whose practice is large, especially if it be surgical on the one hand or mainly obstetrical on the other, to add to his sphere of usefulness (sic) and to retain for himself the *special* fees which his conscience, under other circumstances, would suggest his surrendering to another, whose whole time and work were devoted to the acquiring of special knowledge.

These are the causes why there are so *many* specialists in gynecology and at the same time practically *so few.*

Now let us consider the scientific effect of these combined causes. The most evident one to-day is *confusion of thought and practice.* If a man's education has been that of general surgeon, the only important part of gynecology to him is that which needs extirpation. It is the quickest, neatest, surest way, and leaves nothing behind to suggest a doubt as to the competence of his special knowledge. If, in deference to old-time prejudice and the importunities of patients he operates upon those very "minor" conditions which are connected with laceration of the cervix and rupture of the pelvic floor, he adopts that operation which is simplest and quickest of execution. The principle upon which the particular method used is founded, anatomically and physiologically, is of little consequence; if the operation fail to benefit the patient, hysterectomy can always be done and—the patient is cured! The *obstetric* gynecologist, on the other hand, deals principally with the vagina and what it contains. If after labor he finds the cervix or posterior vaginal wall torn, he puts in enough stitches to close the wound and then feels that his patient is safe from the *special* gynecologist. The injury to the pelvic fascia, which he does not see, does not trouble him, for his point of view is that of the obstetrician. The result of this want of a *gynecological* standard and consequent confusion of practice induces a rather contemptuous opinion of the specialty in the minds of those who acknowledge their ignorance of it. Not only the general practitioner, but the layman as well, is beginning to think that as a specialty it is somewhat of a humbug and that its so-called professors are actuated by other motives than purely scientific ones. It is not difficult to trace this disposition of the outside mind to the three causes we have enumerated. As to the effect upon the patient the oft-heard tale "Oh! doctor, I have been doctoring for years and have been to five or six specialists who promised to cure me, but I am just as bad as ever," speaks for itself.

What inducement is there to the true specialist to devote himself exclusively to his branch when he sees men all about him boldly attempting the work which they do not understand but which they assure their patients they "can do as well as any specialist"? He knows that not one patient in a hundred will ever come to realize that the hysterectomy which followed repeated failures at plastic work in ignorant hands was totally unjustifiable and might easily have been obviated. He knows that when the proper plastic operation is properly applied, failure is owing *always* to the operator and not to the operation. Unfortunately, patients themselves can not know this, and the operator himself is generally too ignorant or too human to make this unpleasant acknowledgement to himself. The real specialist, therefore, finds nothing but discouragement in the pursuit of special study and is subjected to a ruinous and unjust competition.

As to the general practitioner who dabbles in gynecology and "rushes in where angels fear to tread," the harm which he does, with barely a smattering of the most superficial gynecological knowledge, can not be calculated.

And yet he is hardly to blame for attempting that which he does not in the least understand when there are so-called gyne-cologists who assure him that the whole subject is "quite easy" and that he "is a fool to pay to another a fee which he might keep for himself." If he would only confine the application of this advice to the giver of it, what a boon for many a suffering woman!

No, we can not deny that the general medical man has much excuse for not recognizing the wheat in so much chaff and for putting to practical use the proverb, which fits so aptly our spe-cialty to-day, that "when doctors disagree, disciples are free."

OUR JULY ISSUE.

It is a great satisfaction to be able to state that there is so much original reading matter in this issue as to necessitate our cutting down the matter culled from our exchanges. This is for the most part to be regretted, as it has always been our aim to offer our readers a synopsis of the current literature embraced in the scope of the QUARTERLY.

However, with the wealth of original matter contained within these covers the clippings will not be greatly missed. We must say a word for the typographical excellence of this issue with the number of illustrations which speak for themselves.

Necrology.

MEMOIR OF DOCTOR ORRIN DERBY TODD, OF EMINENCE, KENTUCKY.

BY LYMAN BEECHER TODD, M. D.,

LEXINGTON, KY.

Doctor Orrin Derby Todd, son of James Mulherin and Mary Porter Todd, was born in the beautiful county of Shelby on the twenty-fourth day of April, eighteen hundred and forty-one, and there his useful and honorable life was passed, among a noble race of people whom he faithfully served, and who in return gave him in fullest measure their unbounded confidence and their devoted love.

His ancestry were Scotch-Irish. His great-grandfather, a Scotch dissenter, to escape persecution from the Established Church, left Scotland, and settled near Belfast in the north of Ireland, where he died. His grandfather subsequently came to America and made his home in Pennsylvania, where he engaged in farming, and at intervals taught school until the commencement of the War of Independence, when he enlisted in the army and served under Washington, participating in many hard-fought battles of the Revolution. He married Jane Buchanan, who was the aunt of President James Buchanan. His grandfather with some of the Buchanans emigrated, after the war, to South Carolina, and soon thereafter crossed the mountains and settled near Nashville, where, of their family of eight children, six were born in the Block House Fort—and there his grandfather Buchanan was shot and killed by the Indians while reading his Bible at the gate of the fort.

It was in this fort that Doctor Orrin Derby Todd's father, James Mulherin Todd, was born on the seventeenth day of July, 1795. When the second war with England began he frequently served as a scout in Indian expeditions, and subsequently volun-

ORRIN D. TODD, M. D.

teered under General McArthur, going to the relief of Fort
Meigs. He afterward was a volunteer in the expedition to
Green Bay. Soon afterward declaration of peace terminated his
military career. Then he came to Kentucky and settled on a
farm in Shelby County. In politics he was an ardent Henry
Clay Whig until the dissolution of that party. During the Rebellion he was firmly and unconditionally attached to the cause
of the Union. From early life Doctor Todd's father was a member of the Presbyterian Church, always esteemed as one of the
most useful, upright, and honorable citizens of Shelby County.
He married, November 18, 1823, Miss Mary Porter, only
daughter of his first and firm Kentucky friend, William Porter.
He died at the age of eighty-two years, and his mother at
eighty-one. They raised a large family, and after the death of
Doctor Todd—so universally lamented—there survive two sons
and two daughters, James Mulherin Todd, of Shelby County;
John Buchanan Todd, of Lexington; Mrs. W. C. Calloway, of
Shelby County, and Mrs. W. T. King, of Henry County, Ken-
tucky.

Doctor Orrin Derby Todd received a liberal education and
commenced the study of medicine with the late excellent physician, Dr. Hugh Rodman, of Frankfort. In 1862 he entered the
famous Jefferson Medical College, of Philadelphia, where he
attended three full courses of lectures and was graduated in
March, 1865. In the same year he was appointed assistant
surgeon of the Twenty-first Kentucky Union Regiment, and
remained with it until it was mustered out in Louisville
in 1866.

He then began his useful professional life at Eminence, forming a partnership, which continued five years, with Dr. D. N.
Porter, a beloved physician, learned and liberal, who was elected,
in 1868, President of the Kentucky State Medical Society.
Doctor Todd from the start enjoyed the fullest confidence—which
he retained until his life's close—of the entire community, as
well as several adjacent counties, and did an extensive and
lucrative practice, was one of the most skillful and successful
surgeons in his portion of the State, performing many of the
most delicate and difficult operations of that neighborhood. He
was a man of fine personal appearance, of great physical endur-

ance; Doctor Todd was a most generous and noble man, charitable always; of a bright, cheery, and happy disposition; a very
help and comfort in the sick-chamber, a typical friend in need,
generous, loving, and kind; as a man and physician always
charitable to the stranger within his gate.	He had been a mem-
ber of the Kentucky State Medical Society for more than twenty
years, and seldom missed a meeting, had contributed valuable
papers, and had once been first vice-president.	It has been truly
said that at the annual meetings he will be sorely missed, for he
was so sincerely loved and so much deserved that love.	No
banquet gathering could be complete without *our* "Eminence
Doctor Todd," for he was always down on the program to
answer to the toast—"The Ladies," when his genuine love for
fun and wit was overflowing with mirth and pleasantry which
was the delight of all.	No kindlier nor warmer heart ever beat
in the breast of any man.

As during a man's life the most esteemed and valuable of his
possessions is character, thus after he is dead nothing is left of
him that is enduring but character.	And as the just estimate of a
man's character can only be formed by the community in which
he lived—and that is an absolutely impartial tribunal—from
which no appeal is possible, the writer—a devoted friend of
many years of Doctor Todd—gladly introduces here a beautiful
editorial from his town newspaper and other journals of the
State, and eloquent words spoken at his burial, which he believes
to be true portrayal of the high character and usefulness of
Doctor Todd, and the honor done him and the great respect mani-
fested were such as seldom, if ever, have fallen to the lot of any
other physician in Kentucky.

DEATH OF DR. O. D. TODD.

[From The Constitutionalist, Eminence, Ky., Thursday, May 7, 1896.]

Dr. O. D. Todd died at the Norton Infirmary, in Louisville, Monday, at one
o'clock P. M.	The fact of his death was immediately telegraphed to Eminence
and spread all over Henry County before sundown.

Dr. Todd was taken ill about six weeks ago.	For two or three weeks
it was supposed that his illness was not of a serious nature, but about three
weeks ago he was forced to take his bed, and his condition grew gradu-
ally worse.	Saturday, a week ago, he was taken to the Norton Infirmary, where
it was hoped that trained nurses, skillful physicians, and the most modern

medical appliances would all assist in his recovery. However, his condition grew worse, and for several days preceding his death that event was hourly looked for. All that human skill could do was done for him, and he succumbed to uremic poisoning.

Dr. H. H. Grant, Dr. J. A. Larrabee, Dr. J. A. Ouchterlony, Dr. J. M. Mathews, Dick D. Smith, L. B. Helburn, and many other Louisville friends were with Dr. Todd constantly, as well as the members of his immediate family. Every attention that could possibly be given one was his, and he died having before him a living proof of the love and fidelity of both family and friends. The remains were brought to Eminence Tuesday morning on the 9:26 train.

From the hour of six o'clock in the morning, when the remains were taken from Norton Infirmary, in Louisville, the church bells and that of the fire department began to toll and continued to toll until six o'clock in the evening, when the last sad rites had been paid at the grave. In compliance with the request of the Mayor, made known by printed proclamation, all business was suspended, and every business house in Eminence was closed at three o'clock P. M., and remained closed until after six o'clock. In addition, Circuit Court being in session at New Castle, Judge Carroll adjourned court to accommodate members of the bar and jurors who desired to attend. Crowds came from Louisville, Frankfort, and Shelbyville and Shelby County.

The funeral services were conducted by Rev. J. B. Andrews, at the Presbyterian Church, at four o'clock Tuesday afternoon, and the interment was at Eminence Cemetery. There were probably twelve hundred people present at the funeral and burial services. Rev. Wm. Irvine, of Anchorage, paid a splendid tribute to the deceased. After the funeral, the local lodge of Odd Fellows, assisted by the lodges from New Castle, Shelbyville, and other places, conducted the burial. The town was filled with sorrowing, heart-broken people, and the occasion, as a whole, was one to be remembered for many years.

The story of his illness and death is briefly told, and the details are known generally to the public. But the story of his life of devotion to his profession, his extreme conscientiousness in the performance of the duties of that profession, and the daily charities that were welcomed in his busy, self-sacrificing life, no pen can accurately write.

Born within five miles of this town, reared among our own people, educated in medicine at Jefferson Medical College at Philadelphia, served as a surgeon in the army during the last years of the war, settled down at Eminence, and acquired a large and lucrative practice, and died holding, as he had always held, the fullest love, esteem, and confidence of all our people, is a brief synopsis of his career.

As a physician he ranked high among the physcians of Kentucky. He was probably the most popular physician, personally, in the State Medical Society, and the annual reunions were full of mirth and pleasantry to all.

Of Dr. Todd every one was fond. He was a most noble, generous soul, generous when it was to his interest to be otherwise; charitable when the burden would have broken another man's shoulders; full of the sweet breath of life and buoyant in the very nature of things. He was the happy, bright, sunny-natured physician at the bedside of the sick; he pressed the hands of the dying and made comfortable their last hours, while his great heart broke in sympathy and love; he was the typical friend in need, generous and loving to his friend and brother, kind and attentive to the stranger within his gate; and at all times he was just what one would love to have his best, his warmest friend to be.

Full of love and its attendant loves, genuine in the unbroken ties that bound him to his people, a lover of fun, and a great noble man with the heart of a great noble boy—that was Dr. Todd. There has been no public gathering where Dr. Todd was not a welcome visitor. There was no private entertainment where his smiling face did not beam a welcome and a benediction. How sadly he will be missed, let the good people of Eminence and all this section of country say. There was no man among us so sincerely loved, and let us say it now, truthfully, there was no man who so much deserved that love.

Years may come and go, and the great scroll of time may be rolled up and passed into eternity, yet will Dr. O. D. Todd's memory, sweet as the breath of violets by the brookside, "dear as remembered kisses after death," be held in the sacred heart of Love, and live forever. When our people shall gather about their firesides, when the laughter and prattle of the babe shall be a benison to the mother's heart, when rest shall come to relieve the cares which infest the day, and when heart-strings are strained to their utmost tension, his benign spirit shall come to bless and soothe.

He was one of God's noblemen. He has mourned with us, he has tasted of our joys and sorrows—he has been one of us. The great family circle of friends is broken, but he has simply gone before.

"Age can not wither, nor custom stale
His infinite variety."

Nor can time efface the memory of his noble character, his kindly deeds, and his cheerful soul. There is no night of death, but indeed have the stars gone down on a brighter world than this, and yet there is a spirit in that great land beyond the tick of time whose nobility and splendor is only heightened by his temporal life here.

The active pall-bearers were: W. T. Hanks, J. B. Moore, G. W. Young, I. B. Helburn, L. R. Stark, Birney Hunt.

Honorary pall-bearers: W. L. Crabb, W. S. Wilson, W. B. Crabb, J. S. Fuqua, R. T. Jenkins, D. A. Sachs, N. W. Gould, W. O. Moody, A. D. Hudson, W. P. Thorne, Dr. W. A. Jameson, F. M. Karr, J. S. Turner, R. W. Moody, W. L. Hudson, W. A. Holland, J. A. Logan, J. A. Harris, W. T. Williams.

Mayor Turner issued the following proclamation:

"NOTICE.—Owing to the death and funeral of Dr. O. D. Todd, you are requested to close your place of business from the hours of 3 o'clock to 6 o'clock this afternoon. •   J. S. TURNER, Mayor.

"Eminence, Ky., May 5, 1896."

Mr. Howard H. Gratz, editor of the *Kentucky Gazette* at Lexington, on seeing notice of Doctor Todd's death, wrote the following beautiful tribute, which is of especial and impressive interest from the fact that he had seen Doctor Todd only on one occasion, but had remembered for many years his bright intelligence, his loyalty to the dignity and honor of his profession, and his devotion to the interest of suffering humanity, which he here so well expresses:

Dr. Orrin D. Todd, of Eminence, died at the Norton Infirmary, Louisville, Monday afternoon, of a complication of diseases, which began with pneumonia. His remains were taken to Eminence for interment.

Dr. Todd was an eminent physician and a noble, generous-hearted man, beloved by all who knew him.

Some years ago Col. J. Stoddard Johnston, then of Frankfort, was sitting in our office when a gentleman passed by, when Col. Johnston called him in and introduced him to us as " Dr. Todd, of Eminence." Col. Johnston remarked that " Dr. Todd had made a valuable discovery in the line of his profession, but so strong was his feeling in regard to medical ethics that he would not patent it and make a fortune out of it." Dr. Todd remarked " that he considered his discovery very valuable, but he preferred to let the profession have it to making any amount of money out of it," and he proceeded to tell us about it. He said a few months before that day he received an urgent call to the house of a friend, and when he arrived he found his little son of about eight years of age lying in a comatose state, but not in any pain. His stomach was distended, and all he could do was to administer an emetic, which in a short time brought up considerable food and a quantity of cedar berries, and the child soon revived. He left an active cathartic to be administered at bedtime. The next morning a number of tape-worms passed from the child and he was well. Late that afternoon he was called to see his sister, a year or two younger than her brother. She was suffering exactly as her brother had, and her mother said that the little girl had told her in the afternoon that "she was going to eat some cedar berries and have tapes like her brother." She was treated in the same way with like results. There was a negro man in Eminence who all the doctors believed was suffering from tape-worms, and none of them had been able to relieve him. He tried cedar berries on him, and to his astonishment it cured the old darky, who got entirely well. He said, furthermore, he had relieved at least thirty persons in Shelby County by the same treatment, and he wanted the profession to have the valuable discovery, as he believed it to be. This circumstance shows the high character of Dr. Todd, and it is no wonder that his neighbors and friends so sincerely lament his untimely death.

The clinical report of the foregoing identical cases our friend, Dr. L. B. Todd, heard Dr. O. D. Todd make at a meeting of the Kentucky State Medical Society, which with other valuable papers he contributed, caused the State Society to appreciate and lament the death of this eminent physician.

The *Louisville Times* contained the following:

No matter how hardened or resigned we become to the passing of all that is dear and noble in life, there are some generous natures who so endear themselves to us that we can not realize that we must lose them. Such a one was Dr. O. D. Todd, of Eminence, who has just died, and for whom all who knew and loved him wear crepe upon the heart to-day. It would hardly have been possible to find a man who enjoyed life more, and who made it sweeter and brighter for all around him.   He was so genial, so kind of heart and open of hand, that when he came into the sick-room it was like a breath of fresh air, a glimpse of the sun and sky to the weary, despondent invalid.

He was everybody's friend and nobody's enemy.   Generous, chivalrous, and kind, his stainless life was an example to all who knew him, and he was a worthy follower of the Great Physician who has called him from the scene of his labors.

RESOLUTIONS OF RESPECT.

To the Officers and Members of Eminence Lodge No. 140, *Independent Order of Odd Fellows:*

Your committee submit the following resolution on the death of our beloved brother, Past Grand Dr. O. D. Todd, who died May 4, 1896:

He was a true Odd Fellow; his hand and heart were ever ready to respond to the calls of the needy and distressed.   He was honored, respected, and loved by his brethren for his many kind deeds.   As a citizen he was loved by all, honest, kind and just, possessing all those traits of character that make a true man.   As a physician he was widely known for his extensive and successful practice and his great and generous charity.   His memory will live long in the hearts of his brothers and friends.

Life's duty done, he was gently laid away, there to sweetly repose until the call of the Grand Master above.   Peacefully rest, our brother, and may the leaves fall lightly where he sleeps   To his relatives and friends this lodge extends the heartfelt sympathy of every member.   Your committee recommends that a copy of this report be spread upon the records of the lodge and be published in the Eminence *Constitutionalist.*

(Signed)   W. G. SIMPSON, *Chairman,*
W. C. BROWN.
I. B. HELBURN.

ADDRESS AT THE FUNERAL OF DR. O. D. TODD BY HIS FRIEND, REV. WILLIAM IRVINE, OF ANCHORAGE, KY.

As my eye rests upon this large audience of sad and solemn faces gathered from this and other communities to pay respect to the memory of our dead friend, I am most deeply impressed with the irreparable loss we have sustained. I see before me the representatives of all classes and conditions, the old and the young, the rich and the poor, all impelled by a common grief to mingle their tears over the prostrate form of one whom they honored as a physician and loved as a friend.

You have all done well to give this expression of appreciation for a man who ranked among the first in his profession, and who by his eminent abilities and sincere devotion to duty and his truly philanthropic spirit dignified a calling that deserves the highest praise, as it demands for its fulfillment all that is noble and true and gentle and pure in man.

·Dr. Todd was a physician of great skill, and was so recognized among you all.   But the one characteristic that endeared him to you all more perhaps than any other was his intense loyalty to his friends.   And with his great versatility of talents and irrepressible flow of spirits, he touched more people in the varied relations in life than any citizen in our community. His life was in a marked degree an illustration of the proverb of Solomon, "If a man would have friends, he must show himself friendly." Every impulse of his generous nature was to minister to others. He extracted but never intentionally planted a thorn in the breast of any man. He was so thoroughly kind and sympathetic that a great multitude were drawn to him. He made no provisions for enemies and consequently he had none. In this most commendable virtue he has left us an example that we would do well to follow. Time is by no means lost that is devoted to the cultivation and developing of friendships that bring so much of beauty and blessing into our lives.

Believing as we do in the immortality of our spiritual natures with their vast capacities for knowing and loving, may we not hope for a renewal of such friendships? Fidelity to truth and to the memory of my dead friend compels me to say that the greatest mistake of his life was, that for so many years he should have forgotten that "there is a Friend that sticketh closer than a brother." I am glad to say no one was more sensible of this fact at the last. He recognized the need of One who could sympathize and succor when earthly friends could only clasp hands and say " Good-bye."

More than once he expressed the purpose of adding this Friend to his long list of friends, and of making a public avowal of the Saviour. He said, before going to Louisville, "I do not wish to make a secret but an open confession of Christ before the people among whom I have lived, and this I will do as soon as I can go to the church."

As his friend and your friend, I beg you to do at once what he intended to do at the last: Make Christ your friend.

With Our Exchanges.

DISEASES OF THE RECTUM.

BROWN, WARREN, TACOMA, WASH.: PEROXIDE OF HYDRO-
GEN. (*The Medical Sentinel.*)

This is usually made by the decomposition of hydrated peroxide of barium by sulphuric acid. It is employed in the arts
for bleaching. The usual commercial article yields about ten
volumes of oxygen.

Dr. Benjamin Ward Richardson, the famous London physician, who in 1893 received knighthood from Queen Victoria,
first experimented with peroxide of hydrogen in 1857. It was
regarded then as a curiosity, and was soon forgotten. Thirty
years later, Dr. Squibb, of Brooklyn, brought it prominently
before the profession, and since that time it has been used more
and more each year, until its consumption has reached enormous
proportions.

In order to preserve hydrogen peroxide it must be slightly
acid; on this account, a disagreeable irritation and smarting may
be caused by its use on mucous membranes. This can be avoided
by mixing it fresh at the time it is to be used with equal parts of
lime water, or spraying with lime water first.

It effervesces not only with pus, but with blood, serum, mucus,
and cerumen. It is one of our best antiseptics, and it is of the
greatest value in removing septic clots and enveloping fluids
before making application of other drugs.

In appendicitis, the abscess cavity is cleansed with this solution by many operators in preference to any other antiseptic.
Dr. Robert T. Morris, of New York, has laid special stress on the
value of the peroxide in these cases.

Eczema of the anus will rapidly improve if the fissures are
touched twice a day with this solution, then dried gently with
cotton, and a glycerite of lead application made. In nearly
every form of acute eczema in the first and second stages the
peroxide will give us the keenest satisfaction. The regular solution is diluted with two or more parts of water. Hydrogen peroxide is an excellent anti-pruritic, and for this purpose it is widely
used. Of the different preparations of peroxide of hydrogen
Marchand's has been most uniformly satisfactory.

GASTRO-INTESTINAL DISEASE.

MARTINS, ROSTOCK : THE INDICATIONS FOR THE TUBE IN DISEASES OF THE STOMACH. (*Therapeutische Monat. Hefte; Occidental Medical Times.*)

At the recent congress of German physicians and naturalists Martins, of Rostock, specified the uses of the stomach tube in general, viz., to relieve the stomach of its contents, and to secure the action of remedies on the mucous membrane of the stomach. The special indications for the use of the tube were absolute and relative. To the first belong active treatment of diseases of the stomach; to the latter, treatment as determined by the individual experience of the physician. The absolute indications are : (1) Acute poisoning. (2) Stagnation of the gastric contents dependent on stenosis of the pylorus. (3) In all cases of intestinal obstruction from whatever cause. The relative indications are : (1) Carcinoma ventriculi in general. (2) Acute and chronic catarrh of the stomach. (3) Functional affections of the stomach like hypersecretion of the gastric juice, continuous secretion of the gastric juice, etc. The value of stomach lavage is still questionable in neurasthenia gastrica. The writer has seen the suggestive effects of the treatment in such cases. The stomach tube is only one means in the treatment of nervous dyspepsia, and in the experience of the author the most active one. When patients suffering from nervous dyspepsia can be furnished with absolute proof that the food ingested has been digested, it is an important step toward the consummation of a cure. In the discussion Minkowski, of Strasburg, suggested the use of the tube in hematemesis, as it is often the only means of arresting the hemorrhage. He saw no contra-indication for the use of the tube in ulcus ventriculi provided a soft tube is employed. It is important in counting on therapeutic results to determine the time for ridding the stomach of its contents. If, *e. g.*, the stomach is washed in the morning and immediately thereafter food is ingested the latter ferments, and the essential object of lavage proves illusory. Therefore, it is of decided advantage to wash the stomach late at night, thereby giving the organ an opportunity to rest during the night and to regulate its mechanical insufficiency. Quincke, of Kiel, washes the stomach at night three hours after the last repast. In place of the funnel he uses an irrigator, which estimates accurately the in- and outflow of the water. Lenhartz, of Leipzic, referred to the value of lavage in tympanites of the intestines as occurs in typhoid fever, esophago-spasm, and nervous dyspepsia. In the latter affection the results are surprisingly rapid. Leo, of Bonn, suggests the great remedial value of lavage of the stomach during infantile life in acute and chronic affections of the stomach and intestines. Rumpf, of Hamburg, recommended the use of the tube in acute cases of emetocatharsis. It was necessary to individualize the nervous dyspepsia.

KIRBY, N. H., CONCORD, MASS.: A CASE OF ULCER OF THE STOMACH TREATED WITH PROTONUCLEIN. (*The Atlantic Medical Weekly.*)

My experience with protonuclein has been limited, but in all the cases in which I have used it I have been quite successful, especially so in the one which is reported below.

Mr. J., aged twenty-four, farmer by occupation, became one of my patients some two years ago, when he presented the following symptoms: For a long time he had been troubled with inability to retain food. There would be severe pain in the stomach, which would be increased by the presence of food and by pressure over region of stomach. This pain would be relieved by vomiting; sometimes there would be vomiting of blood; bowels constipated, tongue covered with thick coating. I tried several remedies at that time, such as pepsin, bismuth, nitrate of silver, aromatic powder, Carlsbad salts, mustard over epigastrium, etc., together with a strict diet. This course of treatment was followed by temporary improvement, but no real improvement. He finally left me utterly discouraged and came under the care of other physicians, but with apparently no better success. About two months ago he returned to me, very much reduced in flesh, his weight having dropped from 160, his normal weight at the beginning of the trouble, to 124 pounds, and every symptom increased in its severity, and so weak and exhausted was he that he was unable to follow his usual occupation.

I straightway put him upon a strict diet again, and gave him protonuclein, grs. iii, every four hours, to be taken religiously: From the beginning of the administration of protonuclein he began to improve, and gradually to retain food. The pain began to diminish, and he gained fully ten pounds in weight within the first two weeks. His appetite and strength returned, and in fact there was rapid and permanent improvement. At the present time of writing he has ceased taking protonuclein, and when I last saw him he was apparently well.

Book Reviews.

Diagnosis and Treatment of Diseases of the Rectum, Anus, and Contiguous Textures. By S. G. GANT, M. D., Professor of Diseases of the Rectum and Anus, University and Women's Medical Colleges; Lecturer on Intestinal Diseases in the Scarritt Training School for Nurses; Rectal and Anal Surgeon to All Saints, German, and Scarritt's Hospital for Women; and Kansas City, Fort Scott, and Memphis Railroad's Hospitals, to East-side Free Dispensary, and to Children's and Orphans' Home, Kansas City; Member American Medical Association, National Association of Railroad Surgeons, Mississippi Valley Medical, the Missouri Valley Medical, and the Kansas State Medical Associations; of the Kansas City Academy of Medicine, Jackson County, and Kansas City District Medical Societies, etc., etc. With two chapters on "Cancer" and "Colotomy" by HERBERT WILLIAM ALLINGHAM, F. R. C. S., England, Surgeon to the Great Northern Hospital, Assistant Surgeon to St. Mark's Hospital for Diseases of the Rectum, Surgical Tutor to St. George's Hospital, etc., London. Illustrated with sixteen full-page chromo-lithographic plates and one hundred and fifteen wood engravings in the text. Philadelphia: The F. A. Davis Company, Publishers. 1896.

This work has been read with a great deal of pleasure. The author is eminently capable of writing a good book on these subjects, and he has done it. Having devoted a number of years to the practice of rectal diseases, he is enabled to draw from actual practice his observations and deductions, which makes his work more authoritative. Besides the subjects usually treated in works of this kind, Dr. Gant has added a chapter on " Railroading as an Etiological Factor in Rectal Diseases," and one on "Auto-Infection from the Intestinal Canal," both of which can be read with profit. Perhaps the most interesting chapter written by Dr. Gant is that devoted to Fistula. His twelve cautious directions in operating (see p. 991) are admirable and should be carefully observed always. Though mentioning almost all other methods of operating, he gives precedence to the knife, and backs it with the best of reasons. The after-treatment, which, as he says, "is of almost as much importance as the operation itself," is carefully and concisely given. In speaking on the subject " Relation of Pulmonary Tuberculosis to Fistula," the author says "the operation differs slightly from that for ordinary fistula in that it should be performed quickly and in such a way as to have little bleeding." It is well known that wounds in the tuberculous patient bleed freely. Indeed it is one of the diagnostic signs of tuberculosis. Instead, too, of operating hurriedly in this class of patients, they above all others should secure most careful and perfect consideration. Curettement, trimming, etc., should be most thoroughly done, and it takes time to do it, for the reason that all infected tissue should be removed. Again, chloroform, not ether, should be given these patients to avert the danger to the kidneys that the author speaks of.

In speaking of the plastic operation upon the sphincter muscle in incontinence of feces the author would convey the idea of perfect success in the majority of cases when the surgeons who have operated most in this line report bad results.

It is to be regretted that the author has not written more exhaustively of the etiology of *ulceration*. He has made it suffice to classify without going very fully into the pathology. It is of the greatest importance to understand the full nature of an ulceration of the rectum in order to treat it rationally. Then, too, benign ulceration is given as being of frequent occurrence, when contrasted with more serious causes it is insignificant. It is a pleasure to note in this connection that in the chapter on Stricture, when dealing with syphilis as a cause, the following language is used: "Syphilis in the form of gummatous deposits within the rectal walls or from ulceration *heads* the list of causes of stricture of the rectum." All careful and painstaking observers will agree with this statement. Much has been written in denial of this fact, and the author takes his position after an observation of many cases in private and hospital practice. In the chapter on "Benign Stricture," the following words are used: "When the stricture is so high that it can not be felt by the finger, we have to resort to the use of some one of the many kinds of rectal bougies. . . . For purposes of examination the conical or olive-shaped tips fastened to a flexible piece of whalebone are the best." In the first place benign strictures are seldom, *very* seldom, found beyond the reach of the finger, and if they *should* exist, it would be dangerous to explore them with a hard sound of this kind. Students at least should be warned of such danger. In the chapter on Internal Hemorrhoids, it is to be regretted that so much valuable space is taken up describing the so-called "injection plan." Such treatment is discarded by surgeons, and is not only painful and inefficient, but absolutely dangerous.

The author condemns Whitehead's operation for piles, and gives preference to the ligature and clamp and cautery. His own clamp is illustrated, and a number of reasons given why it should be preferred to others.

Chapter XXVI is written by Mr. Herbert Allingham, and is devoted to Cancer of the Rectum, its Etiology, Symptoms, and Treatment. It is most admirably written and deserves careful consideration. As a recapitulation of the subject it will be seen that the author is not an advocate of excision of the rectum for malignant disease. In this day of " rapid " surgery it is refreshing to read : " It is perfectly useless to excise the rectum when the growth is high up in the bowel or at all fixed to the surrounding deeper parts." This is so in keeping with the views so often expressed by the writer that he is glad to copy this expression from one so competent to speak as Mr. Allingham. Chapter XXVII is also written by Mr. Allingham, and is on Colotomy. As we are all so familiar with his views on this subject, it will be sufficient to say that no one has ever written more intelligently of this subject. Whatever may be our objections to some of his premises, certain it is that he makes good defense of them. We have picked out only a few passages for criticism in this book because there are but few to deal with. Every one interested in rectal diseases should buy the work, and they will be fully repaid. The publishers are to be congratulated upon the splendid way in which the book is issued. Typographically it is a model, and the illustrations are first-class.   J. M. M.

The Annual Report of the Supervising Surgeon-General of the Marine Hospital Service for the Years 1894 and 1895. Washington: Government Printing Office. 1895 and 1896

Unfortunately it is not the privilege of many outside of the Service to see such excellent reports of the work done in the Service under Surgeon-General Wyman, and we count it a privilege to have been the recipients of the reports for the past two years.

The report of the Surgeon-General to the Secretary of State contains the reports from the various stations under his control, including photographs of many of the stations. Then follows the financial report and a full report of the fatal cases with necropsies, which is most interesting and instructive reading. Among these is the record of two fatal cases of chloroformization. Then follows a number of interesting papers from the pen of the members of the Marine Hospital Staff which contain many instructive and interesting points, not the least of which is the recommendation of the Surgeon-General that a hospital for consumptives be established by the government "on government reservation in that portion of the United States which combines the proper latitude, altitude, and freedom from moisture," and suggestions as to how this can be brought about. We trust the authorities will heed the suggestions made.

The report of the Hygienic Laboratory of the Marine Hospital Service is very interesting indeed, containing the report of J. J. Kinyoun, P. A. S., M. H. S., detailed to investigate the serum therapy of diphtheria, and the report of the establishing of a plant for the manufacture of the serum by the Marine Hospital authorities.

Dr. Wyman is to be congratulated upon the thoroughness of his work, much of which is so tedious in its detail.

Functional and Organic Diseases of the Stomach. By SIDNEY MARTIN, M. D., F. R. S., F. R. C. P., Assistant Physician and Assistant Professor of Clinical Medicine at University College Hospital; Assistant Physician to the Hospital for Consumptives and Diseases of the Chest, Brompton. With fifty-seven illustrations. Edinburgh and London: Young J. Pentland. Philadelphia: J. B. Lippincott Co. 1895.

This is an extremely attractive book in its make-up with uncut pages,—rather a disadvantage to American readers no doubt,— and gilt edge. There have been no exhaustive treatises upon the stomach published in the last few years, and we consider this one by Dr. Martin the best work in English. It is thorough, concise, and well written, and the author has shown himself to be a master of the subject. As was said of an American book on Practice of Medicine in these columns recently, the more a person reads the more fascinated he becomes, so it is with this book of Dr. Martin's.

The author has followed a very wise course in choosing the order in which the subjects are discussed, first taking up the anatomy of the organ and the normal processes occurring in it; the question of food, its digestibility, and the effects of excess or diminution. Then follows a description of the pathological diseases, and the following classification is made: *Gastric Irritation*, including the simple irritations and most of the forms of nervous dyspepsia or neuroses of the stomach. This is a decided advance, for the old terms of "flatulent and acid dyspepsia, etc.," were vague and indefinite. *Gastric Insufficiency* includes the rest of the forms of functional dyspepsia.

The chapter upon the chemical examination of the stomach contents as a means of diagnosis is very full and up to date.

A feature, however, which is very annoying is the reference in the text to a number of pages in the front or back of the book to illustrate some point, and it is troublesome to have constantly to hunt these up. It would be much better if a foot-note was employed or a brief resumé repeated in the text, and certainly much more convenient to the reader.

There are few criticisms which could be possibly made of the book before us, but there is much to commend; notably is this so of the chapter upon ulcer of the stomach. It is full yet concise; the illustrations are excellent, which, moreover, is true of the others through the book, and it is altogether a very exhaustive treatise.

No library of an up-to-date practitioner will be complete without Dr. Martin's most admirable work, the most thorough English publication upon diseases of the stomach extant.

System of Surgery. Edited by FREDERIC S. DENNIS, M. D., Professor of the Principles and Practice of Surgery, Bellevue Hospital Medical College, Visiting Surgeon to the Bellevue and St. Vincent Hospitals; Consulting Surgeon to the Harlem Hospital and the Montefiore Home, New York; Ex-President of the American Surgical Association; Graduate of the Royal College of Surgeons, London; Member of the German Congress of Surgeons, Berlin. Assisted by JOHN S. BILLINGS, M. D., LL.D., Edinburgh and Harvard, D. C. L., Oxon., Deputy Surgeon-General U. S. A. Volume III. Surgery of the Larynx, Tongue, Teeth, Salivary Glands, Neck and Chest; Diseases and Surgery of the Eye and Ear; Surgical Diseases of the Skin; Surgery of the Genito-Urinary System; Syphilis. Profusely illustrated. Philadelphia: Lea Brothers & Co., Publishers. 1895.

This is the third volume of one of the best works on surgery ever published. As the index indicates, the Surgery of the Larynx and Trachea is given by D. Bryson Delavan, M. D., of

New York; Surgery of the Mouth and Tongue, by Henry H. Mudd, M. D., of St. Louis; Diseases of the Salivary Glands, by Charles B. Porter, M. D., Boston; Surgery of the Neck, by Willard Parker, M. D., New York; Surgical Injuries and Diseases of the Chest, by Frederic S. Dennis, M. D., New York; Diseases of the Eye, by George E. de Schweinitz, M. D., Philadelphia; Operative Surgery of the Eye, by Henry D. Noyes, M. D., New York; Surgery of the Ear, by Gorham Bacon, M. D., New York; Surgical Diseases of the Jaw and Teeth, by Louis McLane Tiffany, M. D., Baltimore; Surgical Diseases of the Skin, by Wm. A. Hardaway, M. D., St. Louis; Surgery of the Genito-Urinary System, by J. William White, M. D., Philadelphia, and Syphilis, by Robert W. Taylor, M. D., New York. Each of these gentlemen is recognized as a master in the department which he treats, and with the supervision of the entire work by Dr. Dennis one can at once understand that this "System of Surgery" can not be surpassed. The work, in a general way, has been reviewed in these columns before, but it is a pleasure to call attention to each succeeding volume. This great work should be in the hands of every physician and surgeon who desires to keep abreast with the latest advances, either in general surgery or the special departments.

The Stomach, its Disorders and How to Cure Them. By J. H. KELLOGG, M. D., Battle Creek, Mich. Illustrated. Modern Medicine Publishing Co., Battle Creek, Mich. 1896.

Dr. Kellogg has written a most readable book, but at the outset we must disagree with the author as to its place in the hands of patients who are afflicted with disorders of the stomach, to whom the book has been dedicated. It is a well-known fact that patients afflicted with disease of the stomach, trivial or severe, are more prone than others to melancholia, etc., imagining all sorts of dire things which may be the matter with them. With a book such as Dr. Kellogg has written for their benefit (?) one can not predict to what extent their imaginings will carry them. In the preface it is stated that "it is not intended that the book is to take the place of the family physician, but rather to aid and abet him in the class of maladies which requires, perhaps more than all others, the thorough co-operation of the patient." But

with the extensive knowledge that can be gleaned from the book
before us the patient will be in the position to dictate to the
physician not only his diagnosis of the case but the treatment
which he has mapped out from his hand-book, and unless concurred in the physician is denounced. If the scope of the work
had been limited to the description of the organs of digestion,
which, however, could have been better done without the many-
colored plate used as a frontispiece, intended to depict these organs,
the classification of foods, the proper modes, etc., of eating,
without going so elaborately into the treatment and the analysis
of the stomach contents, it would have had a place at once, and
could have been recommended by physicians to their patients.

One can hardly agree with all the statements which have been
made in regard to the evils resulting from eating before sleep,
and the nap after eating, or the good to be derived from horse-
back riding and other exercise after eating to promote digestion.

Too much can not be said in commendation of what has been
written about the evils being done by the "orificial surgeons" on
page 282, and it would have been better if a special chapter had
been devoted to that, with special chapter heading.

With the exception of several instances of poor proof-reading
where letters from the wrong font have not been detected, the
typographical work is excellent, but the illustrations for the most
part are crude to an extreme, and convey little to one's imagination.

Infantile Mortality During Childbirth, and Its Prevention. By A.
BROTHERS, B.S., M.D., New York City. Philadelphia: P. Blakiston, Son
& Co. 1896. Price, $1.50.

This is the essay which received this year the William Jenks
Prize offered annually by the College of Physicians of Philadelphia. It is a most excellent work and well deserving of the
honor bestowed upon the author and the essay. It is a thorough review of the subject, and the bibliography given at the
close of each article or subdivision is the most complete we have.
The recital of the findings in forty-seven *post-mortem* examinations made upon newly-born children is most interesting, and the
author has called attention to the neglect of the officials of institutions where these autopsies could be held with the most fre-

quency, in that they do not do their duty at all, a very much neglected field. Many interesting facts would be learned no doubt if more systematic autopsies would be held on all dying at lying-in and foundling institutions. While the book does not contain a great deal of original work, it means a large amount of painstaking work, and Dr. Brothers has given us not only a very readable book but a most thorough and interesting one.

Diets for Infants and Children in Health and Disease. By LOUIS STARR, M. D., Editor of the American Text-book of Diseases of Children. Philadelphia: W. B. Saunders. 1896. Price, $1.25 net.

So little importance is paid as a rule by practitioners to the rigid direction of the feeding of infants in health, not to mention disease, that the little book before us is most timely. The average American needs labor-saving devices, and in this list he will find little to do but fill out the name of the patient, name the amounts of the mixture he wishes to order, and write the directions, all in spaces provided, and the prescription is ready.

The prescriptions provided are for the periods of infancy between birth and seven months, eighth and ninth months, tenth, eleventh, and twelfth months, twelfth to the eighteenth month, eighteenth month to the second and one half year, and the diet of childhood, also diet for various gastro-intestinal diseases.

It is admirably arranged, and where one has not the advantage of the modified milk laboratories and has to rely upon the home modification, these diet lists are indispensable.

Borderland Studies. By GEORGE M. GOULD. A. M., M. D., formerly Editor of the *Medical News.* Philadelphia: P. Blakiston, Son & Co. 1896.

No doctor in America needs an introduction to the facile writings of Dr. Gould, and every one will be glad to learn that in one volume can be found so much that is of interest which has been published in various magazines and periodicals. "Borderland Studies" contains miscellaneous addresses and essays pertaining to medicine and the medical profession, and their relation to general science and thought. To these, as stated by the author in his preface, he has added five not heretofore published, and a number of editorial articles from the *Medical News.* It was with sincere regret that the profession which had identified

Dr. Gould for so long with the *Medical News* learned that he was to give up the editorial management of the *News*, and the book which is before us can not help but have a wide sale, as so much of what he wrote can be found in so small a space.

A Manual of Anatomy. By IRVING S. HAYNES, Ph. B., M. D., New York City. Philadelphia, Pa.: W. B. Saunders. 1896.

This little book is worth many times more than asked for it on account of the many excellent illustrations which abound in it. A method has been pursued not common hitherto in anatomies, viz., the photographing of the various regions and organs from original dissections. These photographs are for the most part reproduced in excellent halftones, and without exception they are admirable pieces of work. No waste of space occurs in the text, in fact the author has erred too much in many places on the side of brevity, the description of the bones and joints being intentionally omitted by the author. As a reference book when a more detailed description of the subject is not wanted we have not seen its equal, and the publishers are to be congratulated on the excellence of their work in reproducing the photographs.

Electricity in Electro-therapeutics. By EDWIN J. HOUSTON, Ph. D. and A. E. KENNELLY, Sc. D. New York: The W. J. Johnston Co. 1896.

This little book is a treatise on the physics of electricity as applied to electro-therapeutics, and is a thorough exposition of that part of the subject. The authors are well qualified to write on electrical subjects, Prof. Houston being president of the American Institute of Electrical Engineers, being a lecturer also on electro-therapeutics in medical schools. Mr. Kennelly is vice-president of the same association mentioned above, and was for many years the principal assistant of Thomas A. Edison. The elements of electricity must be mastered before one can apply it to medicine, and in this book these can be very readily learned, so simply are they treated.

Don'ts for Consumptives, or the Scientific Management of Pulmonary Tuberculosis. By CHARLES WILSON INGRAHAM, M. D., Binghamton, New York. Binghamton: The Call. 1896.

This is an excellent little work and will have a wide sale, for something of its kind is needed to be studied by the tubercular patient in order that he may do the least harm to his fellow-man.

Until cognizance is taken of the spread of tuberculosis and its prevention by State boards and city health authorities, the consumptive must be educated himself as to the true nature of the disease which he has and the precautions necessary to take to prevent its spread by the individual. In the book before us Dr. Ingraham has handled the subject very carefully in a way that will offend the patient the very least that is possible.

A Compend of Diseases of Children. Especially adapted to the Use of Medical Students. Second edition, thoroughly revised, with a colored plate. By MARCUS P. HATFIELD, A. M., M. D., Chicago, Ill. Philadelphia: P. Blakiston, Son & Co. 1896. Price, 80 cents.

For so condensed a work on Diseases of Children this compend is one of the best we have seen. The use of books of this character should be discouraged in medical schools, because it teaches the student to be content with a superficial knowledge of the subject. However, they have their place, and it gives us pleasure to commend this one by Dr. Hatfield. He has covered the field wonderfully well in his limited space, this being especially true of the anatomy and physiology of childhood and the diseases of innutrition.

Obstetric Accidents, Emergencies, and Operations. By L. CH. BOISIL-INIERE, A. M., M. D., LL. D., late Emeritus Professor of Obstetrics in the St. Louis Medical College, St. Louis, Mo. Profusely illustrated. Philadelphia: W. B. Saunders. 1896.

This is one of the latest aspirants for honors in the line of hand-books and compends. It is a thorough review of the subject, including Accidents to the Mother; Obstetric Operations; Accidents to the Child. A number of cases occurring in the practice of the author are recited and add interest to the text; the illustrations are taken for the most part from Dickinson and Winkel, but are well chosen, and the subject is covered as well as is possible in so limited space. The work of the publishers has been well done, the binding being much superior to that usually seen in books of this kind.

Books and Pamphlets Received.

EVOLUTION OF THE MICROSCOPE.—The microscope has been very slowly evolved and is the creation of no one man. In its present form it is, like a living species according to Darwin, the outcome of the survival of the fittest of innumerable variations, the majority of which have been discarded. Indeed, to one interested in microscopes and familiar with the present model, nothing can seem quainter than the old forms, which prevailed during the earlier half of this century and have since become extinct. In the evolution of the microscope two factors have been dominant, the demand for optical improvement and the demand for mechanical convenience. Both of these demands have been well met, so that there appears little left for the future to achieve, until an entirely new direction is opened for further evolution. It need hardly be premised that the optical part is the essential part of a miscroscope. The optical performance of the best microscopes is to-day nearly perfect, having become so very slowly by numerous small improvements. Although magnifying glasses were invented, it is said, in the twelfth century, compound microscopes with achromatic lenses have been in use barely three quarters of a century, while the introduction of homogeneous immersion lenses dates from 1878, and of the perfected apochromatic lenses from 1886.—*From " The Microscopial Study of Living Matter,' by Charles Sedgwick Minot, in North American Review for May.*

A HOT BATH WILL BRING SLEEP.—Suppose a person be tired out by overwork of any kind, to feel nervous, irritable, and worn, to be absolutely certain that bed means only tossing for hours in an unhappy wakefulness. We all know this condition of the body and mind. Turn on the hot water in the bath-room and soak in the hot bath until the drowsy feeling comes, which will be within three minutes; rub yourself briskly with a coarse Turkish towel until the body is perfectly dry, and then go to bed. You will sleep the sleep of the just, and rise in the morning wondering how you could have felt so badly the night before. The bath has saved many a one from a sleepless night, if not from a severe headache the next day.—*Dr. Cyrus Edson in June Ladies' Home Journal.*

A COMPEND OF GYNECOLOGY.—This is a splendidly condensed work by William H. Wells, M.D., with 150 illustrations, Philadelphia, P. Blakiston, Son & Co., 1896, of nearly all the modern writers upon gynecology. The article on "Antisepsis in Gynecology" is particularly good and up to date, as is the article on "Diseases of the External Genital Organs," "The Uterus, Ovaries, and Tubes." The illustrations are splendid reproductions and well chosen. It is a most valuable little work for the student and busy practitioner.

The New Bohemian for May is the first number of that magazine issued by the new management. It comes with a new dress and new cover design, and shows a marked improvement throughout.

Cases Illustrating Various Pathological Stages in Inflammatory Pelvic Disease. By Louis Frank, Louisville. Reprint from the Medical News.

The Necessity of Complete Extirpation of Tumors and the Importance of Rapid Cicatrization of the Wound.   By Frederick Holme Wiggin, New York. Reprinted from the Transactions of the New York State Medical Association.

The Baths of Nauheim in the Treatment of Diseases of the Heart and the Therapeutic Methods of the Doctors Schott.   By William C. Rives, New York City.   Reprint from the New York Medical Journal.

Curettage of the Uterus, History, Indications, and Technique.   By J. W. Ballantyne, M. D., F. R. C. P. E., F. R. S. E., Edinburgh.   Read before the Edinburgh Obstetrical Society.

· Places Rendered Famous by Dr. J. Marion Sims in Montgomery, Ala.   By Edmund Souchon, of New Orleans, La.   Reprint from the New Orleans Medical and Surgical Journal.

Sero-Therapy in the Treatment of Tuberculosis; Report of Cases.   By Paul Paquin, St. Louis, Mo.   Reprint from the Journal of the American Medical Association.

The Prostate, Some of its Acute and Chronic Conditions and Their Treatment.   By L. Bolton Bangs, New York City.   Reprint from the Vermont Medical Monthly.

Congenital Teeth, with Three Illustrative Cases.   By J. W. Ballantyne, M. D., F. R. C. P. E., F. R. S. E., Edinburgh.   Reprinted from the Edinburgh Medical Journal.

Operation for Cataract, with Treatment and Protection of the Eye.   By P. Richard Taylor, Louisville, Ky.   Reprinted from the Louisville Medical Monthly.

Eleventh Annual Report of the Managers and Superintendent of the North Texas Hospital for the Insane at Terrell for the year ending October 31, 1895.

Infantile Intussusception, a Study of One Hundred and Three Cases.   By Frederick Holme Wiggin, New York City.   Reprint from the Medical Record.

Experimental Research Relating to Salivary Secretion and Digestion.   By J. H. Kellogg, Battle Creek, Mich.   Reprinted from Modern Medicine.

Methods of Precision in the Investigation of Disorders of Digestion.   By J. H. Kellogg, Battle Creek, Mich.   Reprinted from Modern Medicine.

City Water Supply and Filtration.   By W. P. White, Health Officer, Louisville, Ky.   Reprinted from the American Practitioner and News.

Tenth Annual Report of the John N. Norton Memorial Infirmary, Third and Oak Streets, Louisville, Ky.   Miss N. Gillette, Superintendent.

Prophylaxis in Infectious Disease in Children.   By A. G. Wollenmann, Ferdinand, Ind.   Reprint from the Medical and Surgical Reporter.

· Excision of Hemorrhoids.   By H. Nevins, San Francisco, Cal.   Reprint from the Pacific Coast Journal of Homeopathy.

Dachrocystitis. By P. Richard Taylor, Louisville. Reprinted from the Journal of the American Medical Association.

Irritable Ulcers of the Rectum. By R. S. Stanley, Memphis, Tenn. Reprinted from the Memphis Medical Monthly.

Brain Surgery for Epilepsy. By Merrill Ricketts, Cincinnati, O. Reprinted from the Cincinnati Lancet-Clinic.

Neuralgia of the Fifth Nerve. By Merrill Ricketts, Cincinnati, O. Reprinted from the Cincinnati Lancet-Clinic.

Treatment of Puerperal Sepsis. By Louis Frank, Louisville, Ky. Reprint from the American Journal of Obstetrics.

Modern Surgery of Serous Cavities. By Merrill Ricketts, Cincinnati, O. Reprinted from the Railway Surgeon.

Physician and Layman. By Arthur R. Reynolds, Chicago, Ill. Reprinted from the Chicago Medical Recorder.

Urotropin. By J. A. Flexner, Louisville, Ky. Reprinted from the American Practitioner and News.

Ileus. By John B. Murphy, Chicago, Ill. Reprinted from the Journal of the American Medical Association.

Petroleum in Pulmonary Affections. By E. P. Jones. Reprint from the New England Medical Monthly.

Gonorrhea of the Rectum. By J. A. Murray, of Clearfield, Pa. Reprinted from the Medical News.

Rupture of the Left Lateral Ventricle. By Merrill Ricketts, Cincinnati, O. Reprinted from Medicine.

The Use of Electricity in Dermatology and Genito-Urinary Surgery. By W. R. Blue, Louisville, Ky.

Tendon Grafting. By Samuel E. Milliken, New York, N. Y. Reprinted from the Medical Record.

Subphrenic Abscess. By Carl Beck, New York City. Reprint from the Medical Record.

A Portrait Catalogue of Books on Medicine, Dentistry, Pharmacy, and Chemistry. P. Blakiston, Son & Co., Philadelphia.

Diseases of the Blood. By Alexander Rixa. Reprint from Practical Medicine.

Merck's 1896 Index. An Encyclopedia for the Physician and the Pharmacist.

Notes and Comments. By Orpheus Evarts, Cincinnati, Ohio.

Aseptolin. By Cyrus Edson.

Notes and Queries.

RELATIONS OF MEDICAL EXAMINING BOARDS TO THE STATE, TO THE SCHOOLS, AND TO EACH OTHER.—Dr. William Warren Potter, of Buffalo, President of the National Confederation of State Medical Examining and Licensing Boards, chose this title as the subject of his annual address at the sixth conference of this body, held at Atlanta, May 4, 1896.

He said there were three conditions in medical educational reform on which all progressive physicians could agree, namely, (1) There must be a better standard of preliminaries for entrance to the study of medicine. (2) That four years is little time enough for medical collegiate training; and, (3) That separate examination by a State Board of Examiners, none of whom is a teacher in a medical college, is a prerequisite for license to practice medicine. It is understood that such examination can be accorded only to a candidate presenting a diploma from a legally registered school.

He did not favor a National Examining Board, as has been proposed, but instead thought all the States should be encouraged to establish a common minimum level of requirements, below which a physician should not be permitted to practice, then a State license would possess equal value in all the States.

In regard to reciprocity of licensure, Dr. Potter thought it pertinent for those States having equal standards in all respects to agree to 'this exchange of interstate courtesy by official indorsement of licenses, but that other questions were of greater moment just now than reciprocity. Until all standards were equalized and the lowest carried up to the level of the highest, reciprocity would be manifestly unfair.

He urged that the States employ in their medical public offices none but licensed physicians. This, he affirmed, would tend to stimulate a pride in the State license and strengthen the hands of the boards.

He denied that there was antagonism between the schools and the boards, as has been asserted. He said that both were working on parallel lines to accomplish the same purpose; that there could not possibly be any conflict between them, and that they were not enemies, but friends.

The medical journals of standing, from one end of the country to the other, he affirmed, were rendering great aid to the cause of reform in medical education, and the times were propitious.

He concluded by urging united effort by the friends of medical education, saying that "the reproach cast upon us through a refusal to recognize our diplomas in Europe can not be overcome until we rise in our might and wage a relentless war against ignorance that shall not cease until an American State license is recognized as a passport to good professional standing in every civilized country in the world."

THE June number of the *Buffalo Medical Journal* is unique in that it is the result entirely of woman's work. The chivalric editor of the Journal, Dr. Potter, gave the Journal over to the fair women doctors, and the June issue is the result of their own efforts and reflects great credit on their ability. Dr. Maud Josephine Frye was the managing editor. Drs. Electa B. Whipple, Ida C. Bender, and Helene Kuhlman, editors in chief, and the following associate editors: Drs. Jane W. Carroll, Jane N. Frear, Lillian Craig Randall, and Evangeline Carroll. The laudable efforts of these editors should be richly rewarded by a large demand for the June issue.

PARKE, DAVIS & Co. have recently opened a branch house in New Orleans and another in Baltimore, to satisfy the rapidly augmenting demand for their pharmaceutical preparations. Their '96 price list, now being distributed, is another indication of the incredible vastness of this institution, comprising twenty-nine distinct lines of preparations and over six thousand different products. This list, by the bye, is so exhaustive and convenient for reference—so admirably printed and cleverly arranged—that no physician should be without one, and we counsel our readers to write to Parke, Davis & Co. for a copy.

THE WESTERN MEDICAL REVIEW, a monthly journal of medicine and surgery, is the name of a new medical journal which has been received among our exchanges. It is printed by the Western Medical Review Co., at Lincoln, Nebraska, and is edited by George H. Simmons, M. D. The first number, dated May 20th, contains 24 pages of reading matter and no advertisements, though it is stated editorially that reputable advertisements will be received, but not placed among the reading matter. We wish the journal a prosperous career.

THE INTERCOLONIAL MEDICAL JOURNAL, of Australia, is the result of the amalgamation of the *Australian Medical Journal* and the *Intercolonial Quarterly Journal of Medicine and Surgery.* The contents of the new journal will embrace original articles, clinical reports, leading articles, reports of proceedings of societies, periscopes and reviews. The list of editors embraces names from New South Wales, New Zealand, South Australia, Queensland, Tasmania, West Australia, and Victoria. Stillwell & Co., of Melbourne, are the publishers.

THE DREVET MANUFACTURING Co., of New York, have secured an injunction against Dr. A. P. Beach, of Seville, Ohio, to prevent his using the name of " Glycozone " in any preparation of his manufacture. The name of "Glycozone " is protected by the Drevet Manufacturing Co., and this application to the U. S. Circuit Court is to prevent the infringements of their rights.

P. BLAKISTON, SON & Co. have issued a portrait catalogue of books on medicine, pharmacy, dentistry, microscopy, hygiene, nursing, and allied subjects, and in the contents make the liberal offer to send books on approval to any address in the United States. The collection of photographs of prominent authors is not only of considerable interest but of some value.

NOTICE TO CONTRIBUTORS.

Articles and letters for publication, books and articles for review, and communications to the editors, advertisements, or subscriptions, should be addressed to EDITORS OF MATHEWS' MEDICAL QUARTERLY, BOX 434, LOUISVILLE, KY.

All necessary illustrations will be furnished free of expense to authors where they send black and white drawings—or negatives—with their MSS.

All articles for publication in the QUARTERLY will be considered only with the distinct understanding that they are contributed to it exclusively.

Alterations in proof-sheets are charged at the rate of 60 cents per hour, which is the printer's rate to the journal.

Reprints may be had at printer's rate if request is made upon proof when it is returned.

All manuscript must be received by the first of the month preceding its publication.

Remittance may be made by check, money order, draft, or registered letter.

The Editors are not responsible for the views of contributors.

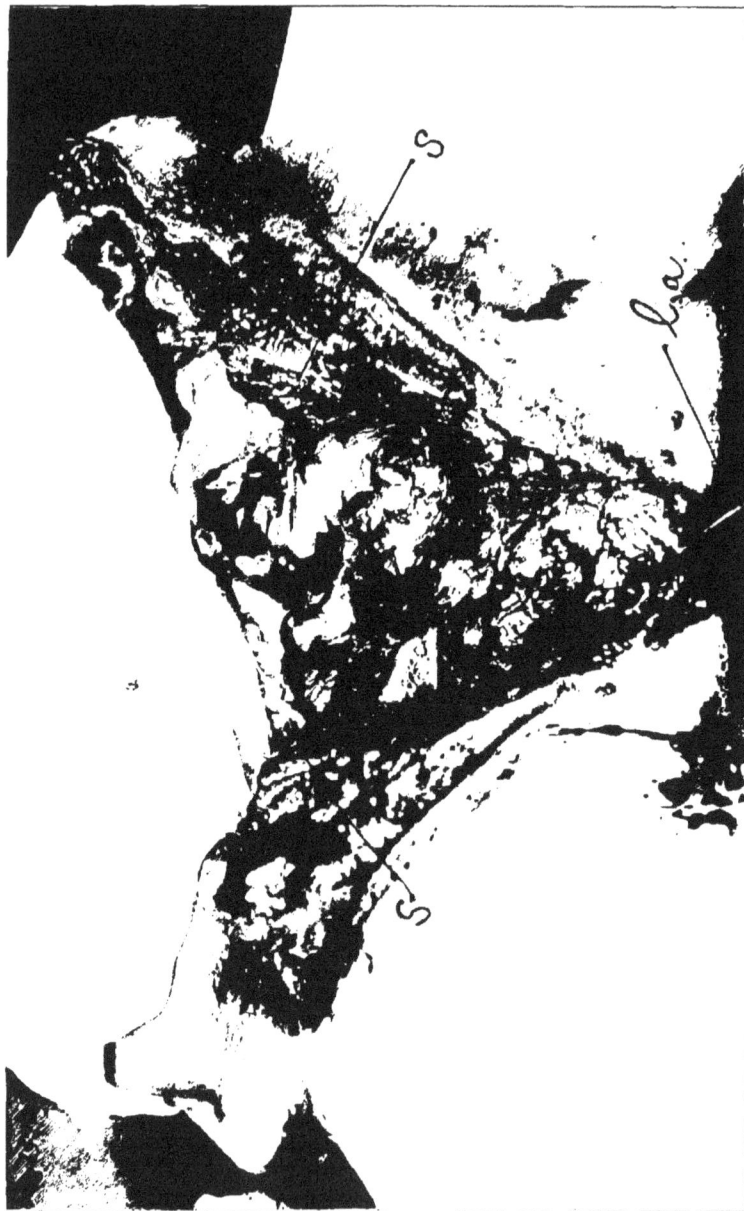

Fig. XVIII. Specimen F—Photograph of female cadaver showing, after laparo-symphysiotomy and removal of bladder, uterus, and adnexa, the upper rectum and sigmoid packed with scybala : s.s, symphysis pubes ; l.a, ligature at anus.

MATHEWS'
MEDICAL QUARTERLY.
"ALIS VOLAT PROPRIIS."

| Vol. III. | OCTOBER, 1896. | No. 4. |

Original Contributions.

THE ADDRESS ON SURGERY.*

BY H. H. GRANT, M. D.,

LOUISVILLE, KY.

When, a few months ago, I was honored with the notice to prepare this address, I thought to recapitulate the important contributions to surgery during the interval since our last meeting. But in the mean time other addresses have so covered that field that I have chosen rather to indicate a narrower but not less important line. Some common ground must be sought on which differing opinions may be reconciled. After Kocher has successfully extirpated a thousand goitres he is not sure it would not have been better to feed them with the thyroid extract instead; though the pendulum of appendicitis has swung very far in both directions, we are not yet even surgically in accord, while our medical friends are by no means satisfied; herniotomy has two or three new methods, not yet a year old, which to most of us promise little better than the former ones; castration for enlarged prostate has yet to prove its acceptability; celiotomy for typhoid ulcer, done successfully two or three times, is yet a successful accident, while other abdominal as well as pelvic and cranial lesions are subject to bitter debate and opposite conclusions. Yet these are at the most but minor differences. The principles are established; more certain every day do we become of our progress.

* Read at the annual meeting of the Mississippi Valley Medical Association, St. Paul, September 14, 1896.

When, twenty years ago, I began the study of medicine, it was no unusual experience to hear apology from the professor as well as the medical orator for the imperfect science. Even in those days of developing theories of antisepsis and of new pathology, while asepsis was yet a latent fancy, the hope of scientific accuracy and definite principles was retiring and modest in comparison with the recent wonderful realization. Doubtless many of us failed to appreciate the final effective agent in these advances. We forget how, in dismal laboratories, many men, who never come to the operating-table or sick-bed side, have with most trying toil and often many dangers, for insignificant reward, elaborated the principles which in the past ten years enabled medicine to take the rank of science and made surgery a marvel to the learned and wise. Without this aid we all recognize the art of surgery must have long lingered in that dark valley where Ferguson, Cooper, Post, Pancoast, Mott, and a host of others almost as great, left it. The knowledge of these men in all but the new pathology is rarely equaled by one of the following generation, and their operative skill is unsurpassed. Much of the great the surgeon has accomplished originated in an humble and unpretentious department of our calling, often unknown to the laity and unappreciated by the masses of the profession. Though for a time the surgeon with the brilliant demonstration and miraculous cures beckoned with excelsior banner the physician brother to follow faster, within a year or two that brother comes dangerously near. The serums and antitoxines, long doubted and derided, have at length so completely made good most claims that not only are the medically treated diseases more scientifically and successfully handled, but not a few formerly intrusted to surgery, with but indifferent results, are reclaimed by these new agents with promise and advantage.

But, while in many of its departments medicine shows great advancement and is still marvelously progressing, the goal is yet far away. In surgery, however, the limit along some lines is appearing. But little improvement comparatively is to be hoped for, either in technique or operative skill, beyond what has been achieved. Not that thorough surgical cleanliness, perfect asepsis, and an ideal technique have yet been presented as a common or ready attainment; not indeed that its perfection has ever yet been attained, still the results which such perfection will guarantee are constantly achieved, and the measures which fail for want of such guarantee can hope for little more with its aid in individual cases. I would not be misunderstood, that any but the most thorough of known methods is a reliance in this chief of all securities, but that in a few hands it is near enough perfect to fearlessly trust. In simplicity consists its safety. Paraphernalia of every kind suggests a lack of personal resource as a formality, it invites confusion and failure. Incompetency and ignorance rely on aid from outside measures, and in the emergencies of life and death incompetency has no place, nor can any support defend it from criminal disaster. I am led to speak thus of the limits of progress, because, as we know, all or a part of every organ in the human body without which the others can support life has been successfully removed, sometimes with good results, sometimes with bad, often with no appreciable influence. Operative surgery has well-nigh proved that it needs but the recognition of disease to successfully remove it, if life can be supported without the diseased structure. The dangers of enterectomy, of gastrectomy, thoracoplasty, extirpation of brain tumors, or malignant diseases of internal viscera, lie not in the operative procedures but in the condition produced by the disease itself. These conditions at the time of operation are often essential and not subject to correction by surgery. Many of them, however, could be anticipated by earlier diagnosis.

Years ago I heard a beginner in medical practice limit his wish to the Genius of Fortune, to an ability to determine when he was consulted by a patient just what was the pathological condition, just what treatment should be instituted, and what surely would be the result. He did not seem to realize that such knowledge could be ever attained under any circumstances. Perhaps, indeed, in the perfection he had in mind it can not; it is the acme of medical science. But I submit to you, that when the first condition of that wish is realized the second and third may be almost as surely applied by present surgery as it may ever be. In other words, advance in surgery must be in determining accurately and at their origin pathological conditions. No surgeon in this audience but has said to himself—to his friends, to his patients—if I had foreknown what I have now discovered the result would have been different; or if I had known yesterday what I find out to-day was the case then, then would not have been this disastrous delay. We reproach the general practitioner, that he calls us too late in appendicitis, intestinal obstruction, and many rapid advancing acute lesions. Should we not realize the fault is too often our own, that a want of confidence in our own power of diagnosis encourages the disposition of our confrère and the family to wait till evening, till to-morrow for developments, and the floodtide in that affair of man or woman is omitted to the disaster of the voyage of that life. Not ten times in a hundred is the deaf ear of patient or friend turned to a hopeful and confident assurance of the surgeon. Rarely does he fail to secure an operation if he has the courage to demand it.

For my part, I have generally found the physician was only too ready to put the responsibility upon me, if I was prepared to give assurances of good results from my suggestion, and that the plainer the case the readier was his compliance. Can we blame him that he objects to the opening of the belly of the patient merely to see if there is any thing wrong inside of it? Death waits upon delay in many of these conditions, and it requires no seer to appreciate that the rescue from these galloping toxemias, which advance hour by hour to essential and irremediable destruction, is early recognition by an unwavering diagnosis, and immediate eradication by irreproachable technique. A distinguished American surgeon, in a recent address on the limits of surgery, refers with some asperity to the "*furor operativus.*" Doubtless the rebuke is to some extent deserved, but I observe an equally fatal danger on the other side. There is no other term in surgery so often misapplied as conservatism. Behind it, as the strong bulwark of approval and countenance, shirk cowardice and ignorance. Here, I dare not, too often waits upon I think I should, and omitted, or slovenly and almost criminally employed, the opportunity for health and life is lost. I do not mean to decry conservatism, but to point out that it differs from incompetency. There is a measure of imperfection in the best of us. The law of relativity applies here as elsewhere.

Along the line that is the most hopeful prospect of improvement lies the consideration of the highest interest. No member of the laity, be his human sympathy and unselfishness ever so great, has the welfare and prosperity of the suffering so at heart as the medical adviser who is in charge. No aim in the life of the true physician is so dear to him as the ways and means to the relief of physical ill. Indifference to results or harshness of method are cruel calumnies not infrequently alleged against us. We, best of all, know how shamefully false they are. There is no enthusiasm of patriotism, no zeal of politics, no fervor of religion, so sincere and unselfish as the concern the true physician brings to the investigation and relief of the helpless who appeal to him. Those of the guild, who in their souls have received the spirit of the Hippocratic thought—and to the credit of its individual members as well as the glory of the profession these comprise the great bulk of it—know no thought of self-interest nor sordid fee between them and their charge. The underlying principle of all our efforts, individual and associated, is the relief of suffering and the averting of death. Among ourselves we need no mention of defense, and though an indiscriminating and uninstructed public, helped by a sensational and somewhat jealous press, doubts our sympathy and feeling, or accuses us of a reckless disregard of life in pursuit of novelty and dare-devilism in our operations, though we constantly hear the friends and neighbors of our patients condemn our unselfish efforts, though our home newspapers misrepresent our want of success where failure was almost inevitable, and distort our motives in our discussion of public measures, quiz and ridicule us when suits for malpractice are instituted, which just judges throw out of court—I say, though these and a hundred more of like kind are our daily reward, I should not hint them in this body were it not that the chief evil falls not on us but on the cause in which we are enlisted. The soldiers in the heat of critical warfare are stricken not with sympathy for the riddled regiment or the wounded general, but with fear less such disaster interrupt common success.

So all this feeling brings surgery into discredit, deters many an operable case from accepting relief until it is too late, or drives it to quacks and incompetent conservators, to delay and perhaps sacrifice lives that skill and knowledge could surely rescue. A surgical friend in a recent unpublished paper says,

" Pandering to the sentimental feeling of the people on the part
of the profession simply tends to keep up the idea among the laity
that surgeons have no feeling, that all human sympathy has been
drilled out of them, that they would rather see blood than eat, and
that they are indifferent about killing people. This calls forth
such expressions as ' The butcher !' 'I will die before I will submit to the knife,' etc., and they do die often for a lack of a proper
knowledge of the surgeon and his ways. They should know that
surgery is conservative, that it is humane, and that procrastina-
tion is often fatal. How often are we introduced to a case when
it is too late to do any good, where if the patient had understood
the necessity for the operation the limb or life might have been
saved. Proof of the harm produced by ignorance on the part of
the people of surgical matters is met with daily in a surgeon's
work, and illustrations of it are abundant, which those not working in surgery often fail to appreciate."

These are among the incidental burdens of our vocation.
They are to be overcome finally only by mastery of the subject.
Apology and explanation, here as elsewhere, are confessions of
failure, and criticisms, just and unjust, must ever act as spur and
whip to urge us to ride over all obstacles until the race is won.

At the outset of these remarks was conveyed the idea that the
future of successful surgery lay in accurate diagnosis added to
our now highly perfected pathology and technique. We hear not
infrequent intimation that the allurements of specialism and the
attractiveness of surgery have created, out of the great hot-house
of medical colleges and surgical clinics, a host of young operators,
who rush in where only skill and experience can hope to safely
tread. It is not their uncommon boast too, that the conditions
of disease, in the presence of which twenty-five years ago men
whose names helped make the medical history of the world stood
helpless, are now wiped away in half an hour by mere boys. It
is measurably true. But it is also true, that it was on the fair and
tender backs of these mere boys that fell the lash of that long
whip which we heard echo *"furor operativus"* at Atlanta but a
few months since. While we can not but be glad to recognize
and fully credit skill and ability without respect to age, we must
not fall into the error of marking but one measure of a heap of
good or evil and approximating all the balance. Whence grew
the fame of Gross and Agnew, Hodgen and Eve, Gunn and

Dawson, together with many more of this and other countries, whose names the pages of surgical history, however partial or incomplete, will show so long as medicine and surgery are written of? The answer is, that a wide knowledge, both historical and experimental, of anatomy, progress of disease and promise of treatment; a knowledge a long life of study and observation only earned, enabled them usually to see under the coverings of skin and muscle, not as through a glass darkly, but face to face with disease.

When improvement in diagnosis shall have so far advanced it from where these able minds laid it down, as pathology and technique have advanced operative work, we will almost if not actually see the gangrenous appendix, the telescoped bowel, the incipient internal carcinoma, as we now see the carbuncles and warts upon the surface. I have heard it told of one of the first (if not the first) diagnostician and operator in abdominal and pelvic surgery in this country, that, on being asked what he expected to find when he opened the abdomen then preparing, he answered, " Any one of fourteen things." Very bad if true, and that it was relatively true we all know, and its correction demands all our efforts, physicians' as well as surgeons', in the interest of the common weal.

Other things being equal, the greatest aid to diagnosis is experience. We hear at times of intuitive diagnosis. It is but the repetition of symptoms seen before, or read of, or previously thought out. Of course the aids are all known to you. Among the later ones, the exploring trocar of Warren and the micro-scope are still too limited and too unsatisfactory to predicate positive conclusions on in many instances, or else bring only unimportant corroboration. The highest interest now centers in the diagnostic uses of the Rontgen, or X rays, or skiagraphy, as the art is called. Its practical value has already been attested in a number of instances, some highly interesting. It is yet in its infancy; but when we remember what growth electricity has attained in other departments, the promise of that infancy is almost without limit. We all know of the method, that it is the transmission of electric light through bodies opaque to ordinary rays, and the production of a shadow indicating the presence of more opaque substances within these bodies. It is well known that certain substances are less penetrable than others, and consequently produce more distinct shadows. Thus metallic substances imbedded in the tissues, or lying in cavities and passages of the body, being impenetrable to these rays, are indicated by these shadows that they leave among the more translucent soft tissues. Fractures and dislocations of bones are similarly indicated, and thus constitute a safer guide for operative steps. Very recently Willard removed a lost intubation tube located in this way. But a month or so ago White photographed a jack stone lodged eight days in the esophagus and successfully operated. Still later Wood, of Philadelphia, performed a similar feat; another Philadelphia surgeon extracted a bullet lodged against the vertebræ and took it out with good result. Such successes are in the highest degree important. Up to this date failure has attended efforts to photograph pathological changes in the soft tissue and viscera satisfactorily. But great improvement is confidently expected.

Pinard reports photographing a product in utero, in a frozen specimen, however, in which the outlines of the fetus as well as the uterine walls and membranes could be well distinguished. Keen, in a rather extended treatise just published, states that the period of exposure determines the intensity of the shadow; that the soft tissues both normal and pathological are represented in the earlier moments of the exposure, but afterward the lines fade away and only the more opaque bodies leave permanent shadows. He thinks there is good reason to expect, as improvement in method and dexterity increases, this graduated plan of varying the period of exposure may disclose with accuracy many pathological changes in the viscera and soft tissues. In other words, the skiagraph now indicates bullets and other foreign bodies in the tissues, dislocated and fractured bones, even though the tissues be greatly swollen; shows the condition of set fractures through undisturbed dressings, and perhaps discovers even tuberculosis and abscess in bones and joints. But beside this, we may confidently expect to see the living child in utero, hidden pathological lesions of the viscera and organs, calculi in the bladder, kidney, and ureter.

Wooden foreign bodies, as splinters, and toys that have been swallowed, offer too little resistance to light for successful photography by present known methods. The bony case of the skull, thorax, and pelvis interferes with some manipulations in these regions. No results are obtained in the brain, and Keen thinks perhaps successful photography will not there be possible. It seems too much to hope that these means open up to us the certainty of diagnosis of cancer of the pylorus and larynx in their operable stage ; gastric and typhoid ulcers before perforation ; appendicitis as a threatening rather than a destroying monster ; volvulus, intussusception, internal strangulation, before sepsis has begun ; lesions in the pulmonary organs as well as the vertebræ in incipiency—yet in great part it can be confidently expected.

But however this may be, no one doorway can open to a royal road to success in surgery. Too much depends upon personal quality. There is no moment, from the time the patient is first seen until the case is terminated, that the discretion and judgment of the surgeon, in what may seem even minor details, is not the key to the situation. What I would assure our able ally, the general practitioner, is that progress in late years has made possible to skill and knowledge in surgery the relief, if promptly applied—and this promptness is a part of skill and knowledge — of most lesions to which promise of relief approaches ; and that our common wish, to secure the highest good, justifies ·the intrusting to the judicious counselor the care of such surgical conditions as he can confidently define—secure that it is the choice of misfortunes. What I would urge upon the surgeon is a recognition of the fact that many of our differences are due to the faulty interpretation of symptoms, and that the mistake which occasions different results from practically the same operation occurs in the diagnosis and not in the technique, and that as the more effectually we clear up the question of diagnosis, by the old means or by new ones, the nearer we agree in the principles that control our surgery, and the more uniform will be our success. What I would have both physician and surgeon believe with conviction, is that there can be no clash of interest between us in this common cause—that together our efforts at enlightenment of the public, and the rescue of our profession from the criticism and sneers it too often encounters, should be in common directed toward the accuracy of knowledge—that we are mutually laborers in one great field, with one object only ; therefore what each finds for his hand to do, let him do it with his might.

DISEASES OF THE RECTUM.

NEW EVIDENCE THAT THE RECTAL VALVE IS AN ANATOMICAL FACT.

BY THOMAS CHARLES MARTIN, M. D.,

CLEVELAND, OHIO.

[Illustrated and written for MATHEWS' MEDICAL QUARTERLY.]

I submit these photographic reproductions as documentary proof of a fact.* Clinical observation convinces me of the necessity of recognition of the semilunar valves as the chief feature of the rectum proper. The unsatisfactory methods of tactile examination by hand or by sound, and the impossibility heretofore of a complete inspection of this organ, as well as the illogical manner in which necroscopic investigation has been conducted, very easily explain how there could have occurred so protracted and eloquent a dispute concerning the anatomy of the rectal ampulla.

The practical surgeon feels the necessity of knowing positively whether there are or are not in the rectum normal obstructions of the nature of these under discussion. If it be proven that these obstructions exist, it will then be imperative that he abandon a method of diagnosis almost universally practiced, and it will also be necessary that he modify his methods of treatment of certain lesions; possibly, too, a new point of view may be afforded from which to study the congenital malformations of the rectum.

A quarrel which is so archaic, so involved, and in which there is such multiplicity of contradiction concerning a matter of scientific interest, can not with proper understanding and with perfect fairness be resumed, in however peaceful a spirit it be undertaken, without the preliminary exact quotation of the expressed opinion of the distinguished workers in this field. I have therefore taken the liberty of making the following quotations and of emphasizing by means of italics such parts as are disqualified or disputed by the evidence which I have found.

*I am indebted to Dr. C. M. Thruston for very great assistance in the preparation of the specimens and for the drawing of the microscopical appearance of the valve. Mr. Theodore Endean made the photographs.

Houston,* 1830, writes : " In the natural state the tube of the
gut does not form, as is usually conceived, one smooth uninterrupted passage, devoid of any obstacles that might impede the
entrance of bougies; it is, on the contrary, made uneven in
several places by certain valvular projections of its internal
membrane, which, standing across the passage, must frequently
render the introduction of such instruments a matter of considerable difficulty.   Cloquet and some other anatomical writers
have made a cursory allusion to this condition of the membrane ;
but all the authors who have treated of diseases of the rectum
appear to have wholly overlooked it.

" The valves exist equally in the young and in the aged, in the
male and in the female; but in different individuals there will be
found some varieties as to their number and position.   Three is
the average number, though sometimes four, and sometimes only
two are present in a marked degree.   The position of the largest
and most regular valve is about three inches from the anus,
opposite to the base of the bladder.   The fold of next most frequent existence is placed at the upper end of the rectum.   The
third in order occupies a position midway between these, and the
fourth, or that most rarely present, is attached to the side of the
gut, about one inch above the anus.

" The form of the valves is semilunar ; their convex borders
are fixed to the sides of the rectum, occupying in their attachments from one third to one half of the circumference of the gut.
Their surfaces are sometimes horizontal, but more usually they
have a slightly oblique aspect, and their concave, floating margins, which are defined and sharp, are generally directed a little
upward.   The breadth of the valves about their middle varies
from a half to three quarters of an inch and upward in the distended state of the gut.   Their angles become narrow, and disappear gradually in the neighboring membrane.   *Their structure
consists in a duplicature of the mucous membrane, inclosing between
its laminæ some cellular tissue, with a few circular muscular fibers.
The only method by which the condition of these valves in the distended state of the rectum can be displayed is that of filling and
hardening the gut with spirit previous to being disturbed from its
lateral connections.*   By the ordinary procedure of distending it

°Dublin Hospital Reports, 1830, Vol. V, page 158.   Hodges & Smith, College-Green,
Dublin.

after removal from the body the valves are made to disappear. Their presence may likewise be ascertained in the empty state, if looked for soon after death, and before the tonic contraction of the gut has subsided.

" They will be found to overlap each other so effectually as to require considerable maneuver in conducting a bougie or the finger along the cavity of the intestine."

Chadwick,* 1878, discussing this subject, says: "Hyrtl, in his treatise on Topographical Anatomy, devotes three pages to the consideration of what he designates as the *Sphincter Ani Tertius*. From his description the only inference is that Hyrtl has generally found a bundle of muscular fibers so encircling the rectum as to exercise the function of a sphincter, at least when the other sphincters are for some reason inoperative. On inflating recta, however, in accordance with the direction given by him, it is rather surprising to discover that no such annular constrictions appear. At the point of the rectum designated by him is, nevertheless, observable a semicircular constriction of the rectum confined to the anterior wall; corresponding to this, but an inch or more higher up, is always seen a second semicircular constriction affecting the posterior wall only. If, now, the rectum be cut open, and its mucous membrane dissected off, as directed by Hyrtl, each of these two constrictions may be demonstrated to consist, as he says the ' third sphincter' does, of an agglomeration of the circular muscular fibers of the rectum. I am able to show you seven recta taken from dissecting-room subjects, from which we dissected off the mucous membrane after cutting them open longitudinally. In all of these you can not fail to find corroboration of my statements in the presence of two distinct masses of circular fibers each encircling about half the circumference of the canal.

" If, now, a mass of feces be supposed to advance through the rectum following the sinuosities, it is evident that these bundles of fibers, when not in active contraction, would present scarcely any obstacle to its progress. It is further noticeable that these partial constrictions of the canal differ only in degree from the constrictions visible in the higher segments.

" At about two and a half inches from the anus the finger

*Transactions of the American Gynecological Society, 1878, Vol. II, page 43. Houghton, Osgood & Company, Cambridge.

Fig. 1. Specimen A—Dorsal posture. Anterior inferior view of cast filled rectum and lower sigmoid.

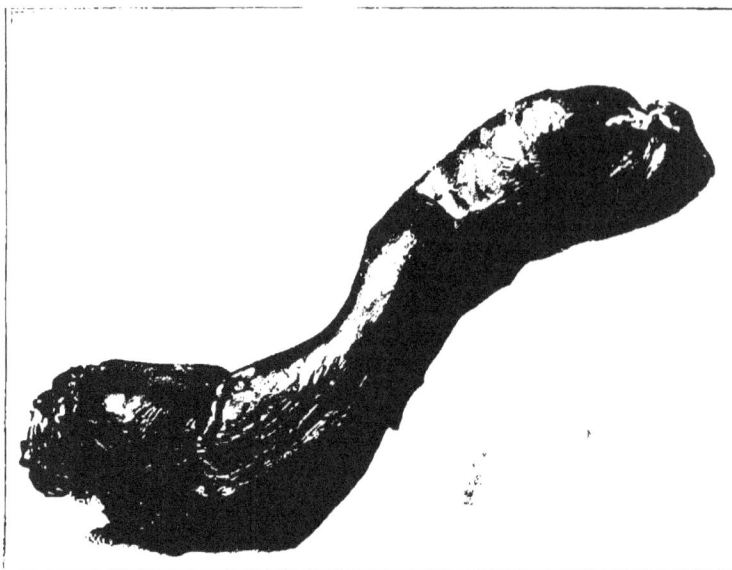

Fig. II. Specimen A—Dorsal posture. Left lateral view of paraffin filled rectum and lower sigmoid.

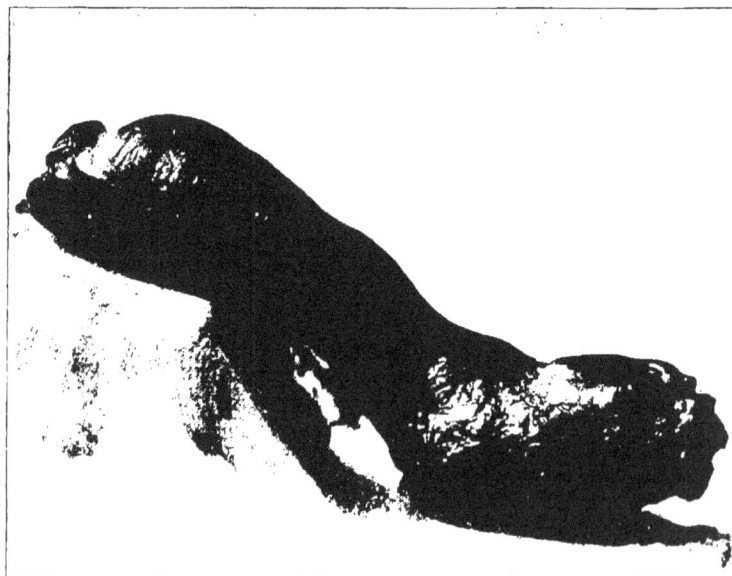

Fig. III. Specimen A—Dorsal posture. Right lateral view of cast filled rectum and lower sigmoid.

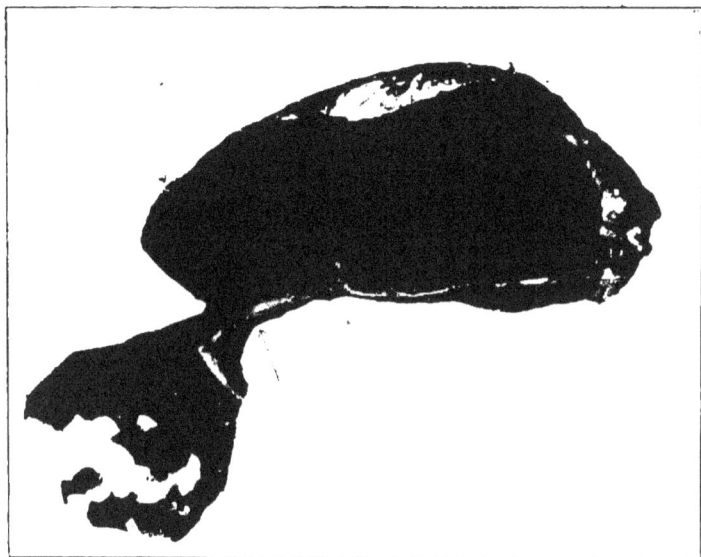

Fig. IV. Specimen A—Knee-chest posture. Right lateral half—interior view of rectum, sigmoid out of focus.

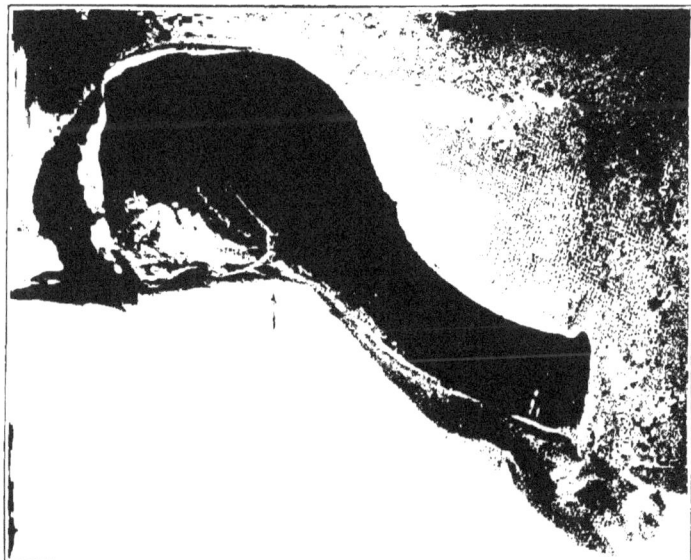

Fig. VI. Specimen A—Knee-chest posture. Left lateral half—interior view. Upper rectum and sigmoid out of focus.

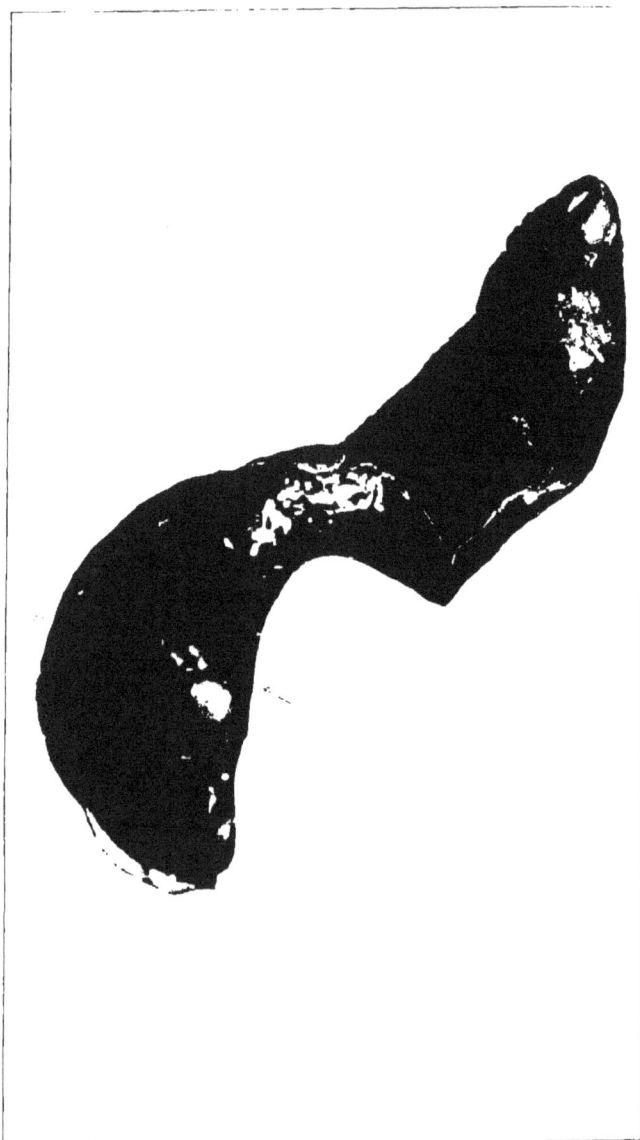

Fig. V. Specimen A—Knee-chest posture. Left lateral half—interior view. Anal region somewhat out of focus.

encounters a confused mass of folds through which the continuance of the canal can only be discovered by considerable burrowing. Here an annular constriction, diminishing the lumen by about one half, seems to be felt.

" If, now, the rectum be distended with water, the finger will almost invariably detect, in place of the lax folds, what still seems to be an annular constriction, but which a more careful examination will show to be composed of two distinct semicircular bands slightly overlapping each other, the posterior being somewhat higher than the anterior."

Chadwick continues : " Being familiar with the views of Nelaton, Hyrtl, and others, I at first sought to assign to this apparent constriction of the rectum sphincteric functions, but soon had to relinquish that idea, for the exploration of very many *recta in the living failed to reveal a single one in which the lumen of the supposed sphincter, when quiescent, had a smaller diameter than three quarters of an inch, while in the majority it was over an inch.*

" These anatomical and clinical observations all tend to indicate that the term ' *Third Sphincter Ani,*' *applied by Hyrtl to these constricting bands, is a misnomer,* and to show that they are simply a part of the general circular layer of muscles, whose function is to dilate before and contract behind the scybala, thereby propelling them on their way and not retarding them."

Chadwick closes by saying : " Having seemingly elucidated the true function of the 'Third Sphincter Ani,' and proved by the above observations that it should more properly be termed a *Detrusor Fœcium,* if deserving of any special appellation, my attention was next directed to the action of the internal sphincter."

Kelsey,* 1893, in a somewhat exhaustive discussion of this subject, says : " It is now about half a century since Nelaton first described the third sphincter muscle, and, in spite of all that has been written concerning it since that time, it is only a few years since Van Buren characterized it as an organ to which anatomy and physiology had been equally unsuccessful in assigning either certainty of location or certainty of function. For the original description of the muscle by Nelaton we are indebted to Velpeau, who writes that he has verified the existence of a sort of sphinc-

*Diseases of the Rectum and Anus, 1893, page 26. Wm. Wood & Company, New York.

ter of the rectum, lately discovered by Nelaton, and goes on to say that it is *a muscular ring situated about four inches above the anus*, just in the place where retractions of the rectum are most often found. *If, after turning the rectum so that its mucous surface is external*, it is moderately distended by inflation, the muscles will be seen to be made up of fibers collected into bundles.

" Sappey admits its frequent existence, and locates it at the level of the base of the prostate, in the middle portion of the rectum, six, seven, eight, or sometimes nine centimeters from the anus. It never completely surrounds the rectum, but only one half or two thirds its circumference ; and it appears to him to be caused by a grouping of the circular muscular fibers. Its breadth is one centimeter, and its thickness two or three millimeters. Situated sometimes in front, sometimes behind, and again laterally or antero-laterally, it is constant in nothing except its direction, perpendicular to the axis of the bowel. In place of one he has sometimes found two bands at opposite points and different levels, and in one specimen there were three. Henle adopts Sappey's description in the main. Petrequin found the *muscle irregularly oblique*, less marked in the front wall than in the back, and consisting of a collection of weak bands of fibers."

After these quotations Kelsey comments as follows: " *The third sphincter muscle and the valves of mucous membrane in the rectum are not, as might be supposed, one and the same thing, though it is true that they have become almost hopelessly confounded in surgical and anatomical literature*, and are often spoken of as identical. The valves of the rectum (*we use the word simply as expressing the folds of mucous membrane*) were first described by Houston at about the same time that Nelaton described the superior sphincter; and it is worth remembering that the two authors *were writing about two entirely different things, and two things which stood in no necessary relation to each other, so far as we may judge from their descriptions*.

" According to this first and clearest of all descriptions,* for the whole article (Houston's) is written with a force and clearness of style which have perhaps had an undue weight in disarming criticism as to the facts, the valves exist in all persons, but vary much in different individuals as to location and number."

*Kelsey. page 28.

Kelsey quotes Houston's description (which has already been quoted by me), and in contravention says: "The palpably weak points in Houston's article were very soon pointed out by O'Bierne (1833) in a work of marked and almost amusing originality. O'Bierne seems rather to regret that he is unable to accept Houston's statements as to an anatomical condition which would account so fully and so easily for the physiological emptiness of the rectum and fullness of the sigmoid flexure on which his (O'Bierne's) own views depend ; but nevertheless he sets himself to the task of demolishing them with great vigor and considerable success. Although he believed the rectum to be normally empty, except just at the time of defecation, he believes that condition to depend upon the anatomical arrangement of the sigmoid flexure, joined with the narrowing of *the upper end of the rectum, which is entirely independent of any folds of mucous membrane.* He not only denies the existence of any such folds, but states flatly that Houston is altogether incorrect in his statement that Cloquet or any other anatomist before his (Houston's) time makes even the slightest allusion to them. He (O'Bierne) believes the folds to have been produced by the method of making the preparations, distending and hardening all parts with spirit before making the incision, and asserts that this method is any thing but natural, and nothing more or less than an attempt to exhibit natural appearances by placing the parts in an unnatural situation—such a situation, indeed, as is not known to be necessary for the exhibition of the valvulæ conniventes or any other valve of the body. *He* (O'Bierne) *meets the statement, that by the ordinary procedure of distending the rectum after removal from the body the valves are made to disappear, by the question, why, if such valves really exist, and if muscular fibers enter into their structure, they should not be discoverable at any time after death, or in any state of the intestine—a question very difficult of solution."*

Kelsey, continuing further, says: " Four years later (1837) the voice of a New York surgeon was raised against these folds, and in almost the same language as O'Bierne's, though from an entirely different standpoint. Bushe declares that he has never, in the living body, been able to detect any valve of such firmness, and capable of exerting any such influence upon the descent of the feces as Houston describes, though he has frequently met with accidental folds produced by the partial contraction of the bowel. He (Bushe) points out that, by *the method of hardening the rectum after distending it with spirit, the accidental folds are rendered permanent by the induration resulting from the action of the alcohol; and that, by the method of inflating and drying, the projections resembling valves are produced by the angles formed by the setting of the intestine during the process of desiccation.*"

Kohlrausch locates one important fold, the plica transversalis recti, at the same point that Houston locates the most constant of the valves, projecting well from the right side of the bowel, forming a little more than a semicircle and running farther on the anterior than on the posterior wall. Kohlrausch says that this fold is known as the sphincter ani tertius, though he does not think that the anatomical conditions justify the title, *as the circular muscular fibers do not enter into the structure and are not developed more here than elsewhere.*

Sappey describes the bowel in its empty state as presenting various folds of mucous membrane, having no determinate direction and but slightly marked. Of thirty recta examined he found but three that answer at all to Houston's chief valve or Kohlrausch's plica transversalis recti. *He says that there is no proof that these folds persist when the rectum is full, but that they probably are effaced by distension, and that it is an abuse of language to apply the name valve to them.*

Henle says that there is but one permanent valve, the plica transversalis recti, which is present only in a minority of subjects.

Rosswinkler locates and describes two folds, but locates them differently from several of the other authorities.

The elaborate investigations on the cadaver by Otis led him to say that the rectum consists of large saccular dilatations marked off from each other by intermediate partitions or folds, projecting alternately from left to right, one beyond the other. And, agreeing with Houston, he says that these partitions or folds are semilunar in shape, involve rather more than one half of the circumference of the internal surface, extend a little farther on the anterior than on the posterior wall and project at the center, where they are deepest, from one to two-and-one-half centimeters into the lumen of the bowel. The number of visible folds of this kind found by him was always two or three, two of which

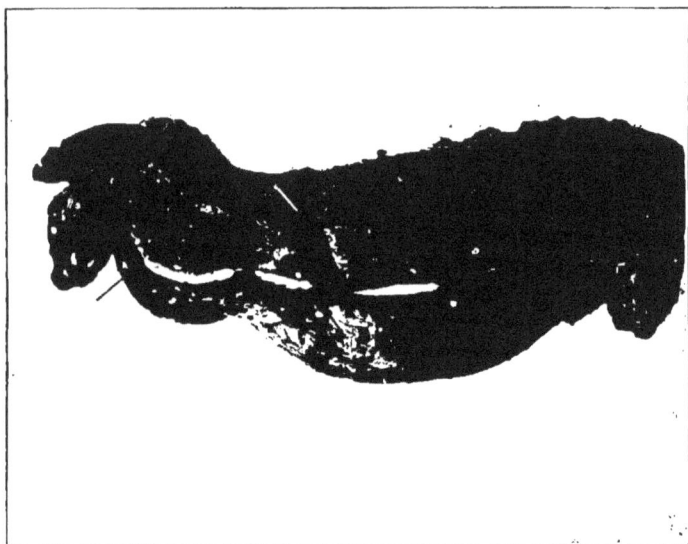

Fig. VIII. Specimen B—Posterior view of rectum cut in bilateral halves, replaced and pinned to its cast.

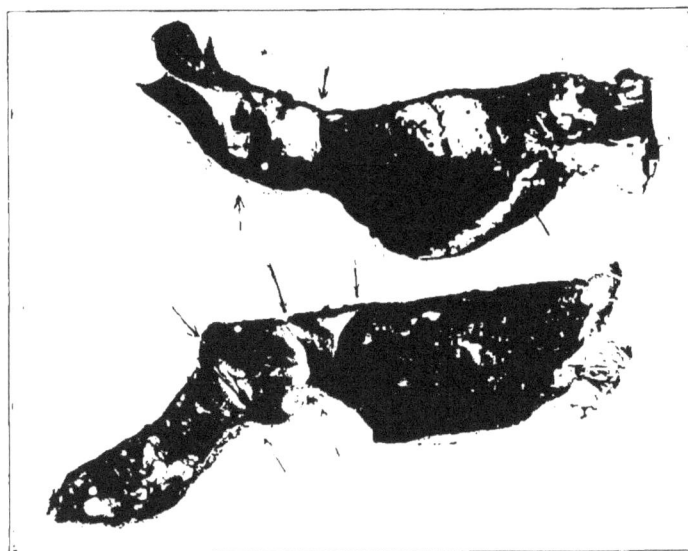

Fig. IX. Specimen B—Right half pictured on the left and the left on the right—very moderate distension.

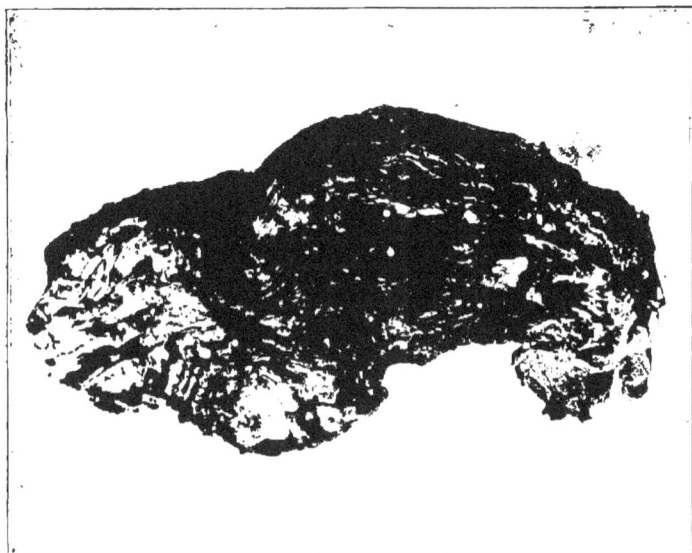

Fig. X. Specimen C—Posterior view—gut cut in median halves and pinned to its cast.

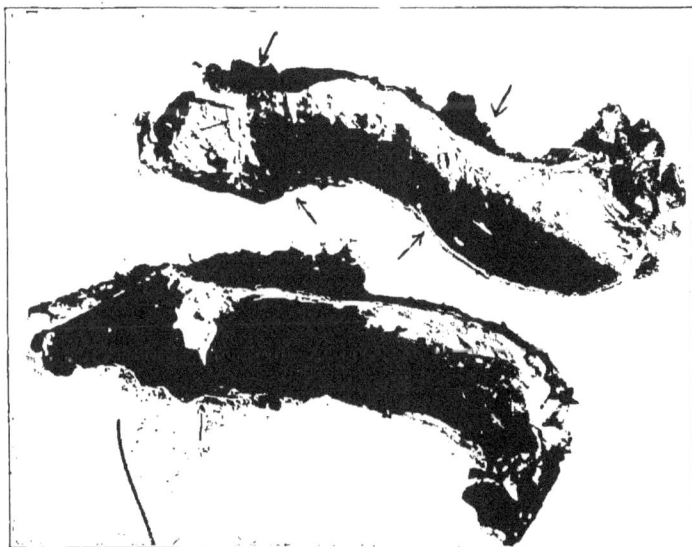

Fig. XI. Specimen C—The right half on the left and the left shown at the right. The uppermost valve shown on the right is a half inch higher than its fellow.

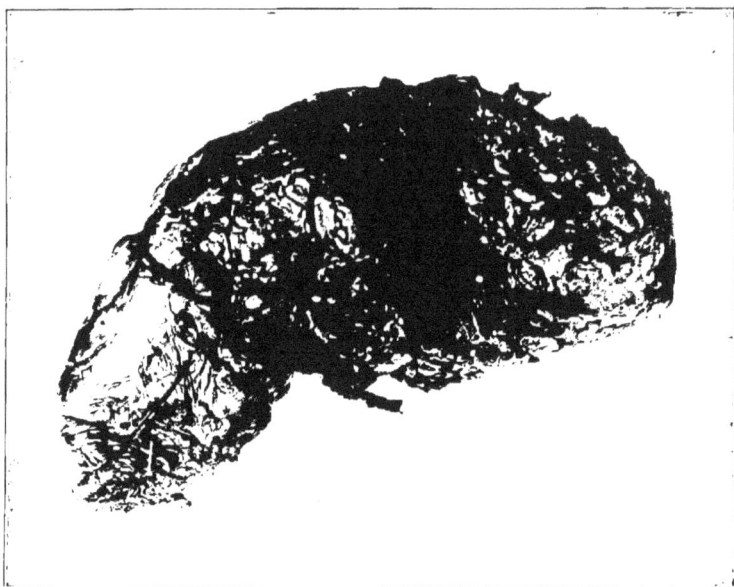

Fig. XII. Specimen D—Posterior view.

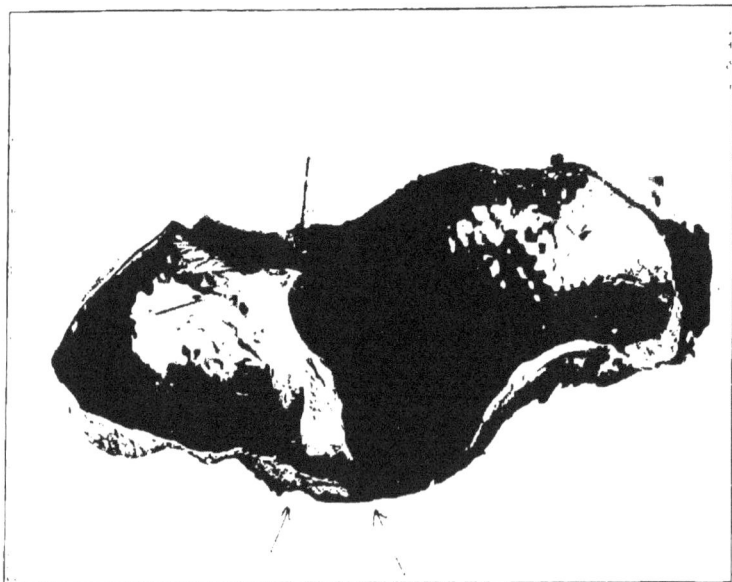

Fig. XIII. Specimen D—Left half.

were constant, the other variable. He also locates these valves as did Houston, and continues : " The folds described within the bowel are *composed of mucous membrane and bands of circular muscular fiber in greater or less proportions.* The longitudinal fibers do not enter into the construction of the folds."

Kelsey comments on Otis' report as follows : " Excepting this description of the arrangement of the muscular fibers and folds of mucous membrane is more exact and definite than any previously given, and as to this constancy of location my own observation does not lead me to entirely agree, the author's conclusions from his dissections are *not different from those of other writers.*"

Mathews,* 1893, says : " Mr. Houston described some *ineffaceable* folds, which have received the name of Houston's semilunar valves. That the student may have an opportunity of looking for them I will give the location where it is said they can be found." Quotes Houston's statements of their location, adds that it is claimed by Houston that their use is to support the fecal mass, and continues : " I have been thus explicit for the reason that I deny their existence, and if they did exist I would deny that their use is ' to support the fecal mass.' For many years I have searched for these folds and have yet to encounter them."

Mathews discusses the subject, and asks : " Is there a third sphincter muscle ? " and answers that Kelsey, in his work on the Diseases of the Rectum and Anus, page 39, says : " From a study of the literature of this question, and from the results of dissections and experiments which we have been able to make, we are led to the following conclusions : (1) What has been so often and so differently described as a third or superior sphincter-ani muscle is in reality nothing more than a band of areolar muscular fibers of the rectum. (2) This band is not constant in its situation or size, and may be found anywhere over an area of three inches in the upper part of the rectum. (3) The folds of mucous membrane (Houston's valves) which have been associated with these bands of muscular tissue stand in no necessary relation to them, being inconstant and varying much in size and position in different persons. (4) There is nothing in the physiology of the act of defecation, as at present understood, or in the fact of a certain amount of continence of feces after

* Diseases of the Rectum, Anus, and Sigmoid Flexure, page 37. D. Appleton & Co. New York.

23

extirpation of the anus, which necessitates the idea of the existence of a superior sphincter. (5) When a fold of mucous membrane is found which contains muscular tissue, *and is firm enough to act as a barrier to the descent of the feces, the arrangement may fairly be considered an abnormity, and is very apt to produce the usual signs of stricture.*"

Mathews then adds : " The only exception I would make to any of these is to No. 2, which says, ' This band is not constant in its situation or size.' I would beg to amend by saying that the band in many instances is entirely absent. I quite agree with all these conclusions of Kelsey, but would relegate the third or superior sphincter-ani muscle to the company of ' Houston's valves ' and to the ' pockets and papillæ'."

Gant,* 1896, expresses himself as follows on this subject : " Internally the rectum presents three or four transverse folds. According to Houston the largest one is situated three inches (7.62 centimeters) above the anus," etc., quoting Houston ; and in conclusion Gant says, " *The folds become almost obliterated when the bowel is distended.*"

After a very critical study of the methods heretofore employed in the investigation of this subject and of the results which have been obtained, at the expense of considerable time and no inconsiderable amount of labor in personal investigation made by means and methods original in character, I believe I am now able to reconcile the very diverse opinions which are shown by my quotations.

It is not improper to assume that, if the judgment of trained observers be equal, their description of the thing considered will vary in the main, only as does the medium through which the view of each is obtained. Our critical review of the literature on this subject has revealed two important facts, viz., that observers employing like means of investigation adduce almost identical evidence, and that the more nearly the method of one approaches that of the other the more in accord are the conclusions reached. A careful and practical study on the living and dead subjects of the various methods employed by these observers results practically and logically in a conclusion as harmonious as it has heretofore appeared contradictory, and proves that for a

*Diseases of the Rectum and Anus, 1896, page 10. The F. A. Davis Company, Philadelphia.

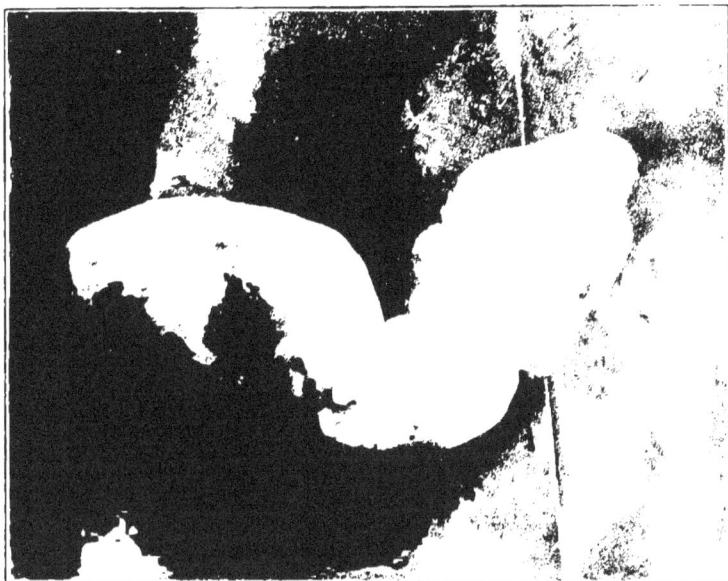

Fig. XIV. Paraffin cast of specimen D—Posterior view showing a single valve mark above with two below.

Fig. XV. Specimen E—Posterior view of a specimen carefully dissected to show the muscular supply to the valve bases.

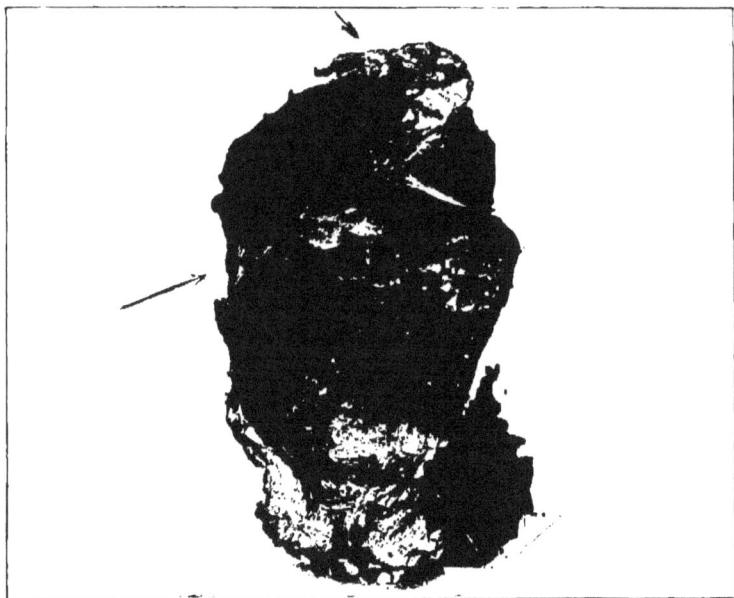

Fig. XVI.   Specimen E—Anterior half.

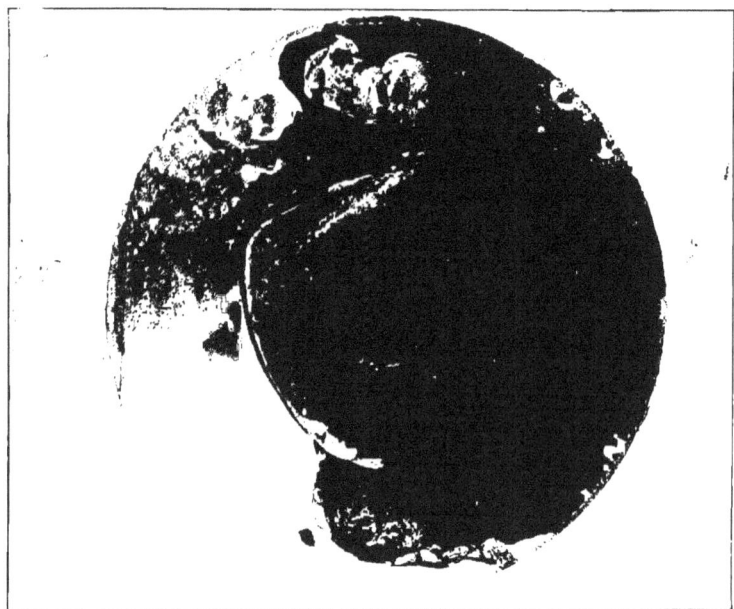

Fig. XVII.   Specimen E—Posterior half occupied by its cast.

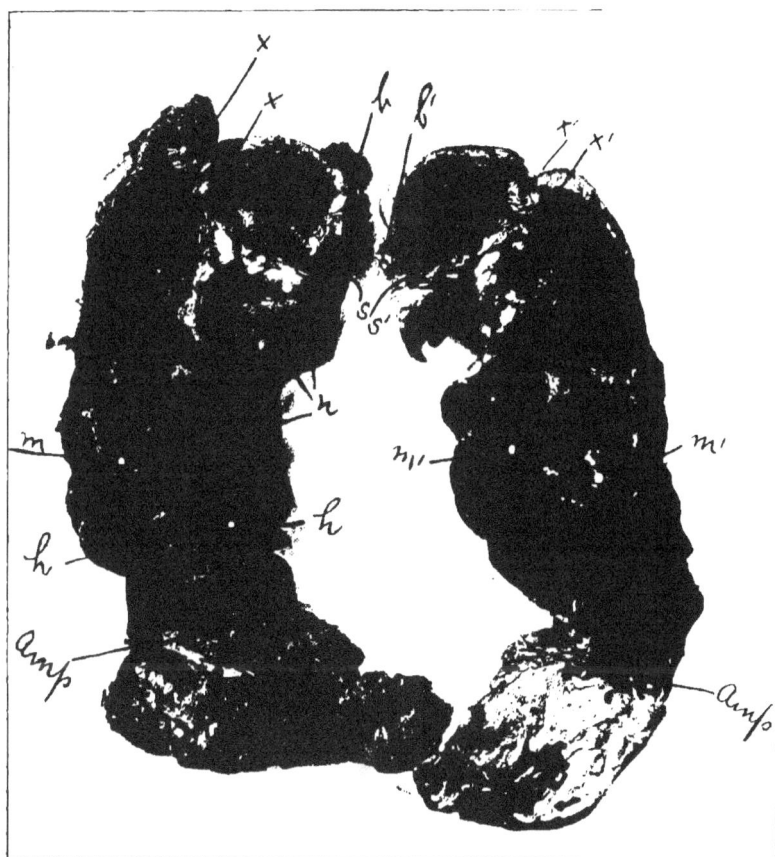

Fig. XIX. Specimen F. Amp., anal end of ampulla; s s. sigmoido-rectal communication; x x. x′ x′. a semilunar valve dividing the upper and middle portions of the rectum; m. the beginning of a semilunar valve on the right of the posterior wall; and m′ m′. the same valve continued over the anterior wall; h h. a small valve not prominent because of the empty state of the gut at this point ; b, b′. sigmoido-rectal valve; n, a portion of mesentery.

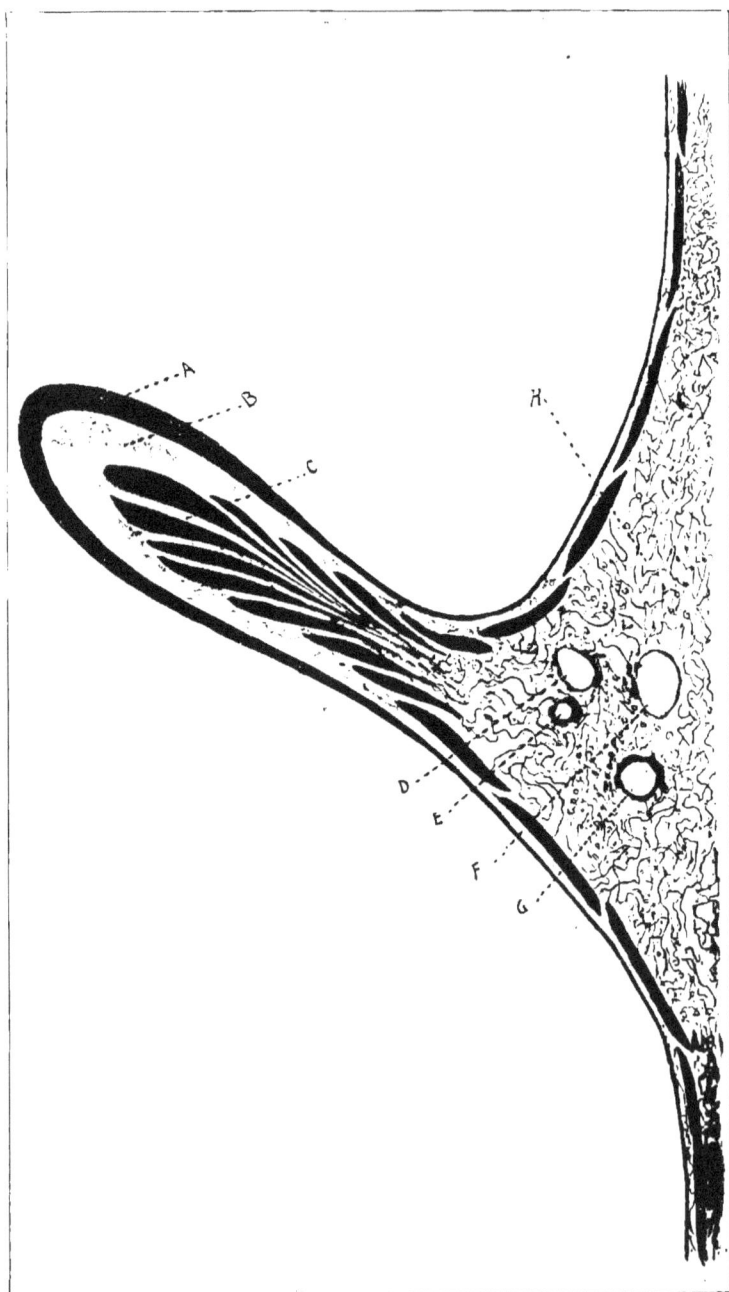

Fig. XX—A semilunar valve magnified fifteen times : A, mucous membrane ; B, fibrous tissue ; C, bundles of circular muscular fibers ; D, vein ; E, artery ; F, vein ; G, artery ; H, areolar and adipose tissue.

period of over sixty-five years these gentlemen have been discussing the same anatomical feature but from quite different points of view.

Houston distended and hardened the rectum *in situ* with spirit. On mesial section of the subject the gut presented valve-like folds unvarying in constancy in all subjects, but in varying number, and in different location in different subjects. He declared their structure to be a duplicature of mucous membrane and bundles of circular muscular fibers. Others, recognizing that in moderate distension the mucous membrane is loosely adherent in the lower rectum, insist that under the conditions employed by Houston the membrane would assume the same appearance in the upper portions, and therefore conclude that these obstructions are accidental folds and not valves ; and, as Houston did not support his statement by attributing to these valves the histological element which anatomists recognize as the essential element of a valve, the opinion of his opponents is seemingly reasonable, but is nevertheless mistaken, as the mucous membrane of the upper rectum is more closely adherent to the muscular wall than is that of the lower.

Hyrtl employed atmospheric distension after removal of the gut and observed an appreciable thickening of the wall of the rectum beneath the mucous membrane, and with apparent good reason assumed this thickening to be muscle only. Under the same manipulations a valve may be made to lose its valvular feature and seem to support this view.

Velpeau supported Nelaton's claim for the superior sphincter by removing the rectum and turning it inside out, so that its mucous membrane was external, and then by inflation demonstrated a marked constriction on the now external surface which was distinctly claimed to be a muscular band. It is very easy to understand how that which is a true valve on the interior of the gut would become a constricting band when the rectum is thus turned.

Dr. Chadwick deserves the very highest praise for having discovered by digital exploration the lowermost of these valves, which he declared to be a detrusor fecium muscle instead of a valve. I find that these valves, when not the seat of disease, frequently elude the finger of average length or are inaccessible

to it; and, as the uppermost is seldom less than five inches from
the anus, this means of determining their presence is not usually
a very practical or satisfactory one, as was very recently proven
in a case in which the lowermost of Houston's valves was mal-
formed into a congenital annular or diaphragmatic stricture with
a central aperture, which although within two inches of the anus
escaped my own digital perception and that of a dozen medical
men in attendance at my clinic and was not discovered until subsequently revealed by proctoscopy.

The method by which I first obtained a satisfactory view of
the semilunar valves in the living subject was described by me
in the July number of MATHEWS' MEDICAL QUARTERLY.

This anatomical feature of the rectum has been macroscopic-
ally examined by many investigators, but the macroscopical
appearance as valves has been described with any degree of
accuracy only by Houston and, with perhaps a little more detail,
by Otis. In the examination of more than a hundred living
subjects I have found evidence to corroborate the testimony of
these two investigators (Houston and Otis), and have compiled
statistics from my cases to recount which would perhaps be
wearisome repetition. If the student will take the pains to
employ my method of proctoscopy he may readily verify what
I have said regarding the appearance of these valves in the
article to which he is referred. The absolute proof which I
present of the existence of these valves is found in the photographic reproductions. The very natural manner of preparing
the specimens for necroscopical examination was also described
in the paper mentioned. After a week's immersion in alcohol
of the paraffin-filled gut the specimen was longitudinally divided
into halves and the interior photographed. The drawing on
page 328 is pictorial testimony of what I believe to be a discovery,
viz., the characteristic element of the semilunar valve of the
rectum is the fibrous, almost tendonous band of tissue which
occupies its free border and underlies the entire surface of the
valve. The fibrous tissue characterizes the valve, the muscular
tissue does not occupy its free border and is not its salient feature,
although it is an important one. The arrangement of the blood-
vessels in its base is evidence of especial provision for its
nutrition.

The photographs demonstrate the presence of the obstructions under discussion; the sketch, which was drawn from the valve while under the microscopic lens, exhibits their character and proves it to be that of *a typical anatomical valve*, and the absence of any other such thickening of the wall in this gut is evidence that the semilunar valves and the so-called plica transversalis, sphincter ani tertius, superior sphincter, and detrusor fæcium muscles are one and the same thing, and *that thing is essentially a valve*, which is most prominent when the gut is most distended.

791 Prospect Street.

AUTHORS QUOTED.—Bushe, Chadwick, Cloquet, Gant, Henle, Hyrtl, Houston, Kohlrausch, Kelsey, Mathews, Nelaton, Otis, O'Bierne. Petrequin, Rosswinkler, Suppey, Velpeau. Van Buren.

THE TREATMENT OF CANCER OF THE RECTUM.*

BY LEWIS H. ADLER, JR., M. D.,

Professor of Diseases of the Rectum, Philadelphia Polyclinic and Post-Graduate College; Surgeon to the Charity Hospital, and to the Out-Patient Department of the Episcopal Hospital.

PHILADELPHIA, PENN.

The question of what surgical procedure is justifiable in any given case of cancer of the rectum is frequently hard to decide, especially from the standpoint of relieving the patient the most with the incurrence of the minimum risk.

The recognized operations are four in number: *Extirpation, Colotomy, Posterior Linear Proctotomy,* and *Curettage.*

Extirpation. The ideal method of treating cancer of the rectum would be by extirpation, as is done in cases of the same disease when the mammary gland is the seat of trouble; but unfortunately it is not often that the rectal neoplasm is discovered in time to permit of the entire removal of the growth and of all glandular involvement, consequently it is my belief that the cases in which this operation is indicated will always be confined to a relatively small number.

Colotomy. On the other hand, colotomy is quite practicable in a large number of instances, and the benefits derived from its

*Read at the meeting of the American Medical Association held at Atlanta, Ga., May, 1896.

performance are thus minutely described by Dr. Chas. B. Kelsey :†
" It relieves pain; does away with the constant tenesmus and discharges from the rectum, which by their exhausting effects are the
immediate cause of death; delays the development of the disease
by preventing the straining and congestion of defecation; prevents absolutely the complication of intestinal obstruction, which
is another cause of death; enables the patient to sleep, eat, and
gain flesh, and often makes him think himself cured in spite of the
plainest prognosis to the contrary. Instead of passing his days and
nights upon the commode, wearing out his life in the effort to
free the bowel from its irritation, he has one or perhaps two solid
fecal evacuations from the groin in twenty-four hours."

The foregoing description of the benefits derived from a
colotomy is no exaggeration. It is the operation to which I
would cheerfully submit were I a sufferer from cancer of the
rectum which had reached the stage of operative interference.
In this connection I would cite the following case which I consider as a typical one for the operation of inguinal colotomy: T.
J. P., a male American, aged fifty-two, a clerk by occupation,
first consulted me in 1893. At that time his history was as follows: Family history: Parents dead, one at age of sixty, of
pneumonia, the other of rheumatism at eighty; oldest brother
died at forty of gastric cancer; patient the youngest of seven
children. Personal history : healthy as a child ; married at forty;
with the exception of lumbago and several attacks of presumed
malaria he has enjoyed good health up to the present illness. He
has been a moderate user of alcohol but an inveterate smoker; no
venereal trouble. For many years has been constipated. Present trouble began about four years before his visit to me. The
pain in his back which he attributed to lumbago became worse.
His bowels became loose. In the morning he would have apparently a normal fecal stool and later in the day one or more
watery passages. The movements were attended by considerable
straining and a sensation as though the bowel had not been entirely emptied. The urine was voided frequently during the day
and he was obliged to empty his bladder several times during
the night. Later on he noticed that his passages, when formed,
were not as thick as they had been, but what caused his anxiety

was the fact that he passed about a tablespoonful of an offensive bloody fluid several times daily, and the movements were accompanied by a severe pain over the sacrum. These symptoms becoming aggravated and the patient being unable to rest comfortably in any position, night or day, together with the fact that he was steadily losing flesh and had no appetite, led him to consult me.

At this time examination revealed the following condition : Varicose veins of both legs, but much worse on the left side. Cachexia not marked. External and internal hemorrhoids noted. A growth was detected about four inches from the anus, which was quite hard and nodular. Its upper limits could not be ascertained.

Consent was given to perform an inguinal colotomy.

Under ether anesthesia, the usual incision was made into the abdominal cavity and the peritoneum was drawn up and attached to the skin by sutures. The colon readily presented, and at the site selected for the artificial anus a loop of it was brought out into the wound and a glass rod four inches long was passed at a right angle to the line of incision under the colon through a puncture made in the mesentery; this was done so as to fix the bowel in the desired position. The intestine was then sutured around the wound to the skin and parietal peritoneum ; seven or eight fine silk sutures being used on either side and the last suture at each angle being placed across from one side to the other. The suturing of the bowel in this manner prevented the protusion of the small intestine and consequent danger of strangulation.*

The dressing consisted of gauze placed between the glass rod and skin so as to prevent undue pressure, and the lower strata of gauze overlying the intestine was soaked in carbolized olive oil, to obviate any subsequent adhesion to the bowel from the lymph filling the meshes of the gauze, which is apt to ensue unless this precaution is taken.

The patient did well after the operation, and convalescence was all that could be desired. The intestine was opened on the third day. In two weeks he was allowed to get out of bed and in another week to walk about generally. By means of a home-

*Fatal results from this accident are on record.

made adominal band and the application of pledgets of old linen
and cotton over the artificial anus, he had and still has no difficulty in controlling the passage of wind and fecal matter. A
truss was obtained for him but was not so comfortable as the
abdominal support mentioned.

He gained flesh, was able to eat and sleep well and attend to
his business. The pain and inconvenience previously suffered
were greatly relieved. During this time opiates were occasionally required. At present writing (three years since the operation), and for a year past, he has suffered considerable distress
from pain over the sacrum and has not been able to work. Opium
has to be used daily. The growth has increased and causes considerable vesical disturbance. It is probable that the patient
may live for six months or a year longer.

In an article read before this Asssociation last year,* I went
into a more extended discussion of the relative merits of extirpation and colotomy in the treatment of cancer of the rectum.

Posterior Linear Proctotomy. I have never attempted to
relieve malignant trouble affecting the rectum by means of a
linear proctotomy. In benign stricture I have found it an excellent plan of treatment when combined with the subsequent
use of bougies. Those surgeons who adopt this method for the
relief of rectal cancer speak highly of its efficiency, some going
so far as to claim that it takes the place of both colotomy and
excision. †

Curettage. In the present paper I desire to call attention to
and to emphasize the value of curettage in those cases of cancer
in which the disease is within the lower three inches of the
rectum and its character of such a nature as to permit of its more
or less complete removal by the curette. In selected cases the
operation is followed by a diminution of pain, bearing down sensations, and discharge, and the lumen of the bowel is enlarged.

Some patients object to having an artificial anus, and refuse
to have a colotomy performed ; others consider the curettage of
the growth a less dangerous operation, and prefer a procedure
which to them is less abhorrent than the idea of an artificial
opening in an abnormal position for the passage of feces. In
some of these cases curettage can be done with decided benefit.

*Transactions Surgical Section of the American Medical Association. 1895, pp. 101-104.

†Mr. Chas. B. Ball, F. R. C. S. I., Diseases of the Rectum and Anus. p. 336.

In certain cases the combined operations of colotomy and curettage will afford the patient much more relief than where one or the other procedure is individually adopted. It is true that only temporary relief is afforded by either curettage or colotomy, but in the majority of cases this is all we can offer the patient under any plan of treatment in vogue at the present time.

During the past two years I have curetted the rectum for the relief of malignant trouble in seven different cases. In only one case was the operation deemed unsatisfactory. This was in a patient of Dr. D. F. Greenwald, of Philadelphia, with whom I saw the patient in consultation. I am indebted to the Doctor for the following history: "Mrs. R. was treated by me (Dr. Greenwald) for a recto-vaginal fistula about five years ago. At that time there was no evidence of contraction of the caliber of the bowel and no other symptoms pointing to the presence of cancer. For a period of three years following the operation for fistula she suffered no discomfort and was able to pursue her usual duties. Two years ago she complained of pain during defecation, which was attended by a discharge of glairy mucus, streaked with blood. Before examination a decided decrease or narrowing of the bowel was discovered about two-and-one-half inches above the anus. Antiseptic douches and the use of rectal bougies daily benefited the case for more than a year. She then complained of bearing down pains, and the countenance assumed the characteristic carcinomatous cachexia. These symptoms continued with greater severity until you saw her. After the curetting operation she was unable to leave her bed, her appetite failed, emaciation followed, and death from exhaustion ensued in two months and three days."

Even in this case, which was a most unfavorable one for any operation, the patient was much relieved of pain, and the two months she lived were certainly passed with more comfort than would have been the case had no operative interference been attempted.

The results following curettage are best described by recording the history of a typical case. J. M., a colored man, aged thirty-three, by occupation an engineer, applied for treatment at our clinic at the Polyclinic Hospital about two years ago. Family history negative. Personal history negative, except a possibility of his having had syphilis; this fact the patient denied, but the records at the Pennsylvania Hospital, Philadelphia, where he had been previously treated, indicate otherwise. Present trouble began some three years previous to his coming under my observation, with bloody discharges from the rectum and a feeling of distress about the anus. He stated that on two occasions during this time he had been admitted into the wards of the Pennsylvania Hospital and had the rectum cut, and that this was followed by the passage of bougies. He dreaded so much the passage of the bougies that on each visit to the hospital he left as soon as he could obtain his discharge.

At the time he came under my care he had a more or less constant bloody discharge from the rectum, considerable pain at stool, and for some time following the movement he was very emaciated. Examination revealed a mass occupying the lower two inches of the bowel, part of the growth protruding through the anus. Abdominal palpation indicated the presence of nodular masses within that cavity, especially marked on the left side above the sigmoid flexure. At first he refused operative interference; six months later he consented to enter the hospital.

Upon admission to the wards of the Polyclinic and after the lapse of several days the patient was prepared for operation, and upon the day selected he was etherized, the sphincters were stretched, and the growth was thoroughly scraped away with a sharp spoon curette. A weak solution of permanganate of potash was employed for douching purposes. In some places scissors had to be used to remove the denser and more indurated portions of the mass. The bleeding at the time and after the completion of the operation was not alarming. The parts were fully dusted with iodoform and the rectum was plugged by means of a good-sized piece of gauze attached to a large rubber drainage-tube, which latter had been closely fitted over a glass tube to keep it patulous; the rubber tube at the same time protecting the glass should it by accident break. Around the tube, but within the folds of the gauze, the packing was placed. The dressing was completed by a fine pad of gauze and cotton and a tailed bandage. Recovery from the operation was perfectly satisfactory, and in a month's time the patient was given employment in the hospital as an orderly. He had not been able to work for over a year previously. The gain in flesh was very marked. For a long period he was given the iodide of potash, but any increase above five grains thrice daily produced nausea; no good effect was noticed from its use. Four months later the patient was in such good health that he married.

A short while after this event the growth in the rectum was noticed returning and the discharge reappeared. Some pain was also complained of and later on defecation was accompanied by considerable distress and difficulty.

Six months after the first curettage he was again subjected to a similar operation. Ten days subsequently he stated that he felt well and was free from pain. Graduated bougies were now used daily.

A month later marked tympanites suddenly appeared and the patient was unable to pass either wind or feces. Various laxatives were employed and several enemas administered before the bowels were made to move, which was not until the third day. Several similar attacks ensued which occasioned considerable trouble before relief was obtained. Finally, the patient became confined to his bed through sheer weakness, and death occurred from exhaustion about nineteen months from the time I first saw the patient.

The autopsy revealed cancerous involvement of the liver and intestine.

In conclusion I would roughly summarize the indications for the operative treatment of rectal cases, thus: Extirpation to be considered in those cases in which the disease admits of the hope of obtaining a permanent cure; colotomy, when the rectum is involved above the lower three inches of the bowel and the disease has produced an appreciable obstruction; curettage, or a posterior linear proctotomy, or the two combined, for those cases in which the disease occupies the lower three inches of the rectum.

1610 Arch Street.

REPORT OF A CASE OF RECTAL PERI-PHLEBITIS TREATED WITH GALVANISM.*

BY DAN'L B. D. BEAVER, M. D.,

Gynecologic and Ophthalmic Surgeon to St. Joseph's Hospital.

READING, PA.

As the title of this paper indicates, it is simply a clinical report of the use of galvanism in a case of peri-phlebitis.

J. R., aged fifty-four, stout, a free liver, developed hemorrhoids several times during the past ten years which I cured with injections of carbolic acid. Two years ago, while fishing and wading a stream in damp, chilly weather, he took cold and came home with an attack of acute rectal peri-phlebitis. The local manifestations of the disease were constant pain, tenderness, swelling with a sense of fullness just inside the anus, and intense suffering with defecation. Upon digital examination of the rectum a hard, painful, cord-like swelling, thick as a finger, was found to extend from the anus upward as high as the finger could reach. Two weeks' rest in bed, with leeches and fomentations locally and anodynes internally, relieved him.

Last fall he again became subject to discomfort in the lower part of the rectum, which in a few months gradually developed into severe pain with and following defecation, for which he called on me about the first of December. The rectum had by this time become so sore that he could not bear the pain incident to passage of feces without first filling the lower bowel with water. He had, for two weeks, been moving the bowels daily with large injections of soapy water, and yet each evacuation was followed by excruciating pain, which compelled him to lie down several hours. After the subsidence of this pain he would feel well and follow his daily routine of work and recreation until the time for the next movement. Then the same round of suffering had again to be endured.

Examination of the rectum revealed a hard, tender cord extending upward three inches in the wall of the bowel on the left posterior side. It was very painful under pressure, incompressible, very slightly movable, and presented an irregular surface.

°Read at the annual meeting of the American Electro-Therapeutic Association, Boston, September 30, 1896.

As he could select the time for defecation and its accompanying pain he went through it in the early hours of evening, and so secured good sleep at night and freedom from pain during the day. Under these circumstances he could not be persuaded to leave his business and get the benefit of rest in bed.

The treatment had to be adapted then to out-door life with exposure to the changes of weather common to our cold season. He was directed to use anodyne and astringent suppositories, and keep the bowels soluble with saline and mercurial aperients. This treatment was continued one month without benefit or even temporary relief. During its continuance it was necessary daily to use half a grain of morphia per rectum after evacuation of the bowels, and take a recumbent position for several hours to relieve pain.

The patient and I having become dissatisfied with the inefficiency of this treatment, I suggested the use of galvanism, although I could not find in the literature at my command any record of its use in this disease.

Galvanism was now applied, cathode in the rectum and anode on the left side of sacrum and buttock, two to three m. for ten minutes, twice a week. After the first dose he felt sore, and consequently somewhat discouraged, but after the second he was elated, saying that it had given him more relief than all that had previously been done.

The improvement now became steadily progressive, so that at the end of two months, and after sixteen applications, the soreness, pain, and swelling had disappeared entirely, and the bowels moved without artificial aid and without discomfort of any kind. The morphia suppositories were required during the first week of the treatment with galvanism, but after that no drugs were used excepting an occasional aperient after the discontinuance of the soap-water injections at the end of the second week.

This was undoubtedly a case of chronic peri-phlebitis, and may possibly have had its origin in small, lingering, morbid foci left over from the previous acute inflammation of the same parts. The tenderness and swelling were in precisely the same position in both attacks. In the first the swelling was more diffused, softer, and with smoother surface than in this, and it was accompanied by fever and anorexia.

I was led to try galvanism in this case by our common knowledge of its resolvent action upon inflammatory indurations in peri-uterine tissue, and by my own observation of its good effect upon the indurated cutaneous and deeper tissue around chronic varicose ulcers of the leg.

The report of this case is offered for record in our transactions because it exhibits the very remarkable effects of electricity in a typical instance of a disease that is regarded by all persons who have had experience in the treatment of disorders of the rectum as annoying and uncertain in its course, slow to disappear under the use of drugs, and prone to recurrence.

GROWTHS OF THE RECTUM; PRESENTATION OF SPECIMENS WITH MICRO-PHOTOGRAPHS, AND REPORT OF CASES.*

BY H. O. WALKER, M. D.,

DETROIT, MICHIGAN.

Gentlemen, the presentation of these specimens and micro-photographs of their structure, together with their histories, will serve as a text for a few brief remarks on growths of the rectum.

Specimen No. 1. This was removed March 6, 1896, from Mrs. R., aged twenty-five. Since the birth of her child, eighteen months previous, she had suffered with rectal tenesmus, hemorrhage, and ichorous discharges from the rectum. When she entered Harper Hospital, March 3d, under the care of Dr. J. H. Carstens, who referred her to me, she was very much emaciated and exsanguinated ; in fact on several occasions she had very nearly bled to death. Rectal examination revealed a very large fungous tumor attached to the left rectal wall, its upper margin being about three inches above the anal orifice. The base of attachment was about two inches in diameter. The tumor was seized and drawn through the anal orifice when it was transfixed well below its base with ligatures and clipped away with scissors.

* Read before the Northern Tri-State Medical Association at Angola, Indiana, July 21, 1896.

Specimen 1.

Specimen 3.

Specimen 2.

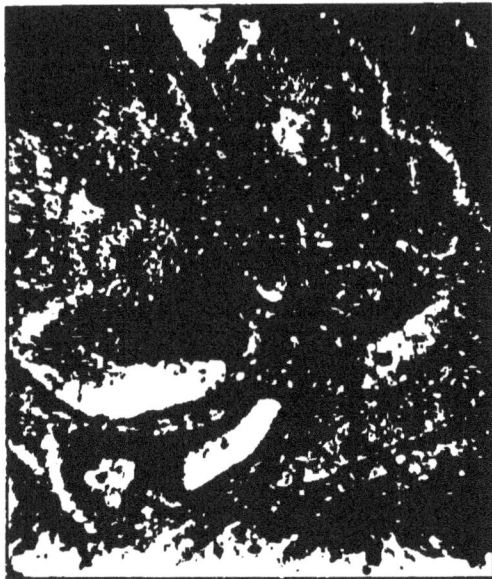

FIG. 1.—Mrs. R.—The section shows tubules irregularly arranged and of varying size, the lumen of which are filled with epithelial cells of more than one layer in depth. Diagnosis—carcinoma (tubular).

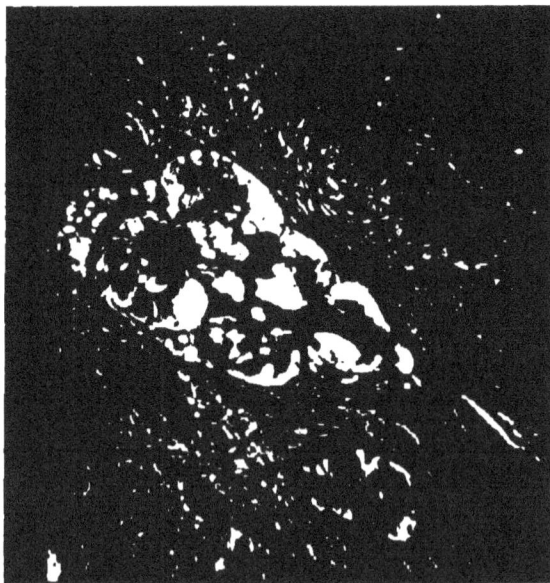

FIG. 2.—Proctor—The specimen on section shows a stroma of fibrous tissue in the spaces of which the epithelial cells are proliferating very rapidly. Here and there are scattered masses of epithelial cells, some of which have undergone colloid degeneration. Diagnosis—colloid carcinoma.

That portion of the rectum to which the growth was attached was also cut away, and afterward approximated with catgut sutures. There were no complications beyond the rectal wall. The hemorrhage was considerable, and it was with difficulty that she was gotten off the table alive. After forty-eight hours she made an uninterrupted recovery, and left the hospital on the sixteenth day. I heard from her three or four weeks ago, when she had gained between thirty and forty pounds in weight with no further evidence of rectal trouble. The micro-photograph, Fig. 1, is described as follows by the microscopist, Dr. Hickey : " The section shows tubules irregularly arranged and of varying size, the lumina of which are filled with epithelial cells of more than one layer in depth. Diagnosis, carcinoma (tubular)." It is quite probable, if the microscopist is correct, that there will be a recurrence of the trouble.

Specimen No. 2. This was removed June 11, 1896, from Mr. J. E. P., aged fifty-three, sent to me by Dr. Macnamara, of Blenheim, Ontario. It was first observed by Dr. M. about six months previously, when it was quite small, just above the internal sphincter. Removal was advised but rejected. When he entered Harper Hospital it involved nearly two thirds of the sphincters on the left, extending well into the ischio-rectal fossa. Its feel was extremely hard. In removing it I was compelled to cut away more than two thirds of the anal orifice, necessarily destroying the office of the sphincters. The recovery from the operation was rapid, and he left for his home on the twelfth day. Although the dissection was well cut from the confines of the growth, its recurrence may be looked for.

The micro-photograph, Fig. 2, is described as follows by the microscopist : " The specimens on section show a stroma of fibrous tissue in the spaces of which the epithelial cells are proliferating very rapidly. Here and there are scattered masses of epithelial cells, some of which have undergone colloid degeneration. Diagnosis, colloid carcinoma.

Specimen 3 was removed from a little girl, aged two years and three weeks, May 22, 1896. The growth was observable at birth but did not produce any inconvenience until about six months ago, when constipation with difficult defecation was apparent. This gradually continued up to May 14th, when I first saw her. Inspection showed a much enlarged buttock of the right side, and digital examination of the rectum revealed a large mass nearly filling the pelvic cavity.

The dissection was difficult, the growth extending up to the brim of the pelvis, and although its removal was tedious there was but little hemorrhage. There was no shock following the operation, and she made a rapid recovery.

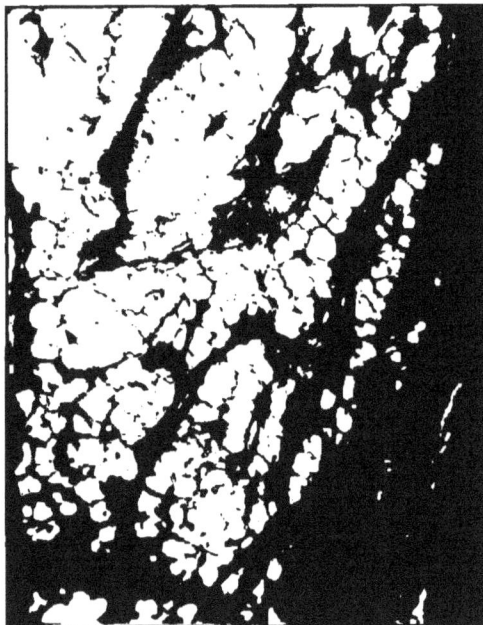

FIG. 3.—Miss G.—This section exhibits fat cells and a dense fibrous tissue. Diagnosis, fibro-lipoma.

Fig. No. 3. This section exhibits fat cells and a dense fibrous tissue. Diagnosis, fibro-lipoma.

Tumors of the rectum, as elsewhere in the body, are either benign or malignant. Benign tumors in contradistinction to malignant tumors of the rectum as a rule have a pedunculated attachment, while the latter have a broad and short base. They are classified as polypi of the soft (adenoid) and hard (fibrous) varieties, and the treatment is excision by any method you choose. Villous growths, although usually non-malignant, are quite liable to become malignant and necessitate a much more radical operation for their removal. Malignant growths of the rectum are in the majority of cases situated at the lower part, and are amenable to excision, and if done early enough are not liable to recur, while those having an origin high up are less liable to a cognizance of their existence until stenosis makes inspection a necessity, and they are then so far advanced that operation by excision is impracticable. Medication locally or internally of a cancer of the rectum is of no avail. It is either excision or a colotomy—excision when confined to the lower rectum and its walls, while advancement into the surrounding tissues or located high up, a colotomy is the only thing to do, which of course is only palliative; but the palliation is so great that it should always be done. Although at first I was averse to making a colotomy, yet my experience with it of late years has been of such a satisfactory character that I do not hesitate to recommend it to a patient with a well-advanced cancer, for I have seen several bedridden, emaciated patients who were suffering extreme torture from its presence pass the remainder of their lives with comparative comfort and without doubt prolonged. Excision that demands a Kraske had better not be done. Better do a colotomy first and a Kraske afterward.

The little girl with the fibro-lipoma scarcely comes within the scope of a rectal tumor, as it did not involve the walls of the rectum, yet its close proximity will be sufficient excuse for its presentation at this time. Its etiology classes it with congenital tumors dependent upon intra-uterine influences, and is not governed in its career as are tumors of postnatal origin.

GASTRO-INTESTINAL DISEASE.

INTESTINAL INDIGESTION.*

BY J. N. UPSHUR, M. D.,

Professor of Practice of Medicine in the Medical College of Virginia, etc.

RICHMOND, VA.

[Written for MATHEWS' MEDICAL QUARTERLY.]

A malady of sufficient importance to be the selected subject for discussion must have something in its nature, causes, and management which indicates a feeling of unrest, or a groping after the light, which urges to the solution, because experience has proven that relief and cure do not come always as the result of earnest clinical work, and that disappointment is not uncommonly our portion in dealing with the affection. I would ask why this is so? Is it not because oftentimes the therapy is merely haphazard and empirical, the result of superficial investigation and consequently inaccurate diagnosis? The nature and the cause of the trouble is not properly appreciated, and consequently there is not a rational therapy. By systematic study of the physiology of digestion, both stomachic and intestinal, at the risk of being too prolix I shall endeavor to make prominent those fundamental underlying principles so essential to properly appreciate the indications for the application of a rational and curative therapy.

It must be remembered that all digestion begins in the mouth; there must be proper mastication and insalivation, that the food may be prepared for perfect solution in the stomach and duodenum of those elements digested by these organs respectively. A failure of any part of this natural process is the initial step in the development of indigestion, whether it be stomachic or intestinal. The character and variety of diet may, nay does, determine whether it be stomachic or intestinal, and a faulty performance of function on the part of the stomach may be the initial point in disturbed function of the bowel. The normal condition of the small intestine is acid immediately below the stomach; the effect of the pan-

*Being subject selected for discussion at the 27th annual session of the Medical Society of Virginia.

creatic and intestinal juices and bile is to make it first neutral
and then alkaline, and this condition obtains along the whole
of the small intestine. In the large intestine the absorptive
function is greater than the secretory. The cecum is rich in
lymphatics—fermentative and putrefactive changes go on in the
large intestine; one of the products of this putrefactive process
is the developed antiseptic substances which kill micro-organisms
(Wernich). We may therefore assume that these substances
limit to a certain extent putrefactive change (Landois & Sterling,
p. 377). All articles digested in the stomach are nitrogenous;
starches, sugars, fats find the condition for their digestion in the
duodenum. The intestinal and pancreatic juices and bile enter
chiefly into the process. The peculiar function of the bile is to
promote absorption by its contact with the mucous membrane of
the bowel, the first step toward assimilation after digestion is
completed. It is also the natural antiseptic and purgative of the
body. The intestinal canal is rich in sympathetic nerve supply,
the nervous influence being the motive stimulant to the performance of function.

I have called attention to these necessary factors in the normal
digestion, because the disorder of any one of the functions may
stand in a causative relation to the development and protracted
continuance of intestinal indigestion. Nor are the functional
derangements which disturb normal digestion in the bowel confined to that viscus, but may, aye often do, lie outside of it.
The fermentative action which finds its culmination in the
bowel often begins in a faulty stomachic digestion. This may be
due to various and varied conditions. The food introduced into
the stomach may be of such a character as to initiate the trouble,
faulty gastric juice, slow peristalsis, thus failing to thoroughly
mix the food with the gastric juice and thereby obtain perfect
solution. The cause may lie within the viscus itself. Catarrhal
mucous lining, dilated stomach, etc., may be the conditions interfering with normal digestion, the essential characteristics of
which are that it must be complete, rapid, and painless. Therefore the function of digestion is a multiple one, the stomach and
intestine mutually dependent upon each other, and when we
find difficulty in the functional action of these organs, it is difficult to say where one ends and the other begins.

24

Definition. We may define intestinal indigestion broadly as *any* disturbance of function or departure from normal action in the process of digestion, attended more or less by pain or discomfort.

Varieties. I would point out as an essential in the management of this trouble the proper appreciation of its existence as a simple functional trouble, or its dependence upon some organic lesion situated either in the intestine itself, in the liver, the pancreas or kidney, thus modifying the constitution of the fluids, the proper character of which is so essential to normal digestion. Therefore we may divide it into *Functional and Organic Intestinal Indigestion.* Further, it may begin in the first and progress to development into the last.

Etiology. It is said that women are oftener affected than men, authorities differing as to the frequency of sex; but overeating and imprudent eating and drinking, irregular living, mental strain from business worry, are causes more common in men. It usually develops between thirty and fifty, but it occurs at all ages and with both sexes. Infants at the breast may be and often are subjects of the malady. The simple affection, infantile colic, which we so lightly regard, is a simple form of intestinal indigestion, due usually to deficient flow of bile and consequent hyperacidity. The development of intestinal dyspepsia in middle life is often the reaping of the wild oats sown in youthful dissipation. Heredity and idiosyncrasy often play a part. Inability in certain families to digest fruit, vegetables, and fats is a part of family history. Heredity does not, however, explain the occurrence of a number of cases in the same family: faulty and improper cooking and irregular hours, defective or impure water, or water containing such mineral properties as render it harsh and irritant.

Whatever tends to impair nutrition resulting in anemias, febrile diseases, rickets, syphilis, indigestion of fats in the strumous diathesis and in consumptive subjects, as well as every thing that impairs functional activity, bad air, deficient exercise, sexual excesses, especially masturbation, depresses function through impaired nerve force. Defective digestion of starches and fats is more commonly the result of excessive mental strain and business worry, because the process is more complex than gastric digestion. Nervous tension incident to the life of physicians, lawyers, and clergymen renders them especially liable to this trouble. The demand upon the sympathetic side of the organism and the expenditure of an unusual amount of nervous force impedes the machine because the motive power is crippled. In children, under the modern system of education, where they are sent to school at too tender an age, so great a demand is made upon the brain by the amount of mental work required, that physical health and development are interfered with, digestive disturbance and faulty assimilation underlying the physical break down. Thus we have a feeble race growing up, costing we can not tell how much, from the standpoint of political economy, by shortening life, inefficient ability for work and a burden upon the State, because of the fact that they become consumers instead of producers.

Wealth and the consequent inactivity from ease and luxury diminishing the demand for food, more of which is taken than the system requires. Sedentary life of book-keepers and clerks, occupations which cause stooping, as in tailors and shoemakers, thus compressing the abdomen at the waist and interfering with the normal peristaltic movements. That iniquity of dress in women, the corset, compressing the abdomen too much, displacing the viscera, and making her an object of pity, because, bowing to the demands of fashion, she ignorantly trammels the body and interferes with its functions for appearance sake. Would that she could learn and act upon the sentiment, "that beauty unadorned is adorned the most," and believe that attempting the physique of a wasp and interfering with proportions is not an addition to the charm of her personality, or a method by which she may attain the greatest of blessings, rude health.

The effect of a hot climate, especially if damp, is to impair the appetite, which is sought to be stimulated by condiments and alcoholic drinks. An amount of food is put into the stomach which it is not competent to digest. The intestine is overwhelmed by a mass of crude material which it can not dispose of, indigestion becomes chronic, alternating at times with attacks of diarrhea or dysentery; it falls to the lot of the liver to dispose of more albuminoid material than it is capable of, uric acid is excessively developed, the liver becomes enlarged, and so a long train of ills is set up. Nor is a hot climate responsible for all this; taking food of a stimulating character, when there is little or no appetite, when the demand of the system does not require it, as for example in fashionable life with those who are in the whirl of society—the eating by courses, too many dishes, etc. On the other hand too little variety in food is bad. Irregular hours and irregular distribution of the time of eating, imperfect mastication and imperfect insalivation interfering with perfect solution of food is a frequent cause. The usual American method of eating in too great haste and allowing no period of repose after eating. The preliminary function must be thoroughly performed and sufficient time allowed to accomplish perfect digestion.

The use of tobacco is also causative. It interferes primarily with the salivary secretion and secondarily with the pancreatic; it also exerts a slow but surely poisonous influence upon the nervous system, it stimulates and then paralyzes the motor nerves of involuntary muscles, also the secretory nerves of glands (Brunton).

Constipation is also causative. Any thing which interferes with general peristalsis, such liver involvement as interferes with the natural outflow of bile, allowing premature and excessive putrefactive changes in the contents of the bowels, excessive acidity in the gastric digestion neutralizing the alkaline changes essential to intestinal digestion. Any disease of the pancreas, catarrh of its duct, or the bowel involving the duct, fatty degeneration, cancer, interfering with the emulsionizing of fats, and causing fatty diarrhea.

Finally, in gastric and intestinal catarrh we find intestinal indigestion more or less prominent, in diseases of heart, lungs or liver, wherever any cause is at work to impede the portal circulation.

Pathology. The nature of the affection and changes to be found are, first, in the interference of function, and secondly, where lesion is present, it is a consequence of long perverted function and the development of organic trouble in some one of the associated organs—stomach, liver, or pancreas. If there be catarrhal lesion of the duodenum, we find the usual gross pathological changes found in catarrh of the stomach or other portions of the intestinal canal.

Symptoms. We are here confronted by a serious difficulty, gastric digestion, dealing as it does with one digestive fluid, is plain in its manifestations; intestinal digestion is more complex,

several fluids entering into the process. The symptoms prominent depend upon the particular fluid deficient, or the way or
avenue through which the faulty nervous stimulus may manifest
itself. We are, therefore, driven to more rigid analysis, and a
more careful and stricter process of exclusion.

It may be acute or chronic, the latter most common. Acute
form may be brought about by imprudence in eating; the matter
entering the intestine can not be physiologically disposed of,
pain, flatulence, borborigmi, fever, coated tongue, headache,
aching in the limbs, rapidly develop and find a culmination in
acute diarrhea. An acute intestinal catarrh being the lesion of
the intestine, if this involves the common bile duct, interfering
with the flow of bile into the intestine, the stools are clay colored, slight jaundice exists, and the urine is loaded with lithates.
Some of the symptoms are reflex, others are due to the absorption into the system of sulphureted hydrogen—thus the lassi-
tude, aching in the limbs, and headache find an explanation. If
the stomach participates we have the symptoms modified. These
attacks are apt to recur at intervals. The intervals become
shorter, until a chronic intestinal indigestion is established.
Food coming from the stomach is not properly prepared for
absorption, decomposition readily takes place, and the prominent
and significant symptom is *pain* in the right hypochondrium, or
about the umbilicus, from one to three hours after eating. The
pain is dull, gives a sensation of weight, not always fixed, and
accompanied by tenderness on pressure. Tympany and borbor-
igmus are present, and the individual experiences a sensation of
fullness and distension. This is caused by the decomposition in
the contents of the bowels. In acute indigestion the rapid formation of gas produces distension and spasm of the gut—acute
colic. When chronic, this is not the case, but the sensation is
one of uneasiness and dull, wearing pain. Constipation is commonly present, due to enervation of the intestinal walls and
overdistension; the dejecta are too dry, sometimes covered with
mucus from the development of chronic intestinal catarrh.
Diarrhea may alternate with the constipation, particles of undigested food may pass off—dark green or black color of stools
indicating excess of or perversion of bile; slate color, its
deficiency; fatty matter, faulty function of the pancreas; appe-

tite is fitful and irregular, not necessarily impaired.   Patient awakes in the morning with a bad taste in the mouth, tongue coated, relaxed, and swollen.

The nervous system is the first to suffer; disturbed function in the intestine, whereby the digestion and assimilation of fats are interfered with, impairs the nutrition of the nervous system, resulting in the depression of spirits, sleepiness, bad dreams, ringing in the ears, vertiginous sensations, pain in the head, confusion of thought, loss of the power of application.   Sometimes epileptiform attacks are traceable to intestinal indigestion, sudden attacks of fainting, collapse, sufficient sometimes to be alarming.   Melancholia sometimes exists.   Cold hands and feet testify to a debilitated circulation ; palpitation is present.   Reflex irritation and impaired nutrition are responsible for heart complication.   Urine high colored, of high specific gravity, loaded with urates, contains a large amount of sediment deposited on cooling.   Sometimes albumin is found ; errors in diet, like eating cheese or pastry, may cause it.   Seminal emissions may occur. Faulty digestion, faulty sexual function, consequent mental depression, reduce the subject to a condition of abject misery.

As the result of malassimilation anemia is an early symptom. Various skin affections, such as eczema, psoriasis, impetigo, etc., supervene to add to the distress of the patient.   Consequent upon disturbed intestinal function is disturbance of the functions of the liver—lithemia results with all of its characteristic symptoms.   Nutrition is still further impaired, emaciation is pronounced, being specially noticeable in those previously stout. Acute intestinal indigestion may terminate in a day or two and the patient be restored perfectly.   Sometimes an obstinate diarrhea is left behind.   Chronic intestinal indigestion, if in children, is apt to persist until the diet is changed.   In adults, if young, phthisis may have its inception here ; in those more advanced, interference with so important a function may result more seriously, and the trouble may last for years. The influence of the intestinal diseases upon the mind alters the whole character of the man.   Every thing seems out of joint; he becomes peevish, fretful, depressed, moody.   It paves the way for some organic trouble, hepatic or renal ; the result of malnutrition closes the scene.

Diagnosis. In the acute form, pain, flatulence, sequence of diarrhea, enables us to determine the nature of the trouble. In the chronic form, the history of the case, the existence of the symptoms, digestive, mental, nervous, and cardiac, evidences of malnutrition, all of these and careful analysis of the symptoms pertaining to liver, pancreas and stomach, will enable us to reach a correct conclusion with little difficulty. It must be borne in mind, however, that the presence of fats in the stools, or a fatty diarrhea, does not always indicate pancreatic disease; it has occurred in ulceration of the duodenum. The existence of glycosuria may be an aid to pointing out pancreatic disease.

Prognosis. Treated promptly and decisively, a cure may be effected, especially in the acute form. The chronic form is not so promising. Of supreme importance is implicit obedience on the part of the patient to orders. Every thing depends upon this, and must be made subordinate to it. This being done, patience and perseverance may and will accomplish much. If the general health has become much impaired, the prognosis is more doubtful. Anemia, debility, coexistent gastric dyspepsia, hypochondria, or a strumous diathesis in children, add much to the gravity of the prognosis. If organic disease of the stomach, liver, pancreas or heart exist, the prognosis is very bad.

Treatment. In the acute form, the first indication is to free the bowel from all irritant matter. I prefer a brisk mercurial purgative for this. Diet should be regulated, pain relieved, if unbearable, at the beginning of the attack, by a hypodermic of morphia. If any intestinal catarrh exists, I know of no remedy superior to the phosphate of soda in doses of thirty grains every two hours, given in a half a teacup of hot water. The tongue cleans promptly under its influence, pain, soreness and flatulence disappear. It should be continued for twenty-four or forty-eight hours. Diet should be simply milk, or some suitable animal broth. In the chronic form the first indication is to put the patient upon a strict but nutritious diet, such a diet as will put the duodenum as much as possible at rest. In the beginning, an exclusive milk diet is best; if it does not suit the patient, it may be given skimmed, or if it still disagrees, or is distasteful, the following preparation which I have found most palatable and beneficial in most cases, to a tumbler of sweet milk,

add the white of one egg whipped up to a light froth and
a wineglassful of lime-water. Such farinaceous food as barley
gruel or oatmeal may be combined with it. Animal broths
may be allowed, or some of the prepared meat essences.
They stimulate the appetite and the secretion of the glands.
When the demand comes for an improved diet, fish boiled,
oysters, sweet bread, the white flesh fowl once a day, soft boiled
eggs, one or two a day. Bread one day old, *sweet and light*,
maccaroni boiled in milk, and rice may be allowed. Coffee is
contra-indicated, and tea should only be allowed when largely
diluted with milk ; cocoa (Phillips' digestible, which contains
extract of pancreatine,) is a very pleasant drink. Butter is to
be given in moderation. Wines are contra-indicated, except in
special conditions—it being determined for each individual upon
its own merits. Mineral waters do good, *by change of air and
scene*, and in those cases where little water is drunk, and as a
means of correcting constipation. Cheese, crabs, lobster, richly
dressed meats, too much fat, pastry, vegetables, are to be avoided.
The patient should be earnestly enjoined as regards the thorough
mastication and insalivation of food, the harmfulness of haste in
eating, the necessity of a sufficient rest before and after eating.
Chewing and smoking should be forbidden as impairing the
character of the saliva and interfering with pancreatic secretion.
Careful examination of the mouth should be made, and if teeth
are imperfect, artificial ones should be substituted for them. All
causes which have combined to bring about the attack should be
removed. If possible, change to a cooler climate ; early hours of
sleep, regularity as to time of and interval between meals
should be earnestly advised. In overworked professional men,
rest from all mental and physical effort should be insisted upon ;
massage and electricity are very useful. The overtaxed child
should be taken from school, and careful effort made for his
physical upbuilding. Sea bathing, properly regulated, is beneficial. If the condition of the patient be so extreme as to
demand it, an artificially digested food may be given, as pepton-
ized milk.

Finally, as to the administration of drugs, sulphuric ether
has been recommended, as it possesses the power of stimulating
the pancreas, given about an hour before meals. Good results

follow the exhibition of pancreatic extract and soda. Gastric digestion should also be cared for, as the more perfectly this function is performed, the better the quality of the substances which go to the duodenum for the completion of digestion. When the liver is at fault, the bichloride of mercury in co. tincture of cinchona is very useful. This organ should also be stimulated to improved action by the administration of benzoate of sodium, or ammonium, or salicylate of sodium, etc. Atony of the intestinal walls, or deficient peristalsis, should be treated by the administration of strychnine given in full doses. I can not too highly recommend its power for good. Quinia in tonic doses is also available, and may be given in combination with it. If flatulence and abdominal distension are present, bismuth subnitrate is indicated in combination with benzosol or salol. In a case recently under my care I found great benefit from large (thirty grains) doses of the hyposulphite of soda. Constipation is to be relieved by enema, or the exhibition of one of the bitter waters. I have found the Rubinat most satisfactory. Fluid extract of cascara is also useful, acting as it does, as a tonic to the intestine and improving the muscular tone. Fowler's solution and tinct. of nux vomica will be found useful; chloride of gold and sodium is a most valuable remedy, its only objection, which may be combated, is that it constipates. This property makes it useful in the diarrheal form. So soon as the patient's condition will admit of it, chalybeate tonics should be administered, backed up by such other remedies as will improve assimilation and nutrition.

210 W. Grace Street.

TREATMENT OF INFLAMMATORY DISEASES OF THE STOMACH.*

BY GUSTAVUS M. BLECH, A. B., M. D.,

DETROIT, MICH.

For several years I have had under my care quite a number of patients afflicted with acute or chronic inflammatory diseases of the gastro-intestinal tract. The records of my clinic (143 of such cases) show that stomach diseases are to my knowledge the most distressing ailments which may afflict human beings. When the stomach is out of order life is a burden and every thing seems to go wrong.

The majority of general practitioners, as far as I could learn, still adhere to the old-fashioned treatment of gastric disorders, and I confess that during the first years of my practicing medicine I have, like others, used remedies which every one of us have prescribed, in order to relieve their patients, and to my great disappointment I never was fortunate enough to cure chronic gastritis by treating the symptoms, although I have occasionally relieved my patients, but only when the disease was not chronic.

You have—as well as myself—prescribed menthol, cocaine, opium, ice, and other remedies to relieve nausea and to stop vomiting; you have cleansed the stomach by lavage and purgatives, and subsequently irritated the lining membrane of that much abused viscus with modern antiseptics; you have called to assistance pepsin and innumerable drugs, but have you cured your patient? No. You have merely lost track of him. The patient did not call again, because the treatment did not do him any good, and frequently because it aggravated his trouble.

So the world goes on and the poor creature afflicted with chronic gastritis goes on suffering more and more. Why did you fail to cure catarrh of the stomach? It is because you merely attempted to relieve the symptoms instead of prescribing remedies to subdue the existing pathological condition, the inflammation of the lining membrane of the stomach, which condition prevents the digestive process from being normal.

* Read before the Mississippi Valley Medical Association at St. Paul, September 16, 1896.

In order to subdue this abnormal inflammatory condition of the wall of the stomach antiseptics are indicated, but you know as well as I do that powerful antiseptics have the same destructive action upon both vegetable cells (germs) and animal cells. Consequently, they will in all cases aggravate the disease.

I am much opposed to the use of strong drugs in my practice on account of sad results which I have witnessed, and I put more stress on harmless, although most powerful antiseptics, than I ever did since I successfully treated hopeless cases of cholera infantum with hydrozone (30 vols. H_2O_2, aqueous solution).

Therefore my method of treatment of all inflammatory diseases of the stomach may be summed up as follows : First, destroy the morbid element which is present in the stomach, so as to thoroughly cleanse the mucous membrane ; second, heal the diseased surface after it has been made aseptic.

As a cleansing agent which acts both mechanically and chemically, I know of nothing as powerful as hydrozone. Therefore I prescribe one tumblerful of lukewarm water containing two per cent. of hydrozone, half an hour or so before meals.

The nascent oxygen which is set free in the stomach by its oxidizing action destroys the morbid element and cleanses the mucous membrane more thoroughly than any thing I know of. This being done, the patient should wait for at least fifteen minutes before taking his meals.

As a healing agent I prescribe one to two teaspoonfuls of glycozone diluted in water to be taken immediately after meals.

The results which I obtained in submitting my patients to the above rational treatment are so gratifying that I do not hesitate to say here that the great majority of cases of stomach disorders may be cured or at least much relieved in a very short time by this treatment, which is already indorsed and used by some of our most skillful practitioners.

On this occasion I wish to state that I cured a well-defined case of gastric ulcer, at least all the characteristic symptoms like circumscribed pain, indigestion, and hematemesis have disappeared for fifteen months under the above treatment, save lavage, which when practiced once caused an alarming hemorrhage. I wrote to the patient, who lives in St. Louis, and he informs me that neither of his symptoms have appeared since I left that city, which was about fifteen months ago. The patient has been instructed to resume the treatment as soon as even the mildest symptoms reappear, but he wrote me that he needed to use no medicine whatever.

While my experience with gastric ulcer is but limited, I could suggest no better treatment ; first, because all usual remedies do not influence the ulcer itself, and second, because I have seen healed the most stubborn cases of ulceration of the cervix and chronic ulcers of the leg under the same method of treatment.

203 East Columbia Street.

THE SIGMOID FLEXURE AND ITS MESO-SIGMOID.

BY BYRON ROBINSON, B. S., M. D.,

Professor of Gynecology in the Post-Graduate Medical School.

CHICAGO, ILL.

[CONTINUED.]

The substance composing the dark inter-endothelial line is of a semi-solid, soft consistence, a jelly-like substance resembling the white of an egg—albuminous. The silver nitrate solution only precipitates the superficial portion of it. The line may be straight, curved, sinuous or serrated, resembling the cranial sutures. Situated on this dark, inter-endothelial line are two structures ; stomata vera and spuria which have served as a bone of contention and as a field for interpretation for over thirty years. Recklinghausen discovered the stomata vera by the use of nitrate of silver solution on the diaphragm in 1861. He placed milk-drops on the extirpated diaphragm and found that the milk-drops disappeared in small whirlpools. By marking the places where the milk-drops disappeared, and then allowing silver nitrate solution to trickle or percolate around the same spot, a peculiar structure came into view which consisted of an aperture surrounded by dark, granular polyhedral cells. This structure, stoma verum, is situated at the common junction of three or more endothelial cells. It is found distributed over almost the entire peritoneum. So far as I can judge the stoma verum is the regulator of peritoneal fluids. The stoma verum forms a direct communication between the peritoneal cavity and the subserous lymph channels. The granular polyhedral cells surrounding the mouth of the stoma verum color dark red with Ag. No. 3 solution on account of their rich possession of albuminous matter. The stoma verum is most abundant on the diaphragm, but is found on the ligamenta peritonei. It is not regular in numbers or distribution in any single place of the meso-sigmoid. No doubt leucocytes can pass out of the stoma verum into the peritoneal cavity and repass into the stoma verum and gain access to the subserous lymph channels. I proved that by injecting Berlin blue in the peritoneal cavity of rabbits and killing them at different periods later. Particles of Berlin blue were found definitely distributed along channel of the subserous lymph channels of the diaphragm. So far I have not found the particles of coloring matter in the subserous lymph channels of the meso-sigmoid, but there is no doubt that the function of the stomata vera is the same on the tunica serosa of the meso-sigmoid as it is on the diaphragmatic serosa. Irregularity in numbers and distribution according to some weighs against the stoma verum as an anatomical structure or a physiological organ. Klein and Recklinghausen and Oedmansson have done excellent work on the stoma verum.

Again, another structure exists on inter-endothelial lines of the tunica serosa of the meso-sigmoid, known as the stoma spurium (stigmata or pseudo-stoma). These structures were especially described by Oedmansson in 1862; Virchow called them lymph corpuscles; Oedmansson called them connective tissue cells. Since that time most investigators consider them connective tissue cells jutting upward between the endothelial cells. The stoma spurium is irregular in its numbers and distribution. It stains quite dark with silver nitrate solution one half per cent. Thus the tunica serosa, the peritoneal endothelia covering the right and left face of the meso-sigmoidea, consists of flat connective tissue plates with inter-endothelial structures. The stoma verum is to regulate peritoneal fluids, to form a direct connection between the peritoneal cavity and the subperitoneal lymph spaces, to allow the exit and ingress of cell leucocytes and blood discs, and to be a source of reproduction for endothelial plates. The stoma spurium is to reproduce endothelial cells. The tunica serosa or peritoneal layer proper, facing the right and left side of the meso-sigmoid, can be stripped off from both its sides as a thin, almost transparent membrane, leaving the real mesentery, the membrana mesenterii propria or the anatomical neuro-vascular visceral pedicle. This middle mesentery consists of nerves, veins, arteries, capillaries, lymph vessels, fibrous and elastic tissue, all held together in definite order, allowing no entanglement or misdirection of function. The membrana mesenterii propria of the meso-sigmoid is a very strong and powerful anatomical structure. It contains dense bundles of white fibrous tissue interspersed by large numbers of elastic fibers which are arranged in mesh and network style. The white fiber bundles are chiefly arranged along the vessels. The membrana mesenterii propria of the meso-sigmoid is the chief and essential support of the S-romanum. It supports the sigmoid loop of bowel, and if not elongated by the dragging of loaded bowel or prolapsed by age it will seldom suffer permanent volvulus, though sixty per cent. of volvulus occurs at this bowel loop. Hence in a consideration of the meso-sigmoid we must consider three layers—(a) the tunica serosa dextra, (b) the tunica serosa sinistra, and (c) the membrana mesenterii propria. Disease attacks first the tunica serosa of either side, but the chief effect results in the membrana mesenterii propria. The chief amount of disease of the meso-sigmoid attacks its left face. A much less amount expends itself on the right face, but mainly in Gruber's fold. The nerve supply of the sigmoid and meso-sigmoid arise from the (a) abdominal brain, (b) from the ganglion of the bifurcation of the abdominal aorta, (c) from the lumbar plexus, and (d) from the lumbar plexus. It is thus seen that the nerve supply of the meso-sigmoid and sigmoid is from both sympathetic and cerebro-spinal, medulated and non-medulated, nerves ; however, the chief source of nerve supply is sympathetic. In the sigmoid we are approaching the last end of the digestive tube, when the sympathetic and spinal nerves combine to usher the parts into the sensations of the external world.

The meso-sigmoid and sigmoid is supplied by the inferior mesenteric artery, whose distinct branches are the colica sinistra,

the meso-sigmoid, and the superior hemorrhoidal.   The branches
of the inferior mesenteric artery possess from two to four series
of arches and arcades.   In quite a number of the subjects exces-

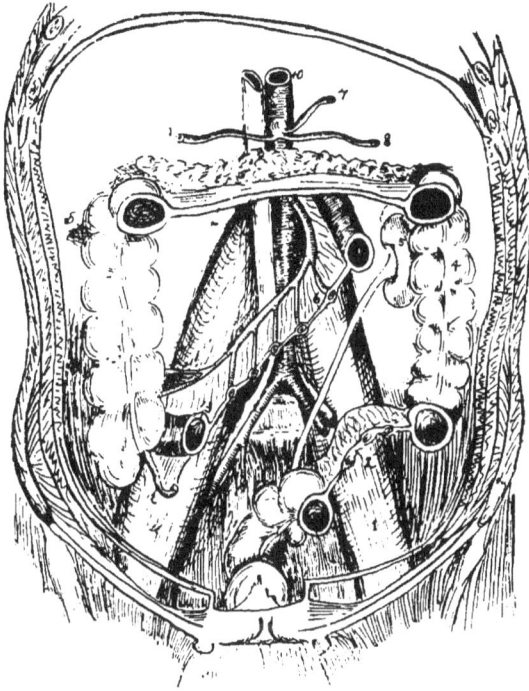

FIG. 5.—(After Byron Robinson) Represents the line of the meso-sigmoid
as it crosses the psoas muscle.   I sketched this diagram to show how prominent
and what an important rôle the psoas plays in relation to the peritoneum and
especially in regard to the meso-sigmoid. 1, psoas muscle; 2, cut edges of
meso-sigmoid; 4, descending colon; 5, kidney; 6, cut edge of mesentery; 13,
ilium; 12, appendix; 11, right psoas; 9, hepatic; 8, splenic, and 7, gastric
arteries; 10, aorta; 14, cut edge of meso-colon transversum; 15, hepatic flexure.
The peritoneum in the region of the pelvis is important to every practicing
physician, especially in the female.   Several years ago I made considerable
investigations in regard to the condition of the peritoneum of the male and
female, and found that in some animals and in man the femoral pelvic peritoneum is the thicker and stronger; probably from ages of experience of peritoneal trauma and repeated infectious invasions.   The diagrams are drawn not
always exact to Nature's lines, but nevertheless they approximate some of its
truths.   (From author's article in *Journal American Medical Association*.)

sive amount of mesentery, or in other words depression in meso-sigmoid, existed with arches or arcades of vessels.   One could push the peritoneum as a deep wide pouch toward either side of the meso-sigmoid between the vascular circles.   In almost every subject possessing such a pouch the pouch itself was of a leath-ery consistence resembling a dried bladder moistened with water. The origin, formation, and condition of such pouches is not explained or commented on in any work at command.   Of course one would look first at the blood supply as the etiologic factor, but why the blood supply should be cut off apparently too much and yet the portion of the meso-sigmoid, i. e., the pouch, still excessive, is not quite clear.

Some ten years ago I began the practical investigation of the viscera in animals and man.   I was enabled to carry out the project to some extent by my appointment to the chair of anat-omy in the Toledo Medical College.   One of the most complicated tasks was to fit the sigmoid to the old traditional description, which is a literary legacy found in almost all English and German anatomies.   Over fifty years ago the anatomist Pittard said, in Toldt's Encyclopedia, that practically the sigmoid loop and rec-tum could not be divided.   Treves revived the same idea.   Long before I had carefully examined three hundred and fifty sigmoid loops, I entirely gave up the idea of attempting to divide the S-romanum.   There is no anatomic or physiologic landmark to divide the sigmoid loop.   Hence I now consider it all one loop, as Pittard suggested in 1845, from the psoas muscle to the point where it loses its serous covering.   To be exact, the sigmoid flex-ure extends from the external border of the left psoas muscle to the point where it loses its complete peritoneal covering or to the third sacral vertebra.

Let us quote Quain's Anatomy, ninth edition, 1882, as especial anatomy of which the English are proud (page 616, Vol. 2): "The sigmoid flexure of the colon, situated in the left iliac fossa, consists of a double bending of the intestine upon itself in the form of a letter S, immediately before it comes in contact with the rectum at the margin of the pelvis opposite to the left sacro-iliac articulatum.   It is attached by a distinct meso-colon to the iliac fossa and is very movable, falling into the pelvis when the blad-der is empty.   It is placed immediately behind the anterior wall of the abdomen, or is concealed only by a few turns of the small intestine. The sigmoid flexure is the smallest part of the colon."

In examining over three hundred and twenty-five adult bodies with reference to the sigmoid flexure, I must say that "Quain" is in error. It is not situated in the left iliac fossa. It does not form a letter S between the psoas and the pelvic brim, and its meso-colon does not arise in the right iliac fossa. The above may appear sweeping assertions against "Quain," but any one can test the same for himself. Only scant general ideas can be gained from an anatomical text-book, and so far as my observation is concerned many are erroneous. The only study of comprehensive value on the S-romanum is that of developmental consideration. Perhaps, when anatomists learn that the meso-sigmoid has eighty per cent. of adhesions on its left face, different views will be held as to its position. Adhesions drag it out of place and dislocate it, which keeps it permanently out of its normal position.

Appendices epiploicæ or appendices fallopiæ or appendices omentula are small sacs or pouches of peritoneum filled with fat and attached along the right and left border of the sigmoid flexure. They vary from very small appendages to sacs from one inch wide to two inches long. These pouches are elongations and folds of the serous blades of the sigmoid flexure. Sometimes two rows of the appendages are very large, close together, and present quite an ornamental appearance. I have frequently found the distal ends of these appendages attached by adhesions to adjacent organs, forming arcs and arcades under which viscera could slide to and fro, endangering strangulation from mechanical causes. Each appendage is supplied with an artery and vein. Their arrangement has curiously, on the transverse colon, a loop of bowel very similar to the sigmoid loop, has only one series of appendices epiploicæ, and also they are relatively much smaller than those of the sigmoid flexure. In dogs, some fish, and other animals, one may observe peritoneal pouches jutting into the peritoneal cavity. Such serous fat pouches, appendages, and folds of peritoneum have been compared to the great omentum— some relations to the blood-vessels. The most numerous and largest appendages of the entire colon are attached to the lateral edges of the sigmoid flexures. The right row of appendages are

the largest (near to the blood-vessels). I have found some appendages entirely on the meso-sigmoid (dislocated). The origin and
original function of the appendages are veiled in obscurity.
They have no function in man, and no doubt are fading out of
existence like the appendix.

It is not generally known that the transverse colon and sigmoid colon are so near alike in their length and the length of
the meso-colons.

In 148 males the average length of the sigmoid is 19 inches;
in 150 cases the transverse colon averages 23½ inches. The
transverse colon is only 4 inches longer than the sigmoid. In
200 cases the average length of the meso-sigmoid was 3½ inches;
in 150 cases the average length of the transverse meso-colon is
4 inches. The transverse meso-colon is only half an inch longer
than the meso-sigmoid. The transverse colon and sigmoid are
quite similar in their aspect. Both have a similar loop and a
similar meso-colon. Both have an acquired mesentery. Their
clinical history is entirely different. The sigmoid is liable to
strictures, cancer, and adhesions far beyond the transverse colon.
The reason of this remarkable fact is due to muscular action.
The sigmoid comes within the range of action of the psoas
muscle. The transverse colon does not come within the range of
action of any muscle, so that the adhesions about it (with the
exception of its kinked ends) are very limited. The sigmoid
has two rows of appendices epiploicæ, while the transverse colon
has only one row. Both have sacculations and equal rows of
ligamenta teniæ coli.

" Quain's " Anatomy, tenth edition, 1896, is just out and lies
before me. It might be interesting reading, and besides a manifestation of English opinion, to quote what these authors say. I
also quote the passage for another reason, which is, that Dr. J.
Symington, Professor of Anatomy in Queen's College, Belfast, is
the author of the part on splanchnology, and his views agree
very much with mine in visceral anatomy. Yet Symington
adopts in my opinion the old traditional, impractical division of
the sigmoid colon at the pelvic brim. The following is Symington's description of the S-romanum :

" The sigmoid colon may be defined as that part of the colon
which is attached to the left iliac fossa from the iliac crest to the

brim of the true pelvis. In front of the crest of the ilium it is
continuous with the descending colon ; from this point it usually
passes downward, forward, and somewhat inward for two or
three inches, approaching the anterior abdominal wall internal
to the anterior superior iliac spine. This part generally lies in
close relation with the fascia in front of the iliacus, and is covered by peritoneum on its anterior and lateral aspects only. The
rest of the sigmoid colon is generally very movable, being pro-

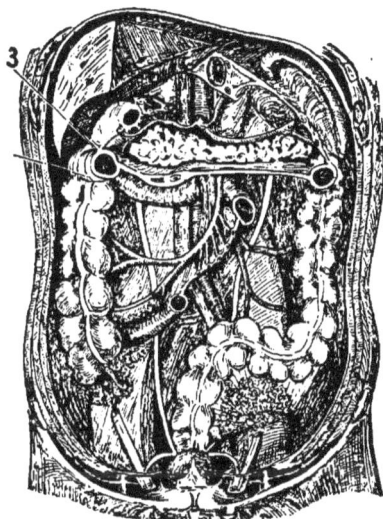

Fig. 6.—(After Byron Robinson). A sketch to illustrate the sigmoid and
meso-sigmoid passing over the psoas muscle. Adhesions appear on the left face
of the meso-sigmoid as it crosses the psoas muscle. Also adhesions appear
around the ceco-appendicular apparatus over the right psoas. (From author's
article in MATHEWS' MEDICAL QUARTERLY, January, 1896.)

vided with a long meso-colon, which is attached transversely in
front of the psoas, and becomes continuous internally near the
bifurcation of the common iliac artery with the meso-rectum.
This portion may be termed the sigmoid loop of the sigmoid
flexure proper. It is very variable in its length and position,
and frequently forms with the first part of the rectum an omega
loop (Treves). In many cases it forms a loop hanging down
into the true pelvis; if the bladder or rectum is distended it is

pushed out of the pelvis and may curve upward as high as the umbilicus, and even in rare cases touch the liver (Treves). Occasionally this loop lies in the iliac fossa, in front and to the outer side of the first part of the sigmoid colon. When its meso-colon is short it simply passes downward and inward across the iliac fossa, usually entirely covered in front by the convolutions of the small intestines. The average length of the meso-colon is about three inches. On turning upward the sigmoid loop and its meso-colon the mouth of a peritoneal pouch is sometimes seen which is called the inter-sigmoid fossa. It is somewhat funnel-shaped, and extends upward a variable distance in the direction of the ureter."

I object to the first part of the description, for the meso-sigmoid may begin to cross the psoas in many cases above the iliac crest, and second, the attachment to the iliac fossa is far from being constant, also the close relation of the upper portion of the S-loop with the fossa in front of the iliacus is certainly not constant. However, the opinions of this excellent anatomist, whose views are gained from modern anatomical methods, are well worthy of notice. But since Dr. Symington has adopted the old impractical, traditional division of the loop at the pelvic brim, it is unnecessary to further criticise, for sooner or later anatomical text-books will adopt the suggestion of Pittard, half a century ago, that it is impracticable to divide the sigmoid loop at all. My experience with some three hundred and fifty adults, some forty children, and about twenty-five early and late fetuses, is that the sigmoid colon or flexure is practically one, an indivisible loop, extending from the external border of the psoas muscle to the point of the rectum where the bowel loses its complete peritoneal covering.

The situation of the sigmoid meso-colon is very variable. But its meso-sigmoid begins at the left border of the psoas, then passes transversely across the psoas, thence toward the median line, and ends at the point where the bowel loses its complete covering. But the loop itself in the adult I have found touching the stomach, liver, and sufficiently high to touch the spleen. In one case the loop would touch the diaphragm; it was elongated into a double-barreled condition with narrow mesentery. Quite a number of times the loop was found in the right iliac fossa. It was occasionally found wound among the small intestines near the second or third lumbar vertebra.

In many cases it forms a loop hanging down into the true pelvis. Occasionally a foot or such a matter would lie on the pelvic floor, but the state of the pelvic organs determine how much S-loop lies in the pelvis. Distension of the uterus, bladder, or rectum forces it upward out of the pelvis. Less frequently a loop lies in the iliac fossa, yet this is often held in this position by adhesions. A distended celum or coils of small intestines might force it upward. It frequently lies in contact with the cecum, transverse colon, or lower end of the duodenum. In one case the spleen filled the pelvis. In short I have frequently seen the sigmoid colon so long that it would touch every viscus in the abdomen. In fetuses and infants the loop is disproportionately long, is not a pelvic organ, and lies in various portions of the abdominal cavity to suit its environments.

The special subject of interest to me in the study of the two hundred carefully recorded adult cases (and some forty fetuses, infants, and children,) is the existence of 85 per cent. of inflammatory adhesions on the left face of the meso-sigmoid in adults. Leaving out all cases of quibble and doubt, we have at least 80 per cent. of old peritoneal adhesions on the left face of the meso-sigmoid where it crosses the psoas muscle. Whence come these adhesions? What are their effect?

The two questions above propounded we will attempt to answer by viewing the facts gained in some 350 to 400 personal autopsies. For the privilege of performing 250 of these autopsies I am indebted to Drs. L. Hecktoen, LaCount, his assistant, A. L. Edwards, and L. J. Mitchell, pathologists to the Cook County Hospital, and by the kindness of the internes.

In the examination of the autopsies we found 72 per cent. of inflammations around the ceco-appendicular apparatus, i. e., over the right psoas muscle. We found at least 65 per cent. of inflammatory deposits about the gall-bladder, i. e., over the right crus of the diaphragm. We found some 45 per cent. of inflammations and peritoneal adhesions where the transverse portion of the duodenum crossed the right crus of the diaphragm and the upper end of the right psoas. We found some 40 per cent. of peritoneal inflammatory deposits in the lesser omental cavity

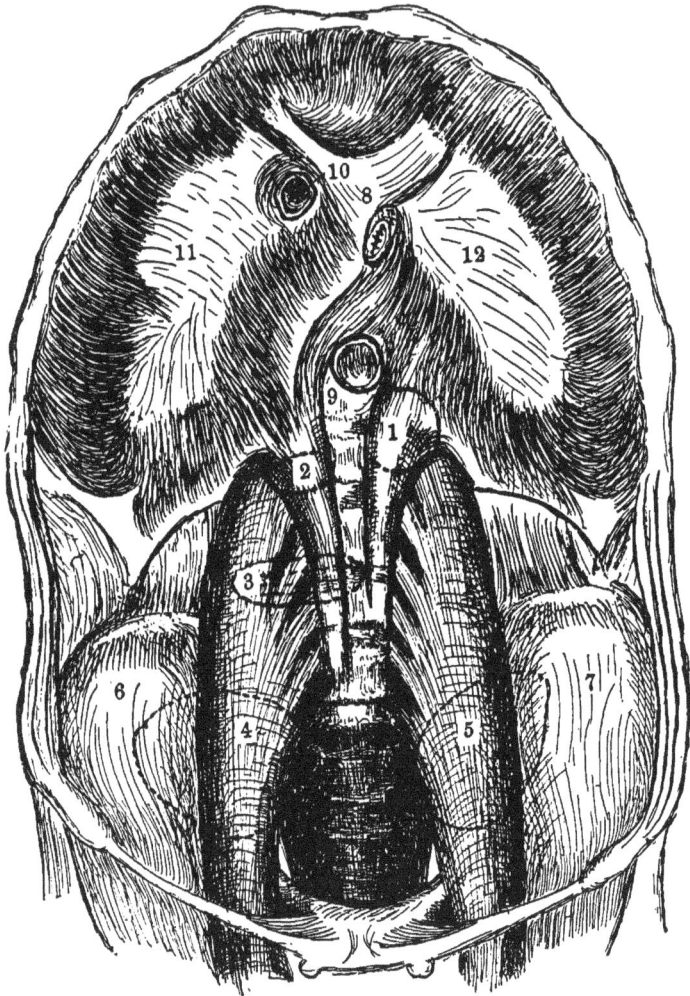

Fig. 1

FIG. 7.—(After Byron Robinson). A sketch drawn to illustrate the local peritonitis over the longest range of action of the psoas, diaphragmatica, and iliac muscles. No. 1 lies in a dotted line over the left crus of the diaphragm, and its local peritonitis is associated with the middle and left end of the lesser omental sac (bussa omentalis minoris), *i. e.*, at and adjacent to the ligamentum gastro-pancreaticum and the spleen. No. 2 indicates with its oval dotted line where the duodenum surmounted by the flexura coli hepatis crosses the right crus of the diaphragm. Its local peritonitis is associated with the right end of the lesser omental cavity and the gall-bladder region. No. 3, with its elliptical dotted line, indicates two points of local peritonitis, viz: (a) where the transverse portion of the duodenum surmounted by the ascending colon passes over the psoas and diaphragmatic muscles. It is at the point where Haller's omentum generally terminates and the anterior concavity of the ascending colon exists ; (b) in the ligamentum phrenico-colicum dextrum induced by the trauma of the right lobe of the liver. Nos. 4 and 5 indicate, with their dotted rings, the local peritonitis at the highest range of action of the psoas and iliac muscles, i. e., around the appendiculo-cecal apparatus and in the meso-sigmoid. Nos. 6 and 7, iliac muscles; Nos. 11 and 12, central tendon of diaphragm ; No. 8, esophagus ; No. 9, aorta, and No. 10, vena cava. (From author's article in *New York Medical Journal*, January 25, 1896.)

at Huschke's foramen, directly over the left crus of the diaphragm. Some 90 per cent. of adhesions exist about the spleen. The first note to make is that all these peritoneal inflammatory deposits exist within the range of muscular action. The peritoneal adhesions are due to the action of the muscles acting or irritating the bowel at times when it contains pathogenic germs or their products, inducing them to migrate to the peritoneum. The vigorous action of the diaphragm on the spleen makes it the most liable to adhesions. The adhesions in the pelvis of woman (80 per cent.) being from a unique cause, viz., infection from the fallopian tubes, should be discussed on different grounds.

Now the peritoneal inflammatory deposits found on the left face of the meso-sigmoid are due to the action of the psoas muscle on the sigmoid loop at times when the bowel contained pathogenic microbes, inducing the microbes or their products to invade the peritoneum adjacent to it. Carefully studying the peritoneum within the range of action of the psoas and diaphragmatica muscle will bring to light various degrees of peritoneal adhesions. Now we know that muscular action alone is not sufficient to induce peritoneal inflammatory deposits. Hence we have interpreted it as due to the muscle irritating the digestive tract when it contains pathogenic microbes.

Peculiar automatic structure or physiologic function throws no light on this subject. The idea of localized peritoneal inflammatory deposits impressed me very much, for one can scarcely examine a single adult without finding them in the usual localities and present in the same localities. In short,

chronic local peritonitis exists in the peritoneum within the
ranges of muscular action which lie in contact with the digestive
tract.   These muscles are the psoas and diaphragmatica.   The
left face of the meso-sigmoid has three localities where chronic
peritonitis or inflammatory peritoneal deposits arise, viz: (a)
Along the left border of the psoas muscle; (b) along the right
border of the psoas muscle (i. e., around the mouth of the
interior-sigmoid fossa), and (c) on the upper or anterior surface
of the psoas muscle (i. e., on its longest range of action).
These are the three chief regions, yet the whole left face may be
covered by inflammatory deposits and so bind the left face to
the parietal peritoneum that one can not measure the meso-sigmoid nor see the intersigmoid fossa which lies buried in this
ancient peritonitis.   The greatest number of adhesions is to be
found on the meso-sigmoid at the point where the external border of the psoas muscle plays its highest range of action, i. e.,
where the sigmoid bowel loop crosses transversely the psoas.
Hence, the origin of the adhesions (85 per cent. of the left face)
of the meso-sigmoid is due to the action of the psoas muscle on
the sigmoid bowel when it contains pathogenic (virulent)
microbes.

We find no adhesions in intra-uterine life, and I have, besides,
carefully observed, in nearly forty infants, that no adhesions
exist in the meso-sigmoid ; i. e., the psoas muscles of these
fetuses and infants have not acted with sufficient vigor on the
bowel to induce migration of microbes or their products because
the subjects had not walked.

In regard to the right face of the meso-sigmoid we find
adhesions immeasurably less.   But adhesions do occur on the
right face, but they are nearly all located on the fold of Gruber,
the ligamentum mesenterico-meso-colon.   The etiology in this
case is the same as for those of the left side, but I am quite convinced that the origin of the microbes in this case is generally
from the small intestines.   The loops of small intestine rest on
Gruber's fold, and when the psoas incites this fold into irritation
by muscular activity infection is liable to migrate to the endothelial covering of the said fold.   The peritoneal adhesions of
Gruber's fold decrease in quality and intensity from its base on
the right face of the meso-sigmoid to its apex at the left side of

the radix mesenterii. The cicatrices on Gruber's fold are frequently stellate in shape, contracted, and hence thickened, raised above the adjacent peritoneum, and of a white color. These cicatrices resemble very much those that occur at the lower end of the mesenterium, due to the action of the right psoas, and both may merge into one another over the sacrum.

What is the effect of the adhesion found about the sigmoid? The second question may not be so easy to dispose of, but the following may be said in regard to it :

1. The adhesions narrow the lumen of the S-romanum.

2. They produce kinks in the S-loop and retard the movement of feces.

3. The acute angles in the bowel are very liable to become traumatized, the mucous membrane abraded, allowing further invasion of infection.

4. The kinks being traumatized may result in stricture.

5. The trauma on the sharp edge of the projecting mucosa may result in malignancy, carcinoma.

6. Abrasion in the kinks and bends of the S-loop may induce pain and dragging sensations.

7. A very important point lies in the fact that any organ like the sigmoid, having a high peristalsis and wide range of motion, i. e., a long pedicle, when fixed, dislocated, may give rise to very much dragging pain. The pain in woman is often attributed to the pelvic organs. In either male or female the interpretation is not always correct.

8. The adhesions prevent volvulus, and no doubt that is the reason that volvulus is so rare in man (one fortieth of all intestinal obstruction).

9. Since the sigmoid flexure is the smallest part of the colon, any narrowing of the lumen by kinking or thickening of its walls enhances very much the possibility of disease. For here the feces are more apt to be hard, rough, and contain sharp foreign bodies capable of traumatizing the mucosa. The wounded mucosa in the S-loop is not in one point, as it is in the flexura coli lienalis or flexura coli hepatis, but extends along the whole range of psoas action. Has carcinoma in the sigmoid loop not a traumatic origin due to its dislocations, compromised lumen and constriction from adhesions? Other portions of the colon free

from adhesions and having equal curves do not acquire malignant diseases. Stricture of the sigmoid is no doubt due to the trauma

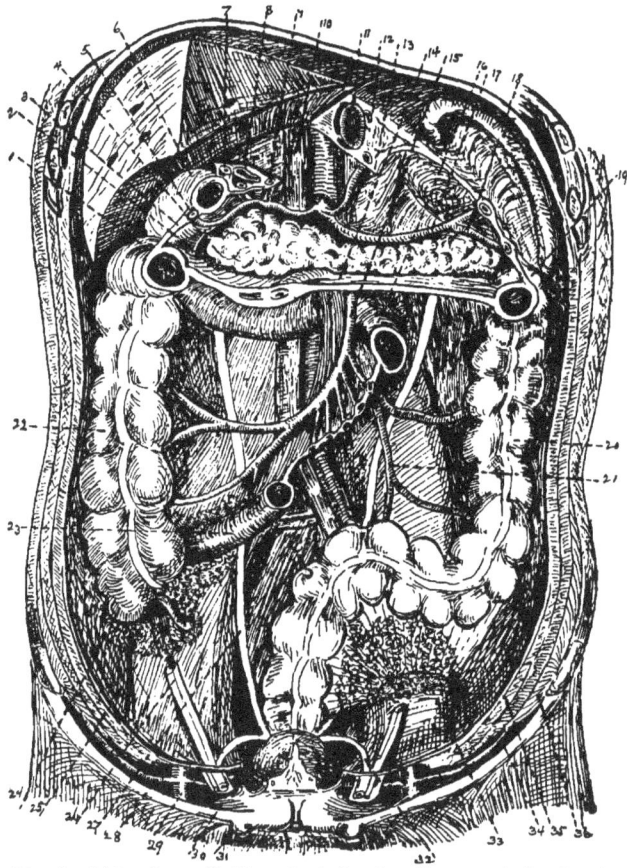

FIG. 8.—(After Byron Robinson). A sketch to illustrate the local peritonitis around the ceco-appendicular apparatus; 23 points to the ilium, with its adjacent adhesions; 24 to the cecum, with its adhesions, and 25 to the appendix, with its adhesions; all over the right psoas and iliac muscles; 33 points to the left face of the meso-sigmoid, with its manifest adhesions adjacent to the left psoas muscle. (From author's article in *New York Medical Journal*, January 25, 1896.)

of the feces on the mucosa, due to a dislocated, kinked, or compromised lumen ; all referring to a meso-sigmoiditis, brought about by the action of the psoas muscle on a bowel containing pathogenic microbes.

10. In laparotomy I found the adhesions on the left face of the meso-sigmoid just as numerous and frequent in proportion as I do in the autopsies. My plan is to break them up and completely free the sigmoid and its meso-sigmoid. The adhesions may reform, but they will reform under the new relations subsequent to the operation on the pelvic organs. So that the new adhesions will be adjusted to the new relations following the operation.

It may be asserted that one can not trace the course of the infectious invasion from the mucosa of the sigmoid colon through its muscular layers to the peritoneum. We may have a diseased mucosa without diseased adjacent musculature or peritoneum. We may have diseased peritoneum without diseased musculature or mucosa.

11. In short, any one of the three structures, mucosa, muscularis, or peritoneum may be healthy or diseased without the others being involved. The same view may be asserted in regard to any portion of the bowel which is usually accompanied by adjacent peritoneal adhesions. One can not trace the infectious disease from the mucosa through the muscularis to the peritoneum in the usual localities as the hepatic flexures, the splenic or appendicular apparatus. No perceptible pathologic conditions can be observed in the layers through which it passes as a pathognomonic sign.

12. With increasing experience in this line, I am more and more convinced that many diseases referred to the rectum have their real seat in the sigmoid, and to be successful we must attack the seat of the disease in the S-loop itself.

13. As we have now learned that operations are required to release adhesions and break them up in (*a*) organs with a high peristalsis and rhythm, as the small intestines, S-loop, tubes, and bladder; and (*b*) organs having a long pedicle or range of action, as the small intestines, the S-loop, and tubes, we may learn that sufficient symptoms arise from adhesions about the sigmoid, manifest by rectal trouble or pelvic pain, to justify opening the abdomen for the purpose of breaking them and restoring the normal mobility of the sigmoid colon. However, the suggestions contained in this article may induce operators to examine their cases

and relieve the adhesions about the sigmoid while the abdomen is open. But it should always be observed that it is not necessary to break up the peritoneal adhesions about fixed and non-peristaltic or non-rhythmic organs. Autopsies prove this fact.

OBSTRUCTION OF THE BOWELS.

BY HAL C. WYMAN, M. SC., M. D.,

Professor of Surgery in Michigan College of Medicine and Surgery.

DETROIT, MICH.

The symptoms of obstruction vary with the part of the intestinal canal obstructed. Vomiting is a constant symptom in obstruction of the small intestine, but is very seldom seen when the lesion is in the large intestine. Pain is a constant symptom, but varies greatly in intensity. Its degree is no measure of the gravity of the case. The abdomen may be distended with gas and intestinal contents, or it may be flat and collapsed. It is said obstruction of the bowels, located in the upper part of the small intestine near the stomach or duodenum, is characterized by a flat abdomen, constipation, no evacuation from the bowels, no movement of intestinal gases—signs infallible but easily overlooked or misapprehended. The physician is often deceived by an apparent thorough evacuation of feces or gas. When enemas have been used, it is easy to misinterpret the significance of an evacuation. Leaky instruments, and improperly charged ones, often fill the intestines with gas, which, subsequently being discharged more or less audibly, leads the physician to think that the continuity of the alimentary canal has been re-established. Feces below the point of obstruction are easily displaced and discharged after an enema, and misapprehension of the situation is the result. A movement of the bowels, to have favorable significance in a case of suspected obstruction, should be complete and contain feces from the whole length of the intestine. There is comfort and satisfaction for the surgeon who recognizes the location and character of the lesion in any case of intestinal obstruction he may be called upon to treat, which is very pleasing in contrast with the doubt and uncertainty attending a case in which character of the stoppage is unknown.

The causes of the obstruction to the continuity of the intestinal stream are several, and in order of their frequency can be enumerated as follows, viz:

1. Localized peritonitis, matting loops of intestines together; leakage of intestinal contents, gas, fluids, and solids from perforated appendix or bowel, and escape of infecting substances from the fallopian tubes, are the most frequent causes of localized peritonitis. Trauma—blows, kicks, stabs, and gunshot wounds are frequent causes of perforation of the intestines. Localized peritonitis sufficient to cause obstruction of the bowels is likely to follow wounds of the urinary or gall-bladder and injuries to any of the viscera contained in the cavity of the abdomen.

2. Obstruction may be due to twisting of the intestine in consequence of changes in the mesentery which give the loop of intestine too much or too little play, and to changes in the muscular action of the intestine in consequence of the irritating character of the contents or perverted nerve influence, neurosis. Notwithstanding all that may be said in favor of a purely neurotic cause of intestinal obstruction, spasm vs. paresis vs. paralysis of the muscular coats of the bowels, the fact remains that surgery recognizes no such case, but always finds some palpable lesion to account for the stoppage of the intestinal stream.

3. Hernia of any part of the intestinal canal which embraces its lumen is likely to stop the movement of the intestinal contents.

4. *Intussusception.* This is said to be a very common cause of stoppage of the bowels in young children. The writer has attended some of these cases, and has observed that the symptoms were distinguished by great pain and restlessness, and that they generally supervened upon a history of diarrhea. In his opinion any case of infantile diarrhea which is suddenly superseded by constipation, vomiting, and pain in the abdomen should be operated upon with the expectation of finding the stoppage due to telescoping of the bowels. The ileo-cecal region is the most common seat of intussusception. In respect to children, notwithstanding the labors of Senn and McBurney—the Livingston and Stanley—the abdomen is still the darkest Africa of surgery.

5. Lesions of the cavity and wall, of the intestine, cicatrices, tumors, and foreign bodies. Under this head is described the cases of obstruction following chronic and acute enteritis and colitis. Typhoid and malarial fevers are not infrequently found in the preliminary history of obstruction of the bowels. Tumors may obstruct any part of the bowels.

I have stated the physical or pathological conditions which may reasonably be looked for in a case of obstruction of the bowels. They suggest the treatment. It is plain that in most of the cases due to local peritonitis little is to be expected from the action of drugs. By the time symptoms are severe enough to require the attendance of a doctor the bowel will be so firmly locked in the embrace of inflammatory products that the only remedy will be found in abdominal section.

There are two kinds of obstruction of the bowels, viz., acute and chronic. In my practice they have occurred with equal frequency, and have called for aggressive treatment. In very rare instances have the symptoms vanished and the patient recovered spontaneously. The first kind of stoppage is ushered in with symptoms which have generally developed suddenly. The illness comes without any warning to the patient. There is generally great pain in some part of the abdomen, and vomiting usually occurs and persists in spite of all medicines and external applications. The temperature may be above or below normal, but the pulse will exhibit a variation from normal which is to the experienced surgeon unmistakable. It strikes the finger with a quick and rhythmic tick-tock, tick-tock, tick-tock vibration, which means that its possessor is dying of fecal infection in consequence of complete cessation of the intestinal current. Sweating profusely, the patient is pale, and presents in his face the approaching symptoms of collapse; kidneys will act vigorously generally, and in fact nature appears finally to break down, and death closes the scene in a supreme effort to eliminate intestinal poisons by all the emunctories of the body except the bowels. The vomit becomes dark and black and grumous in the effort. The mind usually remains clear to the end, and the patient can seldom be brought to a full sense of the approach of death.

Treatment of obstruction of the bowels must be prompt,

vigorous, and incisive. When a reasonable effort with enemas, inversion, and cathartics has been made in a case in which the stoppage is due to intussusception, volvulus, or paresis, or paralysis of the intestine, and the bowels do not move, no time should be lost in preparing the abdomen for an incision which will enable the surgeon to correct the obstruction. Some different opinion may exist as to what a reasonable attempt to move the bowels without resort to surgery is. The writer thinks a reasonable effort has been made when the symptoms persist after inversion of the patient, after a copious enema of some stimulating liquid has been given. Except in those cases in which the patient has taken opium and morphine or other narcotics which arrest peristalsis has he any faith in drugs. In those cases he has known a twentieth of a grain of strychnine and two drops of croton oil in a bread pill to relieve all the symptoms. Because of the frequent and erroneous use of opiates in all painful affections of the bowels, it is well for the surgeon to be alert for obstruction due to the action of these drugs.

In every instance one must heed the danger which comes from delaying. The maxim "Time is a greater healer" does not hold good in cases of obstruction of the bowels. The surgeon who waits will surely have a chance to study his case *post-mortem*. Until quite recently all that was known of the pathology of this affection was learned in the dead-house. Therefore it is not strange that surgeons had little heart to attack it. But now the sum of knowledge of the pathology of the intestinal obstructions is so vast and positive that there is no longer any excuse for delaying surgical treatment after the diagnosis is assured.

After having decided to operate, it is well to know what the resources of operative surgery in these cases are, and how to proceed with them. Section of the abdomen may be made in any part of that cavity. If the seat of the obstruction can be felt by the surgeon's hand, the incision should be made over that part. Before making the section the skin should be made surgically clean and the wound should be kept clean by allowing no infected thing in it. After opening the peritoneum the surgeon's finger should be introduced and the region of the obstruction carefully explored. If there is reason to think an abscess is present and likely to be ruptured, plain, dry, sterile gauze should be pushed around it through the wound so that pus will not come in contact with uninfected viscera. An intussusception can sometimes be corrected with the exploring finger. The bands which bind a volvulus and the adhesions incident to a localized peritonitis can be separated, restoring the movements of the intestinal contents. But sometimes the bowel will rupture in the attempt to free its obstructing adhesions, and then the wound must be sewed so that neither gaseous, liquid, nor solid contents can escape. Two peritoneal surfaces must always be brought together by suture when closing these wounds. When the seat of obstruction is so finely matted together by inflammation or neoplasm that it is impracticable to separate the mass, intestinal anastomosis may be performed : two sound loops of intestine, which can readily be made to touch each other without tension on the sutures or apparatus which may be used to bind them together, are selected, and incised sufficiently to secure a free passage of the contents of one loop of intestine to the other.

Lembert and continued over-and-over sutures, the Murphy button, Senn plates, etc., may be used according to the judgment and experience of the operator. The writer has more faith in the simple and rapid over-and-over suture than in any of the many clamps and devices that are offered. The art of suturing intestines should be learned by practice on the lower animals, the same as fine needle art work is learned by persistent practice on inferior materials. The material contained in the intestine above the seat of obstruction should be evacuated through a section of the intestine which becomes a temporary fistula in all cases in which the surgeon is unable to restore the continuity of the canal by manipulation or anastomosis. In several cases under the writer's care, when operations have been performed for obstruction due to inflammatory deposits about the cecum and appendix, he has found it necessary to reopen the wound, incise the ileum, and allow its contents to escape. Sometimes he has been able with his finger in the cavity of the ileum to stretch the obstructing deposits so that fluids were injected downward through the intestine. In cases that are operated on at the commencement or during the period of collapse, it is a good plan to open the distended intestine at the point of obstruction, remove its contents, and replace them with a mixture of warm normal saline solution as strong coffee. In no other way that is known to the writer is there any hope for cases in which black vomit and other symptoms of collapse have appeared.

The after-treatment in all cases in which operations for relief of obstruction have been performed, should be characterized by rest on the back or side, absence of all opiates. Relieve pain with hot or cold applications to the abdomen; the one which gives relief should be used. Stimulants, brandy, ether, chloroform may be safely used to allay pain. The digestive organs should have rest for at least twelve hours, during which time oatmeal-water or weak gruel may be taken. A wound of intestine which has been properly sutured will be firmly adhered by that time and measures may be safely undertaken to secure two or three thorough movements of the bowels, which are quite essential in preventing post-operative peritonitis. The prognosis is always grave in stoppage of the bowels, but with early diagnosis and prompt surgical treatment there is reasonable ground to expect the patient to recover.

Mathews' Medical Quarterly
"ALIS VOLAT PROPRIIS."

Vol. III.	LOUISVILLE, OCTOBER, 1896.	No. 4.

JOSEPH M. MATHEWS, M. D., - · - · - ·   EDITOR AND PROPRIETOR.
HENRY E. TULEY, M. D., - · - · - ·   ASSOCIATE EDITOR AND MANAGER.

A Journal devoted to Diseases of the Rectum, Gastro-Intestinal Disease, and Rectal and Gastro-Intestinal Surgery.

Articles and letters for publication, books and articles for review, communications to the editors, and advertisements and subscriptions should be addressed to
Editors Mathews' Medical Quarterly, Box 434, Louisville, Ky.

RESPONSIBILITY OF ANESTHETISTS.

Surgeons are at last realizing the importance of having anesthetics administered by experienced and tried physicians when an operation under general anesthesia is necessary. This is right and proper, and it is very much to be hoped that in the near future the medical colleges will give more attention to the teaching of the science and art of giving an anesthetic, the difference between them, the indications and contra-indications for their use, and, in case of accident, remedies and their indications. When this is done, and there are in every town and city several competent anesthetists, there will be fewer deaths as the result of the indiscriminate choice of the anesthetist, probably an invited guest, or the family doctor who perhaps gives or sees an anesthetic administered once in six months or less often.

Accidents from an anesthetic can often be foreseen by an experienced anesthetist and avoided; but often an unfavorable result can not be foreseen in any possible way. Should an accident occur, fatal or otherwise, as the result of the anesthetic directly, the responsibility should be placed where it belongs, not with the operator, but with the anesthetist whose business it is to give it frequently, and who is better prepared to cope with accidents and emergencies than one chosen for that duty at the last minute.

The surgeon's responsibility should cease with the choice of an anesthetist of recognized ability; then, should a fatal accident occur, the anesthetist should hold the responsibility, and the case should not be spoken of as "Surgeon Blank's sad case," as they are in the majority of instances. It is refreshing to see the difference in the operator when he has one whom he can trust in charge of the anesthetic, and one whom he has to be constantly asking the state of the patient.

CHANGE OF NAME.

Having been importuned by many of our friends and readers, the editors have decided to change the name of the QUARTERLY so that in its title on the cover page will appear the true scope of the journal. A notice of the change will be sent as soon as a suitable title has been chosen.

That this change is essential is apparent because of the importance of the contributions contained in each issue, and because there is no other publication in the English language devoted exclusively to the consideration of the gastro-intestinal tract and rectum, the title "MEDICAL QUARTERLY" therefore does not convey to any one the true nature of the publication.

The QUARTERLY has at no time been as prosperous as now, at the close of Volume III. It has filled a place long vacant in medical literature, and under a new name we trust will continue its mission of usefulness.

DISEASES OF THE RECTUM.

HOUSTON, FRANCIS T., DUBLIN: EXCISION OF THE REC-
TUM; A METHOD OF OPERATING. (*British Medical Journal.*)

After detailing two cases the author says: I will now direct
attention to certain points in connection with the operation. In
these cases it is of importance that the peritoneal cavity be not
opened; this may be usually attained by remembering that the
peritoneum usually covers the anterior aspect of the rectum in
the male to within three inches of the internal sphincter, while
in the female you can not be sure of more than two inches,
although there may be much more in either case. At the posterior aspect of the bowel you can count on four or five inches. As
a practical rule I consider that, if the surgeon can by digital examination reach above the malignant growth, the peritoneum
should be safe.

The great objection to excision of the rectum by the usual
operation is the incontinence owing to the removal of the sphincters, and the importance of overcoming this sequela is apparent;
and to point out how this may be attained I will describe the
usual operation, and compare it with the method employed in the
foregoing cases.

The usual operation is commenced by an incision from the
rectum to the coccyx, called by the old writers "the key to the
operation;" then incisions were made round the anus, meeting
in front, the attachment of the levator ani to the rectum was cut
and the bowel with sphincters drawn down; thus most of the
external sphincter and the entire of the internal sphincter were
removed, and incontinence, which was often permanent, resulted.
Further, the remains of the bowel not being usually united to
the skin, owing to the frequency with which the sutures tore
through, the raw surface thus left to granulate was a frequent
source of septic infection and subsequent stricture.

The operation I employ is commenced by an incision from
the rectum to the coccyx, as in the former operation; this being

in the median line, severs the attachment of the external sphincter and levator ani muscles to the ano-coccygeal ligament; the
rectum is now separated from the surrounding cellular tissue, and
clip forceps applied at the commencement of the ampulla, below
which the bowel is divided; the attachments of the levator ani and
sphincter muscles to the bowel are not thus interfered with, and
hemorrhage is restrained. The rectum being now completely
divided above the internal sphincter, there is not much difficulty
in drawing it downward and backward through the median incision, its separation from the prostate and base of bladder being
facilitated by having to draw it backward. Clip forceps are
now applied above the diseased portion, which is removed by a
scissors. The upper portion of the bowel is now fixed by deep
sutures attaching it, about an inch above its cut extremity, to the
recto-vesical fascia and levator ani muscle, thus closing the space
between the levator ani and the peritoneum. In applying these
sutures care should be taken that the mucous coat of the bowel
is not penetrated. A second row of sutures is now applied between the cut extremity of the bowel and that portion which had
been originally left below; these sutures should penetrate the
entire thickness of the bowel in both instances. The object of
the upper row of sutures is to so fix the bowel as to obviate the
danger of the sutures uniting the cut extremities of the bowel
tearing through. The incision from the anus to the coccyx is
now closed, the most anterior suture including the posterior
aspect of the bowel, but not penetrating the mucous coat. The
levator ani and sphincters will thus have their normal relation to
the extremity of the bowel preserved. It might be thought that
the lower portion of the bowel is in danger of sloughing owing
to the superior and middle hemorrhoidal arteries being cut away
from it, but the hemorrhoidal branches of the pudic arteries are
sufficient to maintain its vitality.

BEACH, WM. M., PITTSBURGH: RELATION OF CONSTIPATION
TO DISEASES OF THE RECTUM. (*Pittsburgh Medical Review.*)

It is conceded that the colon is the receptacle of the residue
of the gastro-intestinal digestion, and hence to that organ attention must be directed. The colon is a muscular tube, its muscular coat consisting of external longitudinal fibers and internal
circular fibers, which enter into the construction of the sphincters.

The third sphincter is denied by Mathews, Kelsey, and Allingham, each arguing that if the presence of feces in the rectum is not heeded there is a backward peristalsis sufficient to carry the mass into the sigmoid. If the demands of the sigmoid are heeded by the sphincters, their fibers relax, the anal folds become everted, and defecation is performed. Otherwise the mass is lifted back into the sigmoid to remain till nature again allows it to pass, and each succeeding interval establishes a tolerance of the sigmoid to the fecal mass; it lies there and becomes dry, acts as a foreign body with resulting vascular engorgement of the lower segment of the rectum.

It is evident that the primary cause of constipation is voluntary neglect of the individual, the second or exciting cause is a refractory sphincter together with a damaged sigmoid flexure.

Among the rectal causes of constipation are mentioned, (1) Thickening of the skin and mucous membrane; (2) irritable ulcer or fissure; (3) fistula; (4) hemorrhoids. The colon may present one of the following conditions: (1) Catarrh; (2) ulceration; (3) stricture; (4) malignant disease; (5) polypi.

FREEMAN, LEONARD, DENVER, COL.: VULVO-VAGINAL ANUS. (*The Medical News.*)

Four divisions are made of malformation of the anus: (1) *Atresia ani,* where the anus, or all of that part below the internal sphincter, is absent; (2) *Atresia ani et intestini recti,* the colon ending in a blind pouch; (3) *Atresia recti,* rectum does not connect with anal part of bowel; (4) *Cloaca congenita,* the rectal pouch emptying into some portion of genital or urinary tract.

In embryonal life the rectal and anal portions of the bowel are not formed in a continuous tube, but exist as separate structures, which gradually approach and unite at end of fourth week. At this stage of development the genital and urinary organs empty into the rectal pouch. In the vulvo-vaginal anus the opening is always in the posterior wall and generally low down near the posterior commissure. The opening may be small or large, usually guarded by a sphincter muscle.

The case of a twelve-month-old baby is reported, with an anal opening in the posterior wall of the vagina. A small dimple marked the spot where the normal anus should exist.

Rizzoli's method of operation was done. Sound inserted in vaginal anus as guide, perineal incision made until rectum was reached. Rectum then dissected from surrounding tissues by blunt instruments, cut from vaginal wall by scissors, opening in latter being closed by fine sutures at once. Vaginal anus enlarged and stitched to incision in skin. As internal sphincter had been preserved no incontinence resulted four months afterward.

DRENNEN, C. T., HOT SPRINGS, ARK.: THE ETIOLOGY AND TREATMENT OF PRURITUS ANI.

In discussing the etiology and treatment of *pruritus ani* it will be necessary to take notice of the different diseases which serve as exciting or predisposing causes. Kelsey believes it to be merely a symptom of some local or constitutional disease such as eczema, constipation, or many other diseases to which he cites our attention, but goes on further and admits that it will frequently be found where no cause for its existence can be ascertained. Now we do not believe with Kelsey that no cause can be found in these cases, but to the contrary. No matter what the exciting or predisposing cause or causes may have been there is a cause *sui generis* that operates in all these cases, which is sufficient to make it a disease *per se*. I have seen it persist in its most maddening form, goading the sufferer to the very point of sheer desperation, thereby forcing him to scratch and dig with unrelenting fury until temporary relief would come long after constipation, the exciting cause, had been removed. To illustrate, constipation, where the fecal matter is pressing down the hemorrhoidal blood-vessel, causing congestion, which in turn produces irritation of the terminal nerves, itching, scratching, and consequent thickening of the skin, which in its nature may be either acute, subacute, or chronic, will be the result. Mathews, to whom we are most indebted for the hints thrown out in the method of treatment, says that *pruritus ani* is the most intractable of all the diseases of the anus or rectum, and deserving of our earnest consideration and attention, and I am sure that those of you who have seen this disease will most heartily concur in that statement.

Now as to treatment, a few hints only can be given before going to what I consider the radical treatment of *pruritus ani*.

It is absolutely necessary to remove the exciting cause or causes in these cases to make a permanent cure. We grant in the outset that this can not always be accomplished, but these are decidedly the exceptions rather than the rule. Among the exciting causes that are local we will mention pediculi, eczema, erythema (in fat people), thread worms (which are to be found in the radiating folds at the margin of the anus), lack of cleanliness, the disease known as eczema marginatum (which is most easily cured by rubbing well into the parts night and morning for a week or so an ointment containing from ten to thirty grains of chrysophanic acid to the ounce of vaseline), hemorrhoids, fistula, fissures, etc. Some other causes that we would consider as reflex or constitutional are stone in the bladder, chronic inflammation of the deep urethra, stricture of the urethra, pelvic tumors, uterine derangements, functional disorders of the liver, diabetes, constipation, and lastly, but not least, gastro-intestinal disorders, especially that form that is so commonly known as atonic dyspepsia superinduced by smoking, drinking, irregular eating, irregular sleeping, in fact you might say irregular in all the walks of life. The treatment which tends to the best results is a light breakfast, no luncheon, a good dinner, plenty of hot water an hour before and between meals, and following the homely road of correct habits. Many and innumerable are the local and constitutional remedies we have for the relief of this affection, none of which will I burden you with, but refer you to the text-books on the subject. We do not wish to be thrown out upon the broad field of empiricism or be known as a routinist when we suggest one method which is conducive to the best results in all these cases where no exciting cause or causes remain, that of stretching the rectum and following it up with some procedures which we will endeavor to explain in detail. After having anesthetized the patient and placed him in the lithotomy position proceed to divulse the rectum either by using the thumbs, or, if the hand be small, introduce it well into the rectum, beginning with the index finger and following it up with the middle, and so on until the whole hand is within the rectum, gradually closing the hand and withdrawing until the contraction of the sphincter muscle is completely overcome, observing the utmost gentleness and care, lest violence to the tissues be done, taking in all from three to five minutes. After this take a sharp curette and simply remove every vestige of the thickened and parchment-like membrane. When this will have been done you will have a complete change in the condition of the parts; you will have converted a chronic inflammation into that of an acute, whose tendency is always to recovery. You will have removed the contraction of the sphincter muscle, which will admit of a freer circulation of blood through the parts, and taken off entirely the pressure on the terminal filaments of the nerves. In conclusion we would say that we believe in the radical and surgical treatment of *pruritus ani* as a disease *per se*, and that these remarks have been prompted not alone by theoretical investigation or inquiry, but by practical results.

   SMITH, Q. CINCINNATUS, AUSTIN, TEXAS: IMPROVED REC-TAL IRRIGATOR. (*Texas Courier-Record of Medicine.*)

   This instrument is strong, light, simple in construction, easily kept clean, cheap, can be easily and comfortably used by the patient, is self-retaining, and is very efficient for the purpose intended.

   In description of instrument would say: From the base to the tip is 2¼ inches; outside diameter at the constricted part is ⅜ inch; outside diameter at broadest part is 1 inch. At intake, rubber tubing may be attached, long as desired, through which

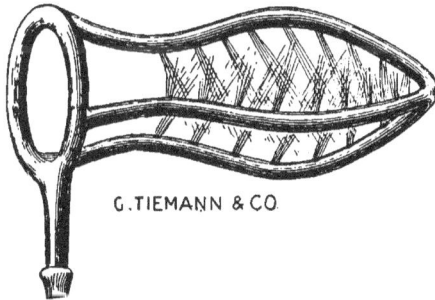

G.TIEMANN & CO

the irrigating fluid is conveyed, passing out at small openings, as shown in cut (nine in each rib), and thence out at the ring in base of instrument. Thus continuous, steady irrigation may be administered as long as desired. As a rule such irrigations—for the treatment of hemorrhoidal conditions, fissures, ulcers, pro-lapsus, etc.—should be continued fifteen to thirty minutes at each seance, and the temperature of the irrigant, hot as patient can bear.

Before administering the medicated irrigant proper, the parts should be well cleansed. The best cleansing agent we have used —of the many we have tried—is Packer tar soap suds, used in plentiful volume, hot as patient can bear. After irrigants have flowed two or three minutes, hot as patient could bear to begin with, their temperature may gradually be considerably increased, to great advantage of the patient. The various solutions that may be advantageously applied in this mode of treatment are too numerous and obvious to mention in detail. The composition of many solutions, well suited to remedy the various and varied conditions of rectal diseases and injuries, will readily occur to the surgeon in charge. But it is well for the beginner to remember the four following points applicable to most cases:

1. First cleanse the parts well.
2. Use solution very warm, or hot as patient can bear.
3. Long seances of irrigation.
4. Use very mild solutions as a rule.

MORAIN: TREATMENT OF ANAL PRURITUS. (*Journal de Médicin de Paris.*)

Try first very warm lotions or irrigations two or three times a day; patient to avoid constipation and always to precede movement by irrigation of oil and to anoint the anus with vaseline. Then try in succession named the following preparations:

Morning and evening a painting of the following glycerole: (*a*) Alum, 4 grams; calomel, 2 grams; glycerine, 20 grams. This pomade: (*b*) Calomel, 4 grams; vaseline, 30 grams. (*c*) Oleate decocaine, $\frac{1}{20}$ part; lanoline pure, 3 parts; vaseline, olive oil, $\bar{a}\bar{a}$, 2 parts. (*d*) Red oxide of mercury, 4 grams; vaseline, 30 grams. (*e*) Introduce into anal orifice a tampon of absorbent cotton soaked in a solution of oxide of zinc to 4 p. 30. (*f*) Cauterize the painful parts with a solution of agnoz, 1-10. (*g*) In quite rebellious cases turn to linear scarifications or to cauterizations with the galvano-cautery.

GASTRO-INTESTINAL DISEASE.

GIRARD, BERNE: CHLOROSALOL IN DIARRHEA.

Chlorosalol is the name of a new preparation which is a derivative, a salicylate, of chlorophenol, of which two isomeric combinations are known. These are orthochlorosalol and parachlorosalol, both appearing as fine crystals, white, insoluble in water, but soluble in alcohol and ether. The ortho derivative has a strong odor while the para is inodorous and insipid, serving the purpose, however, better than the former. The author administers it with success in doses of 2 to 4 grams per day in catarrhal affections of the urinary passages and diseases accompanied by diarrhea. It can also be employed as a dusting powder on some wounds, like salol, and possesses the advantage of producing no local irritation.

HEUBNER, PROF.: THE ANATOMICAL LESIONS OF GASTROENTERITIS AND CHOLERA INFANTUM. (*Universal Medical Journal.*)

Prof. Heubner, recently appointed to the Children's Clinic at Berlin, in a communication on this subject stated that he had examined the intestines of nursing infants dying from other diseases than those of the alimentary tract, and had observed a rapid alteration in the intestinal epithelium, even after a lapse of a few hours. In cases in which the *post-mortem* examination had been made from twelve to twenty-four hours after death, the shape of the cells had entirely disappeared, and a sort of partly desquamated membrane had formed. If the intestine had been contracted at the moment of death, the glands were found lying close together, and the vessels were devoid of blood, causing a certain degree of anemia of the mucous membrane, which occurred after death. If the intestine, on the contrary, had been relaxed, the glands were widely separated and the vessels engorged, causing hyperemia, also cadaveric. These conditions of epithelial desquamation — hyperemia or anemia — Prof. Heubner regarded as normal, no pathological changes of the intestine being found.

In three cases of subacute gastro-enteritis with dyspepsia, examined after death, he found a degeneration of the mucous membrane of the gastric glands, the cells, however, being normal. The epithelium of the large and small intestines was intact, but in the former was considerably degenerated. In cases of cholera infantum the changes were more extensive. The stomach was filled with lumps of mucus; the epithelium of the large intestine, jejunum, and ileum was in a necrosed condition, causing a loss of the power of absorption and secretion; the villi were united, forming a uniform mass. Prof. Heubner did not determine the nature of the micrococci present, although he found no specific bacteria. The lesions did not, in his opinion, account for death, which he was inclined to attribute to an auto-intoxication of intestinal origin, the disease, like other infections, proving fatal before any appreciable anatomical lesions resulted.

CURRIDEN, GEO. A., CHAMBERSBURG, PA.: CHRONIC GASTRITIS OF LONG STANDING WITH PERIODIC ATTACKS OF MIGRAINE. (*Medical Summary.*)

The following report is related by the author:

The history of the case is briefly as follows: Mrs. A., aged fifty-five, since early womanhood has been subject to periodic attacks of migraine at intervals of two, three, or four weeks, but seldom free from them for longer intervals.

An attack comes on by general malaise of usually a day's duration, repugnance of food or drink, marked drowsiness, much depression with request for rest and quiet, followed by complete physical prostration, dull frontal headache, which the least noise or disturbance makes the more intense, invariably accompanied by violent and frequent attacks of vomiting and retching, inability to retain any food or nourishment of any kind, retention of bowels, often cold sweats, pulse somewhat slow and weak and small in volume. This condition lasting usually two days, followed by gradual cessation of symptoms.

During the whole period of usually four or five days' duration she is unable to take nourishment of any kind, remains constantly in bed, and desires only complete rest and quiet. The previous treatment has been so varied and on so many different plans that I refrain from mentioning them.

Two years ago I was able to prevent an attack for over two

months by the use of strychnine in one-twentieth grain doses
t. i. d. with careful diet and artificial digestive.

In May, 1895, I put her on Charles Marchand's "glycozone"
in teaspoonful doses well diluted *t. i. d.*, using this as all other
previous remedies experimentally. She commenced to improve
much in general health, an unusually good appetite, without the
previous distressing symptoms following, a more regular movement of the bowels, freedom from headache, and in every way a
decided improvement; this improvement and enjoyment of good
health lasted during continuation of above treatment for over
three months. Unknown to me she stopped taking the glycozone, thinking herself perfectly well. In a few weeks had a
return attack, milder and devoid of gastric distress. A similar
attack two months later, both of which occurred some weeks
after stopping the above described treatment, and I might say
caused by imprudence in diet.

The conclusion come to in this case is that the headache is
sympathetic, that the stomach becomes acutely inflamed by its
inability to naturally and properly perform its functions, and
responds to the call of nature to unload itself, and thus secure
for a time rest, that the use of glycozone has corrected the existing gastritis, and by so doing has removed the primary cause of
these many years of suffering.

ROGERS, LEONARD, LONDON, ENG.: A RAPID METHOD OF
PERFORMING ENTERECTOMY WITHOUT THE AID OF ANY
SPECIAL APPARATUS. (*British Medical Journal.*)

The method consists in turning back the peritoneal coat off
one end of the small intestine, after resecting a portion of the
gut, suturing the muscular coat thus exposed to the peritoneal
coat of the other end of the intestine, subsequently turning
down the reflected portion of peritoneum over the first row of
sutures, which are thus completely buried, and suturing its deep
fibrous surface to the outer serous surface of the unreflected end
of the intestine. Thus a double sero-fibrous union is obtained
which will unite both quickly and firmly. The inner sutures
are passed through the muscular coat of one end and the muscular and peritoneal coats of the other end of the bowel, which
afford ample material for holding, while the outer sutures

include the peritoneal coats only. The accompanying diagrams
show the method of passing the sutures, and the parts brought
into apposition. The inner wavy lines represent the mucous
membrane, the thick lines the muscular, and the thin straight

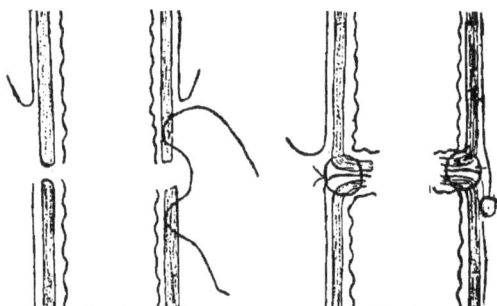

FIG. 1.   FIG. 2.

lines the peritoneal coats of the small intestine. Figure 1 represents on the left-hand side the peritoneal coat turned back from
one end of the bowel, and on the right the method of passing the
inner suture. Figure 2 shows on the left side the parts brought
into apposition on tightening of the inner suture, and on the
right side the second suture is also shown in position, completing
the junction.

If the usual method of numerous interrupted sutures be
employed, the method thus briefly described does not appear to
have any special advantages except the firmness of the union
obtained, and would have the counterbalancing disadvantage of
adding to the length of the operation the time taken in reflecting
the peritoneum, while in the class of cases for which enterectomy is employed a reduction in the time taken in performing it
is the goal of all modern improvements in the operation. The
firmness of the union to be expected from this method, however,
led me to hope that both the inner and outer sutures might be
continuous ones, and thus the length of the operation could be
materially reduced. The question of the possibility of turning
back the peritoneum and the feasibility of obtaining a watertight
junction by means of continuous sutures remained to be decided.

The advantages of this mode of operating are: (1) That it
can be done with the aid of the instruments in a pocket case,

ordinary round sewing needles being used (although curved intestinal needles are to be preferred), and with very little assistance, and is therefore likely to be of special service in military surgery. (2) It can be completed in about half an hour, or only a little longer than the time required with the aid of such special appliances as plates, buttons, and bobbins. (3) The junction is a double sero-fibrous one, and hence, as Greig Smith has shown, will unite the maximum of rapidity and firmness. (4) The mesenteric side can be made very firm by the apposition of the muscular coat of one end to the peritoneum of the other, and the subsequent covering up of this suture by the reflected peritoneum. When special appliances are at hand their use will probably be preferred to the method which I have described, but in country and foreign practice, where these luxuries are seldom likely to be at hand when the occasion arises for their use, I venture to think that my method will be materially quicker and more efficient than the ordinary one of suturing with interrupted Lembert's sutures, in which both Jacobson and Treves are agreed that from twenty-five to forty sutures are required, and even the mesenteric junction gives cause for anxiety.

CRIPPS, HARRISON, LONDON: A METHOD OF TEMPORARILY CLOSING THE OPENING AFTER GASTROSTOMY OR ENTEROTOMY. (*British Medical Journal.*)

Keeping the wound clean after gastrostomy or enterotomy is the chief trouble, for if the gastric juice or fluid feces are allowed to escape the wound is excoriated and the skin for several inches around is kept raw and tender. Dressings or trusses quickly get stopped, and plugs put in the wound unnecessarily enlarge it. The following method is described, having been used by the author in several cases:

A circular disc of sheet rubber thickness of a shilling is cut. Its diameter should be nearly double that of the orifice to be closed. A thread of No. 4 silk is passed by a needle a little to one side of the center of the disc, and is passed back again, coming out a little distance from the original puncture and on the same side. (See Fig. B.) The two ends of the thread are now parallel to each other, and should be left six inches long. The india-rubber disc is now folded into a circular roll and kept in this position by a fine pair of dressing forceps. (Fig. A.) The roll is thus introduced lengthways through the orifice into the interior of the bowel or stomach, and when released from the forceps immediately expands to its original circular form. By drawing up the two strings the disc is lifted up against the

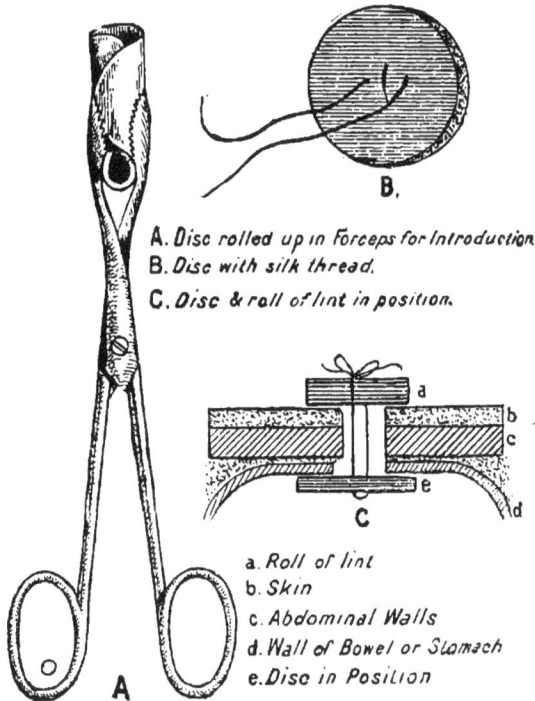

B.

A. *Disc rolled up in Forceps for Introduction*
B. *Disc with silk thread.*
C. *Disc & roll of lint in position.*

a
b
c
e
d

C

a. *Roll of lint*
b. *Skin*
c. *Abdominal Walls*
d. *Wall of Bowel or Stomach*
e. *Disc in Position*

A

mucous surface, and in this way effectually stops any thing coming through. Indeed, the greater the pressure of the fluid within the firmer is the disc pressed against the mucous wall. A little roll of lint the thickness of a pencil and an inch long, stiffened with a pin or a piece of wire down the center, is laid over the external opening between the two strings, which are then tied in a bow across it, with just sufficient firmness to press the roll well down on the skin. (See Fig. C.) It requires some experience in each particular case to find the exact size and thickness of the disc required.

For feeding the patient or obtaining relief all that is necessary is to untie the bow, remove the lint roll, and then, holding the thread lightly, press back with the director the disc into the bowel. It can be at once drawn back again into position when required. I have found this method absolutely effectual in keeping the patients dry and clean.

In a subsequent issue of the same journal a communication was published from Edward Cotterell, in which the following method is described, and which the author claims is superior to the above.

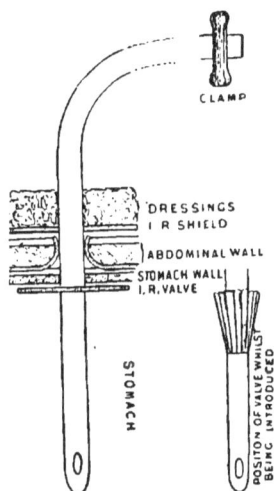

CLAMP

DRESSINGS
I R SHIELD

ABDOMINAL WALL
STOMACH WALL
I.R.VALVE

STOMACH

POSITION OF VALVE WHILST
BEING INTRODUCED

The valve is attached about three inches from the end of a Jacques' catheter. The catheter is introduced, the valve closing like an umbrella and expanding again when in the stomach. A thicker piece of india-rubber is then slipped over the catheter and rests on the skin, being fastened to the catheter by a couple of small safety-pins. The end of the catheter is passed through the dressings and clamped. When feeding the patient it is only necessary to unclamp the projecting end of the catheter and inject the nourishment. There will be no escape of stomach contents.

27

Book Reviews.

An American Text-Book of Applied Therapeutics, for the Use of Practitioners and Students. Edited by J. C. WILSON, M. D., Professor of the Practice of Medicine and of Clinical Medicine in the Jefferson Medical College, etc. Assisted by AUGUSTUS A. ESHNER, M. D., Professor of Clinical Medicine in the Philadelphia Polyclinic, etc. Philadelphia: W. B. Saunders. 1896. Price, cloth, $7.00; sheep and half morocco, $8.00; half russia, $9.00. Sold by subscription only.

This is the latest addition to the series of admirable text-books which have been issued by this firm in the last few years. It compares with the others in excellence and will prove a valuable addition to the series. Among the contributors to this volume may be mentioned Drs. I. E. Atkinson, Baltimore; Sanger Brown, Chicago; the late Wm. C. Dabney, Charlottesville; John C. DaCosta, Philadelphia; F. X. Dercum, Philadelphia; J. T. Eskridge, Denver; F. Forcheimer, Cincinnati; Frederick P. Henry, Philadelphia; A. Laveran, Paris, France; John K. Mitchell, Philadelphia; Wm. P. Northrup, New York; William Osler, Baltimore; Victor C. Vaughan, Ann Arbor; J. C. Wilson, Philadelphia.

The work is a most comprehensive one, considering in regular order, intoxications, infections, diseases due to internal animal parasites, diseases of undetermined origin, and diseases of the several bodily systems, digestive, respiratory, circulatory, renal, nervous, and cutaneous. A consideration of the disorders of pregnancy is also included.

In considering the treatment of diphtheria with antitoxin one is disappointed that Dr. Northrup did not come out more emphatically in favor of the serum treatment instead of contenting himself with quoting from the admirable article of Prof. Wm. H. Welch upon that subject. The antitoxin treatment has come to stay, and we would expect a more decided statement than the author's.

Prof. Wilson's article on the treatment of enteric fever is an admirable one, and it is refreshing to read that "the alleged termination of typhoid fever in the course of a few days as a result of some form of treatment demands the incontrovertible evidence of a large series of cases to establish its correctness." That part of the chapter devoted to the consideration of the hydrotherapy of enteric fever is a most entertaining and instructive one.

The article upon malaria, by Laveran, is very comprehensive and thoroughly up to date.

It is to be regretted that in the text at the beginning of each article the place of residence of the author is not given directly after his name. As it is now, unless one is familiar with the author, he has to look back to the index for this information.

Practical Points in Nursing for Nurses in Private Practice, with an Appendix. By EMILY A. M. STONEY, Graduate of the Training School for Nurses, Lawrence, Mass.; Superintendent of Training School for Nurses, Carney Hospital, South Boston, Mass. Illustrated with seventy-three engravings, nine colored and half-tone plates. Philadelphia: W. B. Saunders. 1896.

It is extremely difficult to know just where to draw the line at what is essential for a trained nurse to know to be most efficient. It seems that in some cases too much knowledge on the part of a nurse is a dangerous thing ; she becomes dictatorial and seems in a measure to try to supplant the physician in the conduct of the case.

However, the nurse who reads the book before us from the beginning to the end will not be likely to allow the information so carefully collected to influence her in any way but for good. The chapters on suggestions to private nurses, etiquette, and dress, are exceptionally fine in general and in detail. With these pages as a guide that she may know to what purpose to put her knowledge to be gained in what follows, a thorough understanding of the book can not but be the greatest help to a nurse.

In some details the book is lacking, as the illustration on page 77 to show the way to administer a hypodermatic injection, the forearm being the least desirable position to make the puncture.

The chapter on obstetrical nursing is fairly good and the glossary in the back of the book is an excellent aid to the studious nurse.

The book should have a good sale, for it will prove very useful to the pupil as well as to the graduate nurse.

400 BOOK REVIEWS.

System of Surgery. Edited by FREDERICK S. DENNIS, M. D, Professor of
the Principles and Practice of Surgery, Bellevue Hospital Medical College, etc. Assisted by JOHN S. BILLINGS, M. D. Volume IV.

In this the fourth volume of this great work on surgery will
be found the most valuable of all the contributions to it. The
subjects, too, are the most interesting to the general surgeon.
The following is the list with the name of author : Tumors, by
Frederick S. Dennis, M. D. ; Hernia, by William T. Bull, M. D. ;
Surgery of the Alimentary Canal from the Pharynx to the
Ileo-Cecal Valve, by Maurice H. Richardson, M. D. ; Appen-
dicitis, by Frank Hartley, M. D. ; Surgical Treatment of Appen-
dicitis, by Charles McBurney, M. D. ; Surgery of the Alimen-
tary Canal from the Ileo-Cecal Valve to the Anus, by Lewis S.
Pilcher, M. D. ; the Surgery of the Liver and Biliary Passages,
by Robert Abbe, M. D. ; Surgical Disorders and Diseases of the
Uterus, by William M. Polk, M. D. ; Surgical Diseases of the
Ovaries and Tubes, by Joseph Taber Johnson, M. D. ; Minor
Gynecological Surgery, by Henry C. Coe, M. D.; Symphyseot-
omy, by William T. Lusk, M. D. ; The Surgery of the Thyroid
Gland, by Robert F. Weir, M. D. ; the Surgical Peculiarities of
the Negro, by Rudolf Matas, M. D. ; Diseases of the Female
Breast, by Frederick S. Dennis, M. D. ; the Use of the Röntgen
or X Rays in Surgery, by W. W. Keen, M. D.

This array is enough to class this work as one of the most
scientific that has ever been issued on surgery. The work as a
whole has been reviewed in this journal, but the fourth volume
needs nothing but praise.

**The Therapeutical Applications of Peroxide of Hydrogen (medicinal),
Glycozone, Hydrozone, and Eye Balsam.** By CHARLES MARCHAND,
Chemist Graduate of the "*Ecole Centrale Des Arts et Manufactures, de
Paris,*" France. Treatment of Diseases Caused by Germs, Bacteria,
Microbes. Eleventh edition. New York.

Besides telling of the therapeutical applications of Glycozone,
Hydrozone, etc., this work embraces articles written by some of
the most prominent men in the medical profession. Indeed, the
book is of as much value as many works devoted to the practice
of medicine. In this day of aseptic and antiseptic surgery the
surgeon must inform himself with the best knowledge attainable,
and this book contains said knowledge. Nearly every variety of

disease falling under the surgeon's care is written about, and especial directions given by the different authors how to use these great agents to effect a cure.

The profession stands indebted to Professor Marchand for the valuable articles from the pens of so many distinguished physicians and surgeons. The book is profusely illustrated, showing rare surgical diseases.

A Treatise on Surgery, by American Authors, for Students and Practitioners of Surgery and Medicine. Edited by ROSWELL PARK, A. M., M. D., Professor of the Principles and Practice of Surgery and of Clinical Surgery in the Medical Department of the University of Buffalo, Buffalo, New York; Member of the Congress of German Surgeons; Fellow of the American Surgical Association; Ex-President Medical Society of the State of New York; Surgeon to the Buffalo General Hospital, etc. Vol. I. With 356 engravings and 20 full-page plates in colors and monochrome. Lea Brothers & Co., Philadelphia and New York, Publishers.

It is always gratifying to welcome an American book if it be one of merit. Especially is it so with the medical profession to welcome a great work on surgery or medicine. The profession then is to be congratulated that in the past few years several works on surgery have been issued that were edited by American surgeons of repute. Standing abreast with the very best of them is the Treatise on Surgery, edited by Roswell Park, M. D. If the volumes to follow equal the first one just issued, it will take rank with any work of the kind published. Among the list of contributors for Vol. I are the following names: William T. Beefield, M. D., Herbert L. Burrell, M. D., Duncan Eve, M. D., John A. Fordyce, M. D., Frederic Gerrich, M. D., William A. Hardaway, M. D., Hobart Amory Hare, M. D., James M. Holloway, M. D., Henry A. Mudd, M. D., Charles B. Nancrede, M. D., Roswell Park, M. D., John Parmeviter, M. D., Joseph Ransohoff, M. D., Chauncey Smith, M. D., Edward Souchon, M. D. These are all familiar names in surgical literature, and give great promise to the work to which they contribute. The contents will embrace every subject pertaining to general surgery. Professor James M. Holloway, of Louisville, writes the chapter on Surgical Injuries and Diseases of the Veins in Vol. I. It is written in a clear, forcible, but plain style, and contains every thing up to date on this important subject. It is truly an American Surgery, truly fit for Americans to follow.

The Diagnosis and Treatment of Diseases of the Rectum. By WILL-
IAM ALLINGHAM, F. R. C. S., Eng.; Ex-member of Council of The Royal
College of Surgeons of England; Late Senior Surgeon to St. Mark's Hospital for Diseases of the Rectum, etc., and HERBERT W. ALLINGHAM,
F. R. C. S., Eng.; Surgeon to the Great Northern Hospital; Assistant-Surgeon to St. George's Hospital; Late Assistant-Surgeon to St. Mark's Hospital. Sixth edition. New York: William Wood & Company.

It is with great pleasure that the profession of this country
will welcome back to print this most excellent work on Diseases
of the Rectum. For many years it had an immense sale in the
States, and it was difficult to understand why it should be allowed
to go out of print. The work is greatly enlarged and appears in
a much more attractive form than before. Rectal Surgery owes
as much if not more to the elder Allingham for its advancement
and elucidation than to any other man. His experience has been
very great in this especial line, and his views have been for many
years quoted by all men writing on this subject.

Herbert Allingham has contributed largely to this work and
thereby enhanced its value greatly. His writings in the past few
years have brought him prominently before the profession, not
only in England but in this country as well. The writer has no
criticism to make of this work, only in the most favorable way.
Suffice it to say that any one interested in rectal surgery must of
necessity purchase Allingham's book, else his fund of information is not replete.

How to Feed Children. By LOUISE E. HOGAN, Philadelphia, Pa. A manual for nurses and mothers. Philadelphia: J. B. Lippincott Company.
1896.

It is a commendable fact that in the last few years so much
has been written in regard to the feeding of children and the
sick, a subject too long neglected. The scientific management
of children's diet has long been neglected and the choice of food
left to the mother and nurse, who, if ignorant, have heeded the
recommendation of some kind neighbor who has perhaps raised
a family on some unheard-of concoction or formula, and professional advice is sought only after the damage of these indiscre-
tions is perhaps beyond remedying.

It is a pleasure, therefore, to see such excellent little works as
the one before us as claimants for recognition at the hands of the

profession and the laity, to both of whom it will be of great service.

The authoress has used the latest writings of eminent pediatricians upon the feeding of children as the basis of the book, and has incorporated many excellent suggestions and recipes. It is well written and will prove of interest and service to many.

A Treatise on Appendicitis. By John B. Deaver, M. D., Surgeon to the German Hospital, Philadelphia. Containing thirty-two full-page plates and other illustrations. Philadelphia: P. Blakiston, Son & Company. 1896.

The medical profession is more or less familiar with the writings of Dr. Deaver on the subject of Appendicitis, but a book containing his views of the various phases of this common disease is welcome.

Briefly, the author's position in regard to the mooted points of this disease is as follows: Pathology he divides into *endo-appendicitis, parietal appendicitis, peri-appendicitis,* and *para-appendicitis,* following the classification of Fowler. Symptoms are divided into *acute,* which include the ulcerative, perforative, and gangrenous varieties, and the *chronic* form. The three cardinal points in the symptomatology of the acute variety are *pain, tenderness,* and *muscular rigidity.*

His position as to the treatment is very positive, stating that "it is a surgical affection and should be treated as such." "The appendix should be removed as soon as the diagnosis is made." Laxatives are strongly recommended, castor oil and the salines being preferred. Opium should not be used, many fatal cases being made so by this drug being used. As to the removal of the sloughing appendix in abscess cases it is stated that abdominal fistulæ are often the result of the stump being left.

If so much trouble had not been taken to print the many colored and variegated plates to illustrate very poorly the various pathological conditions the book would have been much more attractive. The other plates will do very well. In other respects the book is attractively arranged and printed, and it will, we hope, have the effect of establishing the true position of appendicitis, the placing of it among the surgical affections from the start.

A Manual of Obstetrics. By W. A. NEWMAN DORLAND, A. M., M. D., Assistant Demonstrator of Obstetrics, University of Pennsylvania; Instructor of Gynecology in the Philadelphia Polyclinic, etc. With 163 illustrations in the text and six full page plates. Philadelphia: W. B. Saunders. 1896.

Dr. Dorland has crowded into a small space, in a most excellent manner, much that will be of use to the student and the general practitioner who is too busy to look up the subject in a larger and more comprehensive work than the compend before us.

His arrangement of the text is excellent, considering under the head of physiologic obstetrics normal labor and pregnancy from the time of conception to the birth of the child ; under the head of pathologic obstetrics is considered all pathological conditions of the ovum, fetus, and labor ; and the new-born child is considered under a separate chapter.

The chapter on puerperal sepsis is worthy of special mention.

Diet for the Sick. Contributed by MISS E. HIBBARD, Principal of Nurses' Training School, Grace Hospital, Detroit, and MRS. EMMA DRANT, Matron of Michigan College of Medicine Hospital, Detroit. Second edition, enlarged. Limp cloth, 16mo, 100 pages. Price, 25 cents, postpaid. Detroit, Mich.: The Illustrated Medical Journal Co. 1896.

In this little book there is, besides the useful formula for "Sick Dishes," foods and cooling drinks for convalescents, quite complete diet tables for use in anemia, Bright's disease, calculus, cancer, chlorosis, cholera infantum, constipation, consumption, diabetes, diarrhea, dyspepsia, fevers, gout, nervous affections, obesity, phthisis, rheumatism, uterine fibroids. It also gives various nutritive enemas. The physicians can use it to advantage in explaining his orders for suitable dishes for his patient, leaving the book with the nurse.

The Multum in Parvo Reference and Dose Book. By C. HENRI LEONARD, M. A., M. D., Professor of the Medical and Surgical Diseases of Women, Detroit College of Medicine. Flexible leather, 143 pages; price, 75 cents. Detroit: The Illustrated Medical Journal Co., Publishers. 1896.

This is a recent edition of the Dose Book, of which the title page informs us some forty thousand copies have been issued. The present edition is printed on very thin paper, and is bound in red leather, round corners, so as to make it specially light and handy for the pocket ; the weight is not two and a half ounces. Besides the doses of some 3,500 preparations being given, it has numerous tables, such as the solubility of chemicals, pronunciation of medical proper names, poisons and their antidotes, incompatibles, tests for urinary deposits, abbreviations, tables of fees, etc. It will be found a handy pocket companion.

Books and Pamphlets Received.

Clinical Observations upon the Use of Antitoxin in Diphtheria, and the Report of a Personal Investigation of this Treatment in the Principal Fever Hospitals of Europe during the Summer of 1895. By Joseph E. Winters, New York City. Reprint from the Medical Record.

Management and Treatment of Tuberculosis in the Asheville Climate, with the Report of Cases. By James A. Burroughs, Asheville, N. C. Reprint from the North Carolina Medical Journal.

Should Ergot be Used During Parturition and the Subsequent Involution Period? By E. Stuver, Rawlins, Wyoming. Reprinted from the Journal of the American Medical Association.

Reduced Period of Intubation by the Serum Treatment of Laryngeal Diphtheria. By Edwin Rosenthal, Philadelphia. Reprint from the Medical and Surgical Reporter.

The Experience of Several Physicians with the Sero-therapy in Tuberculosis. By Paul Paquin, St. Louis. Reprint from the Journal of the American Medical Association.

Tylosis Palmæ et Plantæ, with the Description of Two Cases, Mother and Daughter. By J. W. Ballantyne, Edinburgh, Eng. Reprinted from Pediatrics.

Hydro-Galvanism of the Urethra. By Robert Newman, New York, N. Y. Reprint from the Transactions of the American Electro-therapeutic Association.

Asexualization for the Limitation of Disease and the Prevention of Crime. By E. Stuver, Rawlins, Wyoming. Reprint from the Ohio Medical Journal.

The Life and Character of Edward Rush Palmer, M. D. By H. A. Cottell, Louisville, Ky. Reprint from the American Practitioner and News.

So-Called Epispadias in a Woman, with an Illustrative Case. By J. W. Ballantyne, Edinburgh. Reprint from the Edinburgh Hospital Reports.

The Use of the Cystoscope in the Diagnosis of Bladder Troubles. By L. J. Krouse, Cincinnati. Reprint from the Cincinnati Lancet-Clinic.

Prevention of Tuberculosis. By E. B. Borland, M. D., Pittsburgh, Pa. Reprint from the Journal of the American Medical Association.

Chancre of the Tonsil and Tongue, with the Report of Four Cases. By T. C. Evans, Louisville, Ky. Reprint from the Medical News.

Chorea. With the Report of a Case Produced by Injury. By E. Stuver, Rawlins, Wyoming. Reprint from the Ohio Medical Journal.

Some Conclusions Drawn from Experiences in Pelvic Surgery. By A. V. L. Brocaw, St. Louis, Mo. Reprint from the Medical Mirror.

Technic of Abdominal Salpingo-Oophorectomy without Pedicle. By T. J.
Watkins, Chicago, Ill. Reprint from the Medical News.

Feeding in Infancy. By Arthur V. Meigs, M. D., Philadelphia, Pa. W.
B. Saunders, publisher. 1896. Price, twenty-five cents.

The Pathfinders. By James T. Jelks, Hot Springs, Ark. Reprint from
the Journal of the American Medical Association.

The Climate of Arizona. By Mark A. Rodgers, Tucson, Arizona. Reprint from the Medical and Surgical Reporter.

The Treatment of Granular Lids. By T. C. Evans, Louisville, Ky. Reprint from the Charlotte Medical Journal.

Neuralgia. By Curran Pope, Louisville. Reprint from the American
Medico-Surgical Bulletin.

Cystitis. By E. Stuver, Rawlins, Wyoming. Reprint from the Ohio Medical Journal.

Milk and Infantile Disease. By Henry Ashby, Manchester, England.

Phthisis as a Neurosis. By H. E. Beebe, Sidney, Ohio.

Notes and Queries.

THE annual meeting of the Mississippi Valley Medical Association was held at St. Paul, Minn., on the 16th, 17th, and 18th of September, and the following officers were elected for the ensuing year: Dr. Thos. Hunt Stucky, of Louisville, President; Dr. C. A. Wheaton, of St. Paul, First Vice-President; Dr. Paul Paquin, of St. Louis, Second Vice-President; Dr. H. W. Loeb, of St. Louis, Secretary; Dr. William N. Wishard, of Indianapolis, Treasurer. The next meeting will be held in Louisville, Ky., the second Tuesday in October, 1897. Dr. H. Horace Grant was selected as the Chairman of the Committee of Arrangements.

For a place to have been selected so inaccessible as St. Paul is to the majority of the members, the attendance was above the average and the meeting an unusually successful one.

DR. THOS. CHAS. MARTIN, of Cleveland, has resigned the Professorship of Genito-Urinary Diseases in the Cleveland Medical College.

TRI-STATE MEDICAL SOCIETY, OF ALABAMA, GEORGIA, AND TENNESSEE.—The eighth annual meeting of this Society will be held in Chattanooga, October 13, 14, and 15, 1896. Doctor, you are invited to be present to contribute in any way you can to the success of this meeting. The railroads will give reduced rates as usual on the certificate plan. The attendance promises to be large and the scientific part of unusual interest as is evidenced by the following list of papers: Convulsions in Children; Report of Cases Treated with Large doses of Morphine, Y. L. Abernathy, Hill City, Tenn.; Treatment of Pus in the Pelvis, W. B. B. Davis, Birmingham; Some Remarks on Syphilis, Willis F. Westmoreland, Atlanta; The Therapy of Antipyretics, P. L. Brouillette, Huntsville, Ala.; A New Splint for Fractures of the Humerus, G. A. Baxter, Chattanooga; Medicine, Hippocratic and Operatic, John P. Stewart, Attalla, Ala.; The Turkish Bath, its Therapeutic Indications, Louise Eleanor Smith, Chattanooga; Puerperal Eclampsia, Soale Harris, Union Springs, Ala.; Puerperal Eclampsia, J. E. George, Rockwood, Tenn.; Humphrey's Operation (Amputation of the Penis) with exhibition of the patient, Cooper Holtsclaw, Chattanooga; Diseases and Treatment of the Accessory Sinuses of the Nose, B. F. Travis, Chattanooga, Geo. S. Brown, Birmingham; A Statistical Report of Some of the More Recent Remedies Used in the Treatment of Tuberculosis and Summary of Recent Preventative Methods of Value, R. H. Hayes, Union Springs, Ala.; Operations for Abscess of the Liver, W. C. Townes, Chattanooga; Cystitis, Report of Cases, D. S. Middleton, Rising Fawn, Ga.; G. C. Savage, Nashville; Paper on Eye Diseases, Aleck Stirling, Atlanta; Paper on Eye Diseases, J. M. Crawford, Atlanta; Report of Case of Bradycardia, W. C. Bilbro, Murfreesboro, Tenn.; Vaginal Hysterectomy for Bilateral Suppurative Processes of the Uterine Adnexa, W. D. Haggard, Nashville; The Woodbridge Treatment of Typhoid Fever, J. W. Duncan, Atlanta; Diseases of the Veru Montanum (Caput Gallinaginis), W. Frank Glenn, Nashville; Microscopical and Chemical Examinations as Aids to Diagnosis, Katherine R. Collins, Atlanta; Some Obstetrical Complications with Report of Cases, R. R. Kime, Atlanta; Treatment of Skin Cancer, C. R. Achison, Nashville.   Frank Trestel Smith, Secretary, Chattanooga, Tenn.

www.ingramcontent.com/pod-product-compliance
Lightning Source LLC
Chambersburg PA
CBHW021347210326
41599CB00011B/788